Exploring Generative AI for Collaborative Robots in Agriculture 6.0

Kishor Kumar Reddy C.
Stanley College of Engineering and Technology for Women, India

Binju Saju
Adi Shankara Institute of Engineering and Technology, India

Asegul Hulus
Association for Computing Machinery, Cyprus

Tatjana Sibalija
Union University, Serbia

Haïfa Nakouri
Quebec University at Chicoutimi, Canada

IGI Global
Scientific Publishing
Publishing Tomorrow's Research Today

Published in the United States of America by
IGI Global Scientific Publishing
701 East Chocolate Avenue
Hershey, PA, 17033, USA
Tel: 717-533-8845
Fax: 717-533-7115
Website: https://www.igi-global.com E-mail: cust@igi-global.com

Library of Congress Cataloging-in-Publication Data

Names: Reddy C., Kishor Kumar, 1990- editor
Title: Exploring generative AI for collaborative robots in agriculture 6.0 / edited by Kishor Kumar Reddy C. [and 4 others].
Description: Hershey, PA : IGI Global Scientific Publishing, 2026. | Includes bibliographical references and index. | Summary: "This book provides a comprehensive understanding of generative AI and its specific applications in agriculture. It explores the design, functionality, and benefits of collaborative robots in farming environments"-- Provided by publisher.
Identifiers: LCCN 2025027593 (print) | LCCN 2025027594 (ebook) | ISBN 9798337342078 hardcover | ISBN 9798337342085 paperback | ISBN 9798337342092 ebook
Subjects: LCSH: Artificial intelligence--Agricultural applications | Agricultural innovations
Classification: LCC S494.5.D3 .E97 2025 (print) | LCC S494.5.D3 (ebook)
LC record available at https://lccn.loc.gov/2025027593
LC ebook record available at https://lccn.loc.gov/2025027594

British Cataloguing in Publication Data
A Cataloguing in Publication record for this book is available from the British Library.

Table of Contents

Preface xv

Chapter 1
Seeds of Intelligence: Generative AI and the Rise of Collaborative Robots in Agriculture 6.0 1
Lingala Thirupathi, Sreenidhi Institute of Science and Technology, India
Mamatha Bontha, Keshav Memorial Engineering College, India
Ravi Aavula, Anurag University, India
K. Sreerama Murthy, Koneru Lakshmaiah Education Foundation, India
Vinay Kumar Nyatha, Sreenidhi Institute of Science and Technology, India
Santhosh Reddy Thuraga, Quantiphi, USA

Chapter 2
A Study on Integrating Generative AI and Collaborative Robots to Provide Smart Agricultural Support 25
Sowmya Pasham, Botho University, Botswana
Jagadeshwari Puttanapura, University of South Alabama, USA
C. Kishor Kumar Reddy, Stanley College of Engineering and Technology for Women, India
Shikha Khullar, Poornima University, India
Jothi Paranthaman, Botho University, Botswana

Chapter 3
AI-Driven Collaborative Robotics in Agriculture: Optimizing Soil and Crop Yield 57
Md. Shoeab Akhter, International University of Business Agriculture and Technology, Bangladesh
Israt Jahan, East West University, Bangladesh
Mahima Tasnim, Underwood University, USA
M. Shohidullah Miah, International University of Business, Agriculture, and Technology, Bangladesh
A. J. M. Sirajul Karim, International University of Business, Agriculture, and Technology, Bangladesh

Chapter 4
Exploring the Technical Foundations and Integration Strategies of Generative AI Within Collaborative Robotics Systems 89
Laya Billinty Varra, Stanley College of Engineering and Technology for Women, India
Jagadeshwari Puttanapura, University of South Alabama, USA
C. Kishor Kumar Reddy, Stanley College of Engineering and Technology for Women, India
Shikha Khullar, Poornima University, India
Ghita Lazrek, Lab LIASSE, ENSA, Université Sidi Mohamed Ben Abdellah, Fez, Morocco
Jothi Paranthaman, Botho University, Botswana

Chapter 5
Overcoming Modern Farming Challenges With Automation Technologies 111
Rupayan Roy, National Institute of Fashion Technology, Kannur, India
Nambiar Arya Rajesh, National Institute of Fashion Technology, Kannur, India
Ayona Padhi, National Institute of Fashion Technology, Kannur, India
Mohana Priya R., National Institute of Fashion Technology, Kannur, India

Chapter 6
From Data to Harvest: How AI and IoT Are Shaping the Future of Farming ... 141
Lingala Thirupathi, Sreenidhi Institute of Science and Technology, India
Suman Dabbu, Osmania University, India
Manvitha Pola, Sreenidhi Institute of Science and Technology, India
Shyam Sunder Pabboju, Mahatma Gandhi Institute of Technology, India
Deva Rajashekar, Anurag University, India
Hari Chandra Prasad Cheerla, Saint Louis University, USA

Chapter 7
Role of AI and IoT in Smart Farming 169
G. Muthupandi, Chennai Institute of Technology, India
N. Rathinavel Subramanian, Chennai Institute of Technology, India
S. Jawagar, Chennai Institute of Technology, India
Angamuthu A., Chennai Institute of Technology, India
S. Pavithra, Chennai Institute of Technology, India

Chapter 8
Generative AI in Practice: From Core Principles to Real-World Use................ 203
Basudha Dewan, Manipal University Jaipur, India
Sabyasachi Bhattacharyya, Barak Valley Engineering College, Residential Girls' Polytechnic, Golaghat, India
Ishu Sharma, Chandigarh Group of Colleges Jhanjeri, Mohali, India
Ehsan Sheybani, University of South Florida, Tampa, USA

Chapter 9
Agriculture 6.0: Advancing Intelligent and Sustainable Agriculture Through the Convergence of IoT and Machine Learning.. 225
Madiha Munawar, Stanley College of Engineering and Technology for Women, India
T. Monika Singh, Annamalai University, India
Shugufta Fatima, Annamalai University, India
R. Madana Mohana, Stanley College of Engineering and Technology for Women, India
Anindya Nag, Northern University of Business and Technology, Khulna, Bangladesh

Chapter 10
Revolutionizing Agriculture and Industries With Smart Solutions: Case Studies on the Transformative Impact of IoT and Machine Learning 257
Laya Billinty Varra, Stanley College of Engineering and Technology for Women, India
T. Monika Singh, Annamalai University, India
Shugufta Fatima, Annamalai University, India
B. V. Ramana Murthy, Stanley College of Engineering and Technology for Women, India
Hitesh Manapragada, California State University, USA

Chapter 11
Enhanced Semantic Segmentation for Identifying Plant Leaf Disease: A Deep Learning Approach 281
N. Sasikaladevi, SASTRA University, India
S. Surekaa, SASTRA University, India
Dhivyabharathi A. J., SASTRA University, India

Chapter 12
Improving Soil Health and Crop Yields With Artificial Intelligence: Innovations, Impacts, and Implementation 313
Ria Ghosh, University of Delhi, India
Rajeev Kumar, Independent Researcher, USA

Chapter 13
Innovations and Challenges in Smart Agriculture: A Case Study on IoT and Machine Learning Applications 341
Jagadeshwari Puttanapura, University of South Alabama, USA
T. Monika Singh, Annamalai University, India
Shugufta Fatima, Annamalai University, India
R. Madana Mohana, Stanley College of Engineering and Technology for Women, India
Anindya Nag, Northern University of Business and Technology, Khulna, Bangladesh

Compilation of References 373

About the Contributors 417

Index 429

Detailed Table of Contents

Preface xv

Chapter 1
Seeds of Intelligence: Generative AI and the Rise of Collaborative Robots in Agriculture 6.0 1
Lingala Thirupathi, Sreenidhi Institute of Science and Technology, India
Mamatha Bontha, Keshav Memorial Engineering College, India
Ravi Aavula, Anurag University, India
K. Sreerama Murthy, Koneru Lakshmaiah Education Foundation, India
Vinay Kumar Nyatha, Sreenidhi Institute of Science and Technology, India
Santhosh Reddy Thuraga, Quantiphi, USA

Agriculture 6.0 marks a transformative era where generative AI and collaborative robots (cobots) reshape farming into an adaptive, intelligent system. Unlike previous technological shifts, this evolution is rooted in co-creation between humans and machines. Generative AI acts as a thinking partner, designing crop plans, predicting pests, and simulating outcomes while cobots execute tasks safely alongside humans, learning and sharing knowledge collectively. Case studies from India, Kenya, and the Netherlands illustrate how these technologies enhance yields, preserve resources, and respect local traditions. Yet challenges remain: data ownership, equitable access, and job displacement demand thoughtful solutions. Looking ahead, regenerative intelligence will enable farms to restore ecosystems while maintaining productivity. Agriculture 6.0 is not just innovation; it's a respectful alliance between technology, nature, and human wisdom, sowing the seeds for a resilient, inclusive, and intelligent future of food.

Chapter 2
A Study on Integrating Generative AI and Collaborative Robots to Provide Smart Agricultural Support 25
Sowmya Pasham, Botho University, Botswana
Jagadeshwari Puttanapura, University of South Alabama, USA
C. Kishor Kumar Reddy, Stanley College of Engineering and Technology for Women, India
Shikha Khullar, Poornima University, India
Jothi Paranthaman, Botho University, Botswana

Advancements in smart farming are being brought on as a result of integrating Generative AI with collaborative robots or cobots. Generative AI models- GANs-, VAEs, transformers- help cobots to be more independent, flexible, and able to take smart decisions on the field. Armed with data from sensors, drones, and IoT devices, AI assists cobots in predicting outcomes, choosing the right actions, and responding within a small window of time to the changing conditions of weather, pests, soil health. For this, it is essential to design a well-established framework that connects both the hardware and software of AI and robots. This practically could be training AI models, managing data flow, or ensuring real-time performance. Another practical application includes crop health monitoring, harvest estimation, and precise spraying, all in favor of saving time, reducing wastage, and maximizing output. On the other hand: limited compute power right there in the fields; the interpretability of AI decision-making; verification of data representativeness and quality constitute a few issues.

Chapter 3
AI-Driven Collaborative Robotics in Agriculture: Optimizing Soil and Crop Yield.. 57
Md. Shoeab Akhter, International University of Business Agriculture and Technology, Bangladesh
Israt Jahan, East West University, Bangladesh
Mahima Tasnim, Underwood University, USA
M. Shohidullah Miah, International University of Business, Agriculture, and Technology, Bangladesh
A. J. M. Sirajul Karim, International University of Business, Agriculture, and Technology, Bangladesh

Emerging trends in precision agriculture are the consequence of the convergence of AI with collaborative robotics, delivering smart and dynamic soil and crop management systems. This chapter covers the theoretical background and application of AI-driven robots in farm environments, with a specific focus on the analytical optimization of soil health and the assessment of crop yield. It emphasizes data-driven AI techniques, including machine learning, deep learning, and sensor data modeling, to simulate interventions by collaborative robots, such as soil spatial mapping, nutrient deficiency detection, predictive analytics for yield estimation, and decision-making support for farmers. The chapter serves as a prelude to forthcoming AI-robot applications and real-world use case scenarios. By bridging human expertise and AI, this chapter moves toward creating resilient, efficient, and sustainable agricultural ecosystems. Furthermore, this approach aligns with global sustainability goals by enhancing resource efficiency and empowering data-driven agricultural policymaking.

Chapter 4
Exploring the Technical Foundations and Integration Strategies of Generative AI Within Collaborative Robotics Systems 89
Laya Billinty Varra, Stanley College of Engineering and Technology for Women, India
Jagadeshwari Puttanapura, University of South Alabama, USA
C. Kishor Kumar Reddy, Stanley College of Engineering and Technology for Women, India
Shikha Khullar, Poornima University, India
Ghita Lazrek, Lab LIASSE, ENSA, Université Sidi Mohamed Ben Abdellah, Fez, Morocco
Jothi Paranthaman, Botho University, Botswana

The fusion of Generative AI with collaborative robots (cobots) represents a new frontier in Agriculture 6.0, by allowing intelligent decisions, adaptive learning, and increased automation. This chapter provides a thorough technical overview of the architectures, methodologies, and implementation approaches of integrating generative AI models (GANs, Transformers, and diffusion models) with cobots for agricultural applications. The focus of this chapter is on real time adaptability, data efficient learning, and multi-modal sensor fusion for cobots and AI generation techniques. The chapter discusses hardware-software co-design, model deployment on edge processing hardware, interoperability issues, and use cases in the field. The information presented is designed to benefit researchers, technologists, and practitioners working to build sustainable, intelligent, and collaborative agricultural environments.

Chapter 5
Overcoming Modern Farming Challenges With Automation Technologies 111
Rupayan Roy, National Institute of Fashion Technology, Kannur, India
Nambiar Arya Rajesh, National Institute of Fashion Technology, Kannur, India
Ayona Padhi, National Institute of Fashion Technology, Kannur, India
Mohana Priya R., National Institute of Fashion Technology, Kannur, India

Modern agriculture faces workforce shortages, climate variability, soil deterioration, pest infestations, and resource inefficiencies. These issues affect global food security, rural livelihoods, and agricultural production. Meanwhile, tremendous advances in automation, AI, the IoT, and robots provide unprecedented prospects to change agriculture. This chapter discusses modern agriculture's major issues and how automation can solve them. The chapter shows how automation might increase farming system efficiency, resource use, and resilience by evaluating worldwide case studies, upcoming technologies, and adoption constraints. It discusses automation's socioeconomic effects and recommends inclusive and sustainable technological integration in agriculture.

Chapter 6
From Data to Harvest: How AI and IoT Are Shaping the Future of Farming ... 141
Lingala Thirupathi, Sreenidhi Institute of Science and Technology, India
Suman Dabbu, Osmania University, India
Manvitha Pola, Sreenidhi Institute of Science and Technology, India
Shyam Sunder Pabboju, Mahatma Gandhi Institute of Technology, India
Deva Rajashekar, Anurag University, India
Hari Chandra Prasad Cheerla, Saint Louis University, USA

The agriculture industry is at a crossroads, faced with rising global food demands, resource constraints, and the pressing effects of climate change. To move forward, farming must evolve, and that evolution is being powered by two transformative technologies: Artificial Intelligence (AI) and the Internet of Things (IoT). Together, they're creating what we now call Smart Farming. This chapter explores how AI and IoT are not only streamlining agricultural processes but are also helping reshape the very foundation of how farming operates. IoT brings the power of constant connectivity and real-time data collection through devices like soil sensors, drones, and weather stations. AI, on the other hand, processes that flood of data to uncover insights, predict outcomes, and drive automated decision making. We'll examine how these technologies are working together to enable precision agriculture, smart irrigation, early pest detection, yield forecasting, and even supply chain optimization. The chapter will also address the challenges that come with digital transformation in agriculture.

Chapter 7
Role of AI and IoT in Smart Farming 169
G. Muthupandi, Chennai Institute of Technology, India
N. Rathinavel Subramanian, Chennai Institute of Technology, India
S. Jawagar, Chennai Institute of Technology, India
Angamuthu A., Chennai Institute of Technology, India
S. Pavithra, Chennai Institute of Technology, India

This chapter explains how Artificial Intelligence (AI) and the Internet of Things (IoT) can change present-day farming practices through smart farming. Dealing with problems like climate change, shortage of resources and labor issues, using AI and IoT could help the agriculture sector provide secure, sustainable and efficient food supplies. This chapter describes how smart farming has advanced, focusing on how IoT devices provide real-time data and AI helps to make future decisions easier. In this chapter, examples and case studies are used to explain how artificial intelligence is changing the methods of farming, covering irrigation, monitoring crops, managing pests and boosting the supply chain. It also covers the difficulties preventing many from using it such as infrastructure, budget and data protection and suggests ways to make it available to more people. In essence, the chapter describes how the future of farming is being affected by the combination of AI and IoT.

Chapter 8
Generative AI in Practice: From Core Principles to Real-World Use................ 203
Basudha Dewan, Manipal University Jaipur, India
Sabyasachi Bhattacharyya, Barak Valley Engineering College, Residential Girls' Polytechnic, Golaghat, India
Ishu Sharma, Chandigarh Group of Colleges Jhanjeri, Mohali, India
Ehsan Sheybani, University of South Florida, Tampa, USA

Generative AI is a subset of machine learning focused on generating new, coherent data samples that mimic a given dataset. This chapter introduces core principles and theoretical foundations of generative modeling, emphasizing probabilistic approaches and models such as GANs, VAEs, and transformers like GPT. It explores design strategies, training techniques, and trade-offs involved in quality and efficiency. Practical applications are discussed across domains including natural language generation, synthetic media, drug discovery, medical imaging, education, and content recommendation. Real-world case studies support the theoretical insights. The chapter also addresses ethical considerations like deepfakes, data privacy, and algorithmic fairness, promoting responsible AI development. Emerging trends such as diffusion models and multimodal generation are also covered, offering a concise yet comprehensive guide for researchers and practitioners in academic, industrial, and cross-disciplinary settings.

Chapter 9
Agriculture 6.0: Advancing Intelligent and Sustainable Agriculture Through the Convergence of IoT and Machine Learning 225
Madiha Munawar, Stanley College of Engineering and Technology for Women, India
T. Monika Singh, Annamalai University, India
Shugufta Fatima, Annamalai University, India
R. Madana Mohana, Stanley College of Engineering and Technology for Women, India
Anindya Nag, Northern University of Business and Technology, Khulna, Bangladesh

This chapter explores the convergence of IoT, Machine Learning (ML), and Generative AI as foundational pillars of Agriculture 6.0. It highlights how IoT enables real-time data collection from agricultural environments, while ML analyzes and optimizes this data for intelligent, autonomous decision-making. Generative AI further enhances capabilities such as crop modeling, yield prediction, and synthetic data generation for precision farming. This integration supports sustainable, efficient, and resilient agricultural systems. The chapter also extends the discussion to broader applications in smart cities, healthcare, environmental monitoring, and industrial automation. Innovations like smart grids, predictive maintenance, and traffic optimization illustrate their wider societal impact. Challenges related to data privacy, ethics, and regulatory needs are addressed, along with future trends such as 5G, edge computing, and federated learning as critical enablers of a more connected and intelligent world.

Chapter 10
Revolutionizing Agriculture and Industries With Smart Solutions: Case Studies on the Transformative Impact of IoT and Machine Learning 257
Laya Billinty Varra, Stanley College of Engineering and Technology for Women, India
T. Monika Singh, Annamalai University, India
Shugufta Fatima, Annamalai University, India
B. V. Ramana Murthy, Stanley College of Engineering and Technology for Women, India
Hitesh Manapragada, California State University, USA

This chapter explores how the convergence of the Internet of Things (IoT) and Machine Learning (ML) is revolutionizing agriculture and a range of industries, including healthcare, smart cities, manufacturing, retail, finance, education, and smart homes. Through real-life case studies, it demonstrates how these technologies are enabling intelligent automation, operational efficiency, real-time monitoring, and predictive decision-making. Special attention is given to agricultural innovations such as precision farming, smart irrigation, and crop health analytics, alongside comparable breakthroughs in other sectors. The chapter also critically examines challenges related to data security, privacy, ethical considerations, and system integration. Furthermore, it addresses the future trajectory of IoT-ML ecosystems, including the role of emerging technologies like edge computing, 5G, and federated learning. Readers will gain valuable insight into how IoT and ML are driving a paradigm shift towards smarter, more connected systems, transforming the way industries operate and evolve.

Chapter 11
Enhanced Semantic Segmentation for Identifying Plant Leaf Disease: A Deep Learning Approach 281
N. Sasikaladevi, SASTRA University, India
S. Surekaa, SASTRA University, India
Dhivyabharathi A. J., SASTRA University, India

Plant diseases are a severe menace to crop health and agricultural output, particularly if detection is slow or incorrect. This chapter proposes a strong, multi-stage deep learning pipeline for leaf disease detection in an automated manner using semantic segmentation and mask-based classification. This method initially improves image quality by using CLAHE and bilateral filtering to enhance the visibility and minimize noise. The output is passed through a UNet architecture based on a ResNet18 encoder to perform segmentation of the infected areas into binary masks that pinpoint the infection. These masks are resized and fed into a specially designed CNN model to classify the disease only based on the structure and shape of the infected area. Through the removal of dependence on raw RGB input, the system enhances robustness to lighting and color variation. The hybrid solution attains the accuracy of 86.83% over a hand-cultivated dataset including Healthy, Powdery and Rust leaf instances. Segmentation with structural mask-based classification improves generalizability in agriculture.

Chapter 12
Improving Soil Health and Crop Yields With Artificial Intelligence: Innovations, Impacts, and Implementation 313
Ria Ghosh, University of Delhi, India
Rajeev Kumar, Independent Researcher, USA

Innovative and sustainable farming methods are desperately needed in light of the growing world's food demand, the depletion of soil resources, and the uncertain effects of climate change. In this chapter, we examine how AI is enabling data-driven, accurate, and adaptive farming practices, thereby transforming crop yield optimization and soil health management. Farmers and agronomists can now continuously monitor soil parameters, anticipate nutrient deficiencies, identify diseases early, and optimize irrigation and fertilization schedules with previously unheard-of precision by combining AI technologies like machine learning, deep learning, computer vision, and the Internet of Things (IoT). This chapter offers a thorough review of AI applications in agricultural production and soil analysis, along with case examples showing quantifiable increases in sustainability and productivity. It talks about how AI and precision agriculture work together, highlighting how intelligent systems may improve environmental results, increase food security, and reduce resource waste.

Chapter 13
Innovations and Challenges in Smart Agriculture: A Case Study on IoT and Machine Learning Applications 341
Jagadeshwari Puttanapura, University of South Alabama, USA
T. Monika Singh, Annamalai University, India
Shugufta Fatima, Annamalai University, India
R. Madana Mohana, Stanley College of Engineering and Technology for Women, India
Anindya Nag, Northern University of Business and Technology, Khulna, Bangladesh

This chapter explores the evolving future of data-driven decision-making through the convergence of the Internet of Things (IoT) and Machine Learning (ML), focusing on applications in smart agriculture. IoT enables the real-time collection of vast data from connected devices, while ML provides the intelligence to interpret and act on this information. Their integration enhances automation, efficiency, and predictive insights in agricultural systems. Core concepts of IoT and ML are presented together, highlighting their role in optimizing crop monitoring, precision irrigation, disease detection, and yield forecasting. A case study illustrates how ML applied to IoT data improves operational performance and decision-making. The chapter also addresses benefits, challenges such as data integration, privacy, and connectivity, and explores emerging trends like edge computing and federated learning. Overall, it offers a holistic view of how IoT and ML are shaping the future of smart agriculture and related domains.

Compilation of References 373

About the Contributors 417

Index 429

Preface

The convergence of Generative AI and Collaborative Robots (Cobots) is revolutionizing the landscape of Agriculture 6.0, marking a significant leap towards autonomous, intelligent, and sustainable farming practices. Generative AI empowers Cobots with advanced decision-making, real-time adaptation, and the ability to generate innovative solutions for complex agricultural tasks. This edited book presents a comprehensive exploration of how Generative AI enhances the capabilities of collaborative robots in various farming operations, from precision planting to smart harvesting. By bringing together cutting-edge research, practical applications, and future trends, this volume serves as a valuable resource for researchers, technologists, and stakeholders committed to shaping the future of intelligent agriculture.

CHAPTER OVERVIEW

Chapter 1 explores the agriculture 6.0 and it marks a turning point where technology and nature come together in a new, smarter way of farming. It's a world where AI and cobots don't replace humans they work alongside them, creating a collaborative, adaptive system that boosts yields and preserves resources. From India to Kenya to the Netherlands, this transformation is already making a difference. Yet, as we embrace this future, we must ensure it's equitable and sustainable for everyone involved.

Chapter 2 delves into the transformative role of integrating Generative AI with collaborative robots (cobots) in the domain of smart farming. By combining models such as GANs, VAEs, and transformers with data collected from sensors, drones, and IoT devices, this chapter explores how AI-powered cobots can make autonomous, context-aware decisions in dynamic agricultural environments. The discussion highlights practical applications including crop health monitoring, harvest estimation, and precision spraying—each contributing to increased productivity and sustainability. At the same time, it addresses challenges such as limited edge compute resources, data quality assurance, and AI interpretability. Through this examination, the chapter aims to provide researchers, engineers, and agri-tech innovators with insights into building robust AI-robot frameworks that balance performance, efficiency, and ethical considerations in modern agriculture.

Chapter 3 explores artificial intelligence (AI) and the powerful integration of collaborative robotics in sustainable agriculture. It also focuses on how these technologies improve soil health, optimize crop yield, and support sustainable farming through data-powered decisions. With real-world examples and practical insights, the chapter aims to guide researchers, neoperculars, and policymakers in the creation of smart, climate-flexible agricultural systems for the future.

Chapter 4 discusses the collaboration of Generative AI and collaborative robots (cobots) to support the next wave of modern farming transformations. The agriculture industry is shifting to smarter, technology-based systems and advancements in models like Generative Adversarial Networks (GAN), Variational Autoencoders (VAE), Transformers and Diffusion Models are creating new ways for cobots to Figure out how to: learn, adjust and make decisions in their newly created ways. These new AI-architecture systems are making decisions and actions more adaptable and efficient - moving beyond detection mechanisms to understand crop health, plan irrigation or navigate to where they are in a field. The chapter explains how these models are built and where they run (inside the cobot or the cloud) in terms of real time decision-making. The chapter also refers to practical examples from crop monitoring to adaptive spraying. This chapter is intended to support researchers and developers of how, in the advent of AI, we can begin thinking about and developing viable, scalable systems that utilize AI and cobots for sustainable farming.

Chapter 5 examines, modern agriculture faces critical challenges including workforce shortages, climate variability, soil deterioration, pest infestations, and resource inefficiencies that threaten global food security and rural livelihoods. However, advances in automation, artificial intelligence, IoT, and robotics offer unprecedented opportunities to transform agricultural practices. This chapter examines these challenges and explores how automation can enhance farming system efficiency, resource utilization, and resilience through global case studies and emerging technologies. Beyond technical solutions, we address the socioeconomic implications of agricultural automation and advocate for inclusive, sustainable technological integration that benefits producers, consumers, and communities dependent on resilient food systems.

Chapter 6 explores a quiet revolution in agriculture, driven by the urgent need to feed more people with fewer resources. At the heart of this transformation are AI and IoT, working together to make farming smarter and more responsive. From monitoring soil health to predicting weather and optimizing harvests, these tools are changing how we grow our food. The chapter highlights their real-world impact on precision farming and beyond. It also takes a closer look at the challenges and opportunities in this digital evolution of agriculture.

Chapter 7 focuses on how the agricultural sector is undergoing a significant transformation with the integration of Artificial Intelligence (AI) and the Internet of Things (IoT). These technologies enable real-time monitoring, precision farming, and data-driven decision-making to improve productivity and sustainability. From predictive crop management to automated irrigation systems, AI and IoT offer solutions to key challenges faced by farmers. This chapter delves into how intelligent technologies are reshaping traditional farming into a more efficient, resilient, and environmentally friendly process. By leveraging AI and IoT, smart farming paves the way for ensuring global food security in an era of rising demand and climate uncertainties.

Chapter 8 explores the world of Generative Artificial Intelligence (AI), how it was theorized, what its most significant architectures are, and what its real applications are. The chapter lists a wide range of uses discussed under natural language processing, drug discovery and personalized education. It also ponders on the ethical issue related to synthetic media, algorithmic bias and data privacy. Future innovation trends are also investigated, and what drives the future will include such emerging trends as diffusion models and multimodal systems. Also written to appeal to students, researchers, and professionals, the chapter is intended to promote responsible and informed interventions into generative AI and effective interactions with it.

Chapter 9 provides a full-fledged view of the combination of IoT, ML, and Generative AI into Agriculture 6.0. It ventures into topics like real-time data, intelligent analysis, and generative capabilities transforming traditional farming into smart and sustainable systems. It then shifts its focus toward cross-industrial development: and smart cities, healthcare, and industrial automation are highlighted. Among these technologies being future enablers are 5G, edge computing, and federated learning. Last but not least, it mentions ethical and regulatory considerations, thus providing a broader overview of this ever-changing digital landscape.

Chapter 10 provides theoretical consideration of the effects of the Internet of Things (IoT) and Machine Learning (ML) in a variety of important fields. Through a selection of specific case studies, it illustrates the potential of the fusion of these technologies to enable intelligent automation, data-driven decision making, and increased operational efficiency across the array of industries discussed in the chapter (i.e., healthcare, smart cities, manufacturing, agriculture, retail, finance, and education). The chapter discusses not only the technical implementations of these smart solutions, and their attendant benefits, but also some of the most significant challenges they present to various stakeholders (i.e., data privacy, cybersecurity, and ethical challenges). Engaging with candidate examples of these technologies from the real world, the chapter provides useful insights for researchers, practitioners, and policymakers, about those changes that IoT and ML are beginning to impact and may impact the development of a more connected, effective, and sustainable future.

Chapter 11 focuses on early detection of plant leaf diseases and how it is vital for improving crop health and yield. Traditional inspection methods are often labor-intensive and prone to errors. This work explores an enhanced deep learning-based semantic segmentation approach for precise disease identification. The proposed method ensures accurate localization of infected regions on plant leaves. Such advancements support automated, efficient, and reliable plant disease management in agriculture.

Chapter 12 discusses how artificial intelligence (AI) is transforming how we perceive and manage soil fertility and crop yields. As the world's agriculture faces the twin challenge of higher food demand and environmental degradation, the convergence of AI technologies presents catalysts for transformative change in precision agriculture, sustainable utilization of resources, and climate resilience. This chapter offers a broad overview of AI uses along the value chain of agriculture, ranging from real-time soil analysis to predictive yield forecasting and robotics-enabled field operations. By reviewing technological progress as well as the challenges of practical implementation, the chapter seeks to give researchers, practitioners, and policymakers concrete knowledge for supporting wiser and more sustainable agricultural systems.

Chapter 13 intends to expose the workings of innovation and challenges being faced in smart agriculture, keeping IoT and ML applications at the forefront. Merging sensor data collection with intelligent decision-making algorithms, this case study presents the transformation of modern agricultural practices by technologies such as precision farming, crop monitoring, predictive maintenance, and yield optimization. These issues include roadblocks for implementation like connectivity, scalability, and data reliability, especially in rural areas. The exploration offers a more grounded view to equip researchers, agritech developers, and policymakers with the necessary insight to develop IoT and ML for a sustainable data-driven agricultural ecosystem in the near future.

C. Kishor Kumar Reddy
Stanley College of Engineering and Technology for Women, India

Binju Saju
Adi Shankara Institute of Engineering and Technology, India

Asegul Hulus
University of Greenwich, UK

Tatjana Sibalija

Union University, Serbia

Haïfa Nakouri
Quebec University at Chicoutimi, Canada

Chapter 1
Seeds of Intelligence:
Generative AI and the Rise of Collaborative Robots in Agriculture 6.0

Lingala Thirupathi
https://orcid.org/0000-0003-0703-3860
Sreenidhi Institute of Science and Technology, India

K. Sreerama Murthy
https://orcid.org/0000-0003-2659-9966
Koneru Lakshmaiah Education Foundation, India

Mamatha Bontha
Keshav Memorial Engineering College, India

Vinay Kumar Nyatha
Sreenidhi Institute of Science and Technology, India

Ravi Aavula
https://orcid.org/0000-0001-8640-1152
Anurag University, India

Santhosh Reddy Thuraga
https://orcid.org/0009-0002-3325-6443
Quantiphi, USA

ABSTRACT

Agriculture 6.0 marks a transformative era where generative AI and collaborative robots (cobots) reshape farming into an adaptive, intelligent system. Unlike previous technological shifts, this evolution is rooted in co-creation between humans and machines. Generative AI acts as a thinking partner, designing crop plans, predicting pests, and simulating outcomes while cobots execute tasks safely alongside humans, learning and sharing knowledge collectively. Case studies from India, Kenya, and the Netherlands illustrate how these technologies enhance yields, preserve resources, and respect local traditions. Yet challenges remain: data ownership, equitable access, and job displacement demand thoughtful solutions. Looking ahead, regenerative intelligence will enable farms to restore ecosystems while maintaining productivity.

DOI: 10.4018/979-8-3373-4207-8.ch001

Agriculture 6.0 is not just innovation; it's a respectful alliance between technology, nature, and human wisdom, sowing the seeds for a resilient, inclusive, and intelligent future of food.

1. INTRODUCTION

Walk through a modern farm today, and you'll still hear the familiar sounds, boots crunching against tilled soil, the rustle of leaves in the wind, the low hum of a tractor in the distance. But beneath these timeless rhythms, something different is unfolding. There's a new presence in the fields, an intelligence that doesn't come from nature or nurture in the traditional sense, but from circuits, code, and data. This change isn't loud or flashy. It doesn't announce itself with chrome and wires or replace the farmer with a robot overlord. Instead, it's subtle like a new partner working quietly in the background, helping to guide decisions, predict outcomes, and handle the heavy lifting with precision and care. Farming is no longer just about muscle or machinery. It's becoming a conversation between human wisdom and machine intelligence. Welcome to Agriculture 6.0, the next major chapter in humanity's relationship with the land. It's not just a technological upgrade; it's a philosophical shift. Where previous agricultural revolutions focused on maximizing production through tools, chemicals, and techniques, this new era is about integration. Intelligence is no longer confined to the farmer's mind or a software dashboard. It's becoming embedded in every seed sown, every crop harvested, every decision made.

What sets Agriculture 6.0 apart is its emphasis on collaboration. This isn't a world where machines replace people. Instead, it's a world where people and machines work together each playing to their strengths. Humans bring intuition, context, and care. Machines bring speed, precision, and tireless effort. And at the center of this collaboration are two transformative technologies: generative artificial intelligence and collaborative robots, or cobots. Generative AI is the brain behind the operation. Trained on vast datasets from weather patterns and soil chemistry to plant genetics and market trends this form of AI doesn't just analyze; it creates. It can generate crop plans, simulate future conditions, and offer novel solutions tailored to the unique needs of each farm. It doesn't dictate decisions but serves as a thinking partner, augmenting the farmer's judgment with deeper, data-driven insight.

Cobots, on the other hand, are the hands in the field. Unlike traditional industrial robots, which operate in isolation, Cobots are designed to work with humans, not separate from them. They're equipped with sensors and adaptive algorithms that allow them to learn, adjust, and even communicate. Whether it's pruning vines in a vineyard or sorting delicate fruit without damage, Cobots are changing what's physically possible on the farm. But what truly defines Agriculture 6.0 isn't the technology itself it's the relationship between the human and the machine. We're witnessing a new kind of harvest, one that doesn't just yield food but produces knowledge, resilience, and adaptability. Farmers aren't surrendering control; they're gaining powerful allies that help them do more with less, reduce uncertainty, and respond dynamically to a changing world.

The impact is already visible. In regions where labor is scarce, Cobots are helping fill the gap. In climates disrupted by erratic weather, generative AI is helping farmers re-strategize in real-time. In communities steeped in agricultural tradition, these tools are being adapted to fit local contexts, enhancing not erasing cultural knowledge. This is a technological revolution that respects roots while encouraging growth. It's important to recognize that this shift didn't happen overnight. It's the result of decades of advancement from the mechanical inventions of the Industrial Age to the precision agriculture techniques of the digital age. What we're seeing now is the convergence of those trajectories, sharpened by artificial intelligence and softened by collaborative design. As we look ahead, the question isn't whether machines will take over farming. The more powerful and urgent question is this: How can we shape these tools to align with our values, our ecosystems, and our communities? Agriculture 6.0 offers the opportunity to re-imagine farming not as a solitary human endeavor, but as a shared act of creation between people, machines, and the living world.

2. RELATED WORKS

(Majumder et al., 2024) uses the adoption of generative AI to visually forecast the potential yield of a crop using stable diffusion and conditional prompts additionally implementing several generative adversarial networks architectures for data generation, augmentation and super-resolution. (Zhao et al., 2024) highlights the potential of advanced generative models in enhancing synthetic dataset creation for fruit object detection and segmentation in agricultural robotics. (Klair et al., 2024) proves valuable in addressing data scarcity issues within the agricultural domain. (Liu et al., 2023) This paper aims to bridge this research gap by examining the current studies and potential applications of digital twins and Generative AI in agriculture.

(Dhavale et al., 2024) The results of this research project demonstrate the potential of cutting-edge AI technology to transform agriculture, making it more accessible, efficient, and environmentally friendly. (Abayaratne et al., 2024) introduces efficient and precise detection and tracking methods for individual lettuce plants. (Venkata et al., 2024) This study follows a similar approach to agriculture, providing employment to approximately 71% of rural Indians.

(Rani et al., 2023) presents a systematic multi-fold survey and analysis of such work focusing on recent AI techniques developed to combat plant diseases. (Thamaraikannan et al., 2023) Deep learning is transforming agriculture by enabling accurate crop prediction and efficient environmental monitoring through advanced imaging and AI techniques. Despite challenges like limited labeled data, evolving models continue to enhance precision and reduce complexity in farming applications. (Shaikh et al., 2022) The authors found that the deep learning algorithms that have been used in recent studies have performed far better than the conventional machine learning algorithms due to recent technological advances that can efficiently process vast amount of data and enable timely intelligent decisions similar to human decisions.

(Kavipriya et al., 2024) This paper investigates the use of Machine Learning and Deep Learning methods to forecast crop yield using remote sensing images. (Hou et al., 2025) proposes an automatic decision-making localization algorithm for robots fused with generative big models in agriculture Internet of Things. (Jockusch et al., 2024) introduces a novel approach method for decentralized land cover and land use classification, utilizing federated learning in conjunction with convolutional neural networks on imagery obtained from Unmanned Aerial Vehicles. (Padmaja et al., 2021) Discussed the distributing of the message over the network using another frequency and timing technique to circumvent the jammers.

(Rizvi et al., 2024) To address the limitations of conventional weed control methods, in this research, Machine Learning and Deep Learning based techniques are proposed to identify and categorize weeds in crops. (Prema et al., 2024) presents a comprehensive approach to plant disease classification using a combination of auto-encoders for dimensionality reduction and a fine-tuned ResNet-9 architecture. (Kolhalkar et al., 2025) Robotic fruit and vegetable harvesting, powered by AI, machine vision, and soft robotics, offers a promising solution to labor shortages and efficiency challenges in agriculture. With continued advancements in detection algorithms and adaptive technologies, these systems are poised to revolutionize harvesting while minimizing fruit damage and operational costs. Detected (Pratapagiri et al., 2021) plant leaf disease using convolutional neural networks.

(Du et al., 2024) The numerical results indicate that our proposed semantic communication framework and resource allocation schemes significantly reduce transmission costs while enhancing the transmission quality of important information in communication services. (Sonmez et al., 2024) proposed an end-to-end workflow for deep learning-based computer vision applications to facilitate the AI model training, model optimization, model deployment, metadata streaming, metadata recording, result filtering/searching and data visualization. Developed (Shailaja et al., 2023) intelligent deep feature based intrusion detection system for network applications. Predicted (Rekha et al., 2021) ensemble machine learning based dengue disease with performance and accuracy elevation patterns.

The purpose of the research is to understand the technological evolution of agriculture over the years and propose a new perspective for the practices that are becoming established in agribusiness. (Neves et al., 2023). (Patil et al., 2024) although in practice farmers face various threats to their economic activity; the "era of agriculture number six dot zero" has already been heralded. (Movahed et al., 2024) the production of healthy and sustainable food seems vital, because nutritional health is one of the lofty goals of the modern industrial revolution. Supply chain cycle improvement of the agriculture industry to increase sustainability and environmental security is one of the concepts that will be important in Industry 6.0. (Kumar et al., 2025) explores the transformative effects of smarter chatbots and text-based prediction, classification model of stress, anxiety, depression among the people.

3. OBJECTIVES

- To explore the integration of generative AI and collaborative robots (Cobots) in transforming agriculture into an adaptive and intelligent system.
- To analyze the real-world applications and impacts of Agriculture 6.0 through global case studies.
- To identify the key challenges such as data ownership, equitable access, and job displacement.

4. METHODOLOGIES

4.1 Integration of Generative AI and Cobots

The key modules in Figure 1 and the process flow with step-by-step integration in Figure 2 are shown below.

Figure 1. Key modules

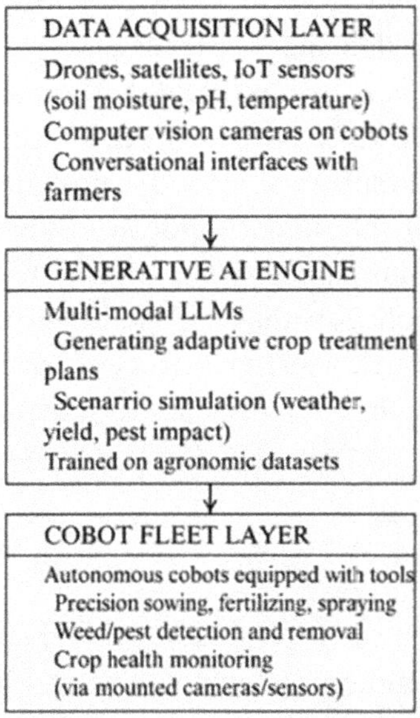

This smart agriculture system operates through four integrated layers, working together to enhance farming efficiency and decision-making. The Data Acquisition Layer collects real-time field data using drones, satellites, IoT sensors (measuring soil moisture, pH, nutrients, temperature), and computer vision cameras mounted on Cobots. This information is processed by the Generative AI Engine, which uses multi-modal large language models trained on agronomic data to generate adaptive crop treatment plans, simulate scenarios such as weather shifts or pest outbreaks, and provide conversational support to farmers. These insights drive the actions of the Cobots Fleet Layer, a group of autonomous robots performing precise tasks like sowing, fertilizing, spraying, weed and pest control, and continuous crop health monitoring. All operational data loops back into the Feedback & Self-Learning Layer, enabling the system to refine its models and strategies continuously, making the AI smarter and more effective with every cycle.

Figure 2. Process flow (step-by-step integration)

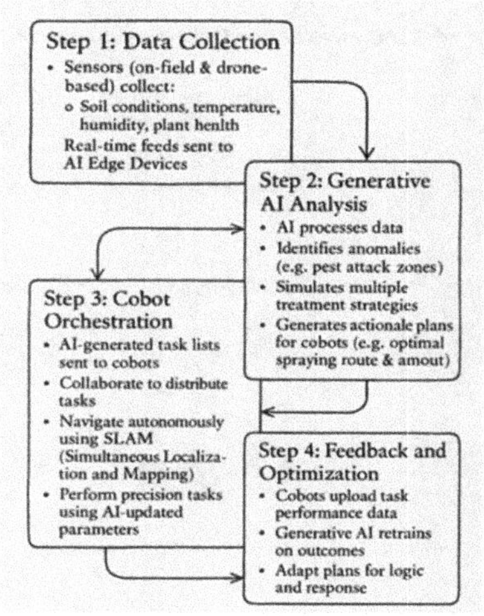

This smart farming system works in four smooth steps. First, sensors on the ground and drones in the air gather real-time data about the soil, weather, and crop health, sending it to AI edge devices. Next, the AI analyzes this data to spot problems like pest outbreaks, simulates different ways to respond, and creates clear action plans along with easy-to-understand summaries for farmers. These instructions go to a team of Cobots that work together, move around the field on their own using mapping tech, and carry out tasks like spraying or fertilizing with high precision. Finally, the Cobots send back performance results, and the AI uses that feedback to get smarter and adjust its plans for the next round, making the system more accurate and efficient over time.

The Figure 3 shows an Integration of Generative AI and Cobots.

Figure 3. Integration of Generative AI and Cobots

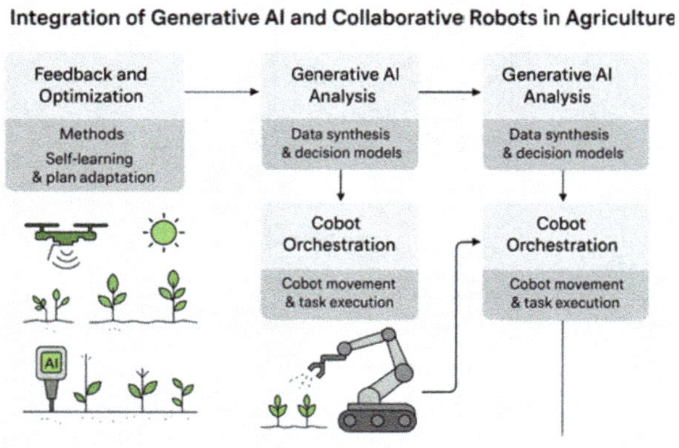

4.1.1 Methods and Technologies

Modern farming is going through a big shift. With the help of some powerful new technologies, farmers can now make more accurate and timely decisions about everything from planting and watering to dealing with weeds and predicting crop yields. Two of the most exciting tools making this possible are multi-modal transformers and diffusion models. Let's break down what they do and how they're being used on real farms today. Imagine you're a farmer managing a big field. You've got data coming in from all over photos from drones flying over the crops, moisture levels from soil sensors, and updates from the weather forecast. In the past, you might have looked at these pieces of information separately, maybe even manually. But now, with the help of transformer technology, all this data can be analyzed together in real time.

These "multi-modal" systems are designed to take in many kinds of data at once images, numbers, text, you name it and put them all together like pieces of a puzzle. For example, if a part of your field looks dry in drone images, and the soil sensors confirm low moisture, while the forecast shows no rain for a few days, the system can suggest watering that specific area. That means no wasted water, no overwatering, and healthier plants.

It's not just about watering. These systems can also help spot early signs of disease or pests by recognizing patterns in the images or by matching past cases with current conditions. And because the system learns over time, it gets better and better at making accurate recommendations. It's like having a digital assistant that understands your entire farm and helps you make smart, timely decisions.

While transformers help you understand what's happening now, diffusion models help predict what might happen next. These models are great at simulating natural processes that change over time, like how crops grow or how weeds spread across a field. Let's say your farm has a few small patches of weeds. A diffusion model can take what it knows about your field wind patterns, past weed behavior, soil type and predict where those weeds are likely to show up next. That gives you a chance to get ahead of the problem, spraying only the areas that are at risk instead of treating the whole field. It's more efficient, more affordable, and much better for the environment. These models can also be used to predict how your crops will grow in the coming weeks. For instance, if you're growing tomatoes, the system can estimate when the fruit will be ripe, based on current weather and how the plants are looking. That helps you plan your harvest more precisely and avoid waste. Another great use for diffusion models is in fighting plant diseases. If a disease starts in one corner of the field, the model can simulate how it might spread and where it could show up next. That lets you treat just the right areas before the disease gets out of control.

The real magic happens when these two technologies work together. Transformers help make sense of the current conditions by bringing together all your different data sources, while diffusion models look ahead and predict what's coming next. Together, they give you a full picture: what's happening now, what's likely to happen later, and what you can do about it. For example, your system might notice that one part of the field is showing early signs of plant stress. It combines that with weather forecasts and sensor readings to figure out why it's happening. Then it uses a diffusion model to estimate how far the problem might spread in the coming days. Based on all of this, it recommends a targeted action maybe some irrigation or a specific nutrient treatment before things get worse.

This kind of smart farming means fewer chemicals, less waste, and better yields. It's also more sustainable, which is great news for both farmers and the planet. All of this technology might sound high-tech, but the goal is actually pretty simple: to help farmers make better decisions. Farming has always been about timing knowing when to plant, when to water, when to harvest. These new tools just give farmers better information, faster, and in a way that's easy to act on. What's especially exciting is that these systems are becoming more affordable and easier to use. You don't need to be a tech expert to benefit from them. Many of these tools are being built with farmer-friendly interfaces, so anyone can use them even those who've never worked with AI before.

As these systems continue to improve, they're going to play a bigger role in everyday farming. Whether it's saving water, reducing herbicide use, or improving crop quality, the benefits are clear. And with climate change and resource shortages becoming more of a concern, smart farming solutions like these are not just helpful they're necessary.

Farming has always been about decisions when to plant, how much to water, where to spray, and when to harvest. But now, as machines and robots start to play a bigger role on the farm, the question becomes: how do we teach these machines to make the right decisions? That's where Reinforcement Learning, or RL, comes in. Think of it as a way to train robots and smart systems to act more like experienced farm workers learning by doing, figuring things out from trial and error, and getting better over time. Imagine you're teaching your dog a new trick. You say "sit," and when your dog does it correctly, you give them a treat. If they don't sit, they don't get the treat. Eventually, your dog learns that "sit" means they should, well, sit because they associate the command with a reward.

Reinforcement Learning works in a similar way, except instead of a dog, we're training a robot or a digital system. And instead of tricks, we're teaching it how to make good choices on the farm. The idea is simple: when the robot does something right say, spraying pesticide only on the infected plants instead of the whole field it gets a virtual "reward." When it makes a mistake like spraying too much or in the wrong spot it doesn't. Over time, the system figures out what actions lead to rewards and which ones don't. Just like your dog, it starts to learn what to do and what to avoid. Let's say we have a robot that's supposed to spray pesticide on a large vegetable farm. The goal is to spray only where it's needed on plants that show signs of pests or disease while skipping healthy areas. This saves chemicals, protects the environment, and keeps crops healthier. At first, the robot doesn't know the best path to take through the field or how to recognize which areas need spraying. So it starts exploring. It tries different routes, sprays different spots, and keeps track of what works and what doesn't.

Each time the robot sprays correctly only on affected plants it gets a reward. When it sprays unnecessarily, or misses a problem spot, it doesn't. Gradually, it starts recognizing patterns. Maybe it learns that shaded areas near irrigation lines are more likely to have fungal problems, or that the far edge of the field usually doesn't need spraying. After a few days or weeks, this robot can cover the entire field efficiently, using the least amount of pesticide, avoiding wasted trips, and skipping unnecessary spraying. It's not just following instructions it's learning how to do the job better each time.

One of the biggest strengths of Reinforcement Learning is that it allows systems to make decisions in real time. That's a big deal on the farm, where things change constantly. Maybe the wind picks up and the robot needs to adjust its spraying direction. Maybe a part of the field is unexpectedly muddy and the robot needs to take a different route. With RL, the robot doesn't just follow a fixed path. It adapts. If it finds that certain actions are working better like choosing a new route to avoid obstacles it can switch strategies on the fly. This kind of flexibility is hard to build into traditional programs, but RL makes it possible by allowing the robot to keep learning as it goes.

Take irrigation, for example. A smart irrigation system can use RL to figure out the best times to water crops. Instead of relying on a timer, the system learns from soil moisture data, weather forecasts, and crop conditions. If watering at night leads to less evaporation and better soil retention, the system starts doing that more often. If overwatering causes root damage, it learns to cut back. Over time, it becomes more efficient just like an experienced farmer would. Or think about greenhouse climate control. Reinforcement Learning can help manage temperature, humidity, and light in the most energy-efficient way possible, all while keeping the plants happy and healthy.

There are lots of smart technologies in farming today, but Reinforcement Learning stands out because it doesn't need constant programming or human supervision. Once it's set up, it can explore, experiment, and learn on its own. It's like giving the system a goal "keep the plants healthy," "save water," "spray only when necessary" and letting it figure out the best way to reach that goal. Reinforcement Learning isn't perfect, and there are a few things to keep in mind. First, it takes time to train. Just like a dog doesn't learn a new trick overnight, a robot or system using RL needs time to try different actions and see what works. This "learning phase" can be slow, especially in farming where the feedback loop is longer.

Second, it requires a good reward system. If the system is rewarded for the wrong actions or not rewarded for the right ones it won't learn properly. Designing smart, meaningful rewards is a key to making RL work well in agriculture. Lastly, it works best when combined with other technologies. RL alone isn't a silver bullet. But when it's paired with sensors, computer vision, satellite data, and human guidance, it becomes a powerful tool in the smart farming toolbox. As more farms adopt automation and smart systems, Reinforcement Learning is expected to play a bigger and bigger role. Imagine Cobots that learn how to harvest without damaging fruit, irrigation systems that adapt to changing weather, or even planting machines that find the best depth and spacing based on soil conditions. These aren't distant dreams they're already being tested in agricultural research labs and pilot programs around the world. The big picture is this: Reinforcement Learning gives machines a way to learn from experience, just like humans do. And on the farm, where conditions are always changing and every season is different, that ability to adapt and improve can make all the difference.

Prompt-tuned Large Language Models (LLMs): These are smart chat bots that understand farmer questions and give helpful, easy-to-understand answers or advice about the farm. For an example: A farmer asks, "Why are some leaves turning yellow?" and the system replies with possible causes and what actions to take. Generative Adversarial Networks (GANs): These create fake but realistic pictures, like images of sick plants, to help train the system when real examples are hard to find. For an example: Creating many images of crop diseases to train a detection system so it can spot diseases even with limited real photos.

ROS (Robot Operating System): This is like the "brain" software that helps robots talk to each other and work together smoothly on the farm. For an example: Robots planting seeds, watering, and harvesting all managed through a common system that coordinates their actions. Swarm Intelligence Algorithms: These are ways to get lots of robots to work as a team, sharing tasks so the whole farm gets covered faster and better. For an example: A group of small robots spread out to inspect every row of crops simultaneously, communicating to avoid overlapping work.

Computer Vision: This technology lets robots "see" what's going on, like spotting weeds or checking if crops look healthy, using cameras and image analysis. For an example: A robot with a camera detecting weeds among crops and removing them automatically. Tele-operation Interfaces: These let farmers control or guide the robots remotely if needed, so they can step in and help or make changes anytime. For an example: A farmer uses a tablet to manually steer a robot around tricky spots or check its status while sitting in the farmhouse.

4.1.2 Use Cases in Practice

The table 1 shows the different Use cases for Generative AI and Cobots Roles in Smart Farming.

Table 1. Use cases for generative AI and Cobots roles in smart farming

Use Case	Generative AI Role	Cobots Role
Precision Seeding	Generate optimal seed placement maps using soil, terrain, and weather data	Execute sowing with centimeter-level accuracy using GPS and sensors
Weed Management	Create weed detection patterns and chemical treatment recipes	Identify and selectively spray or remove weeds in real time
Harvest Prediction	Simulate yield forecasts using weather, soil, and growth data	Scan crops with cameras/sensors to assess readiness and report findings
Disease Control	Generate disease risk maps and treatment plans	Apply targeted treatments and monitor affected crop areas

- **Precision Seeding:** AI figures out exactly where to place each seed to grow best, and the robot plants them at those spots.
- **Weed Management:** AI learns to recognize weeds and recommends the best way to remove them. The robot finds and removes or sprays only the weeds—saving chemicals and protecting crops.

- **Harvest Prediction:** AI predicts how much you'll harvest and when. The robot checks how ripe the crops are and sends updates.
- **Disease Control:** AI spots patterns of disease and plans how to stop it. The robot goes to those places and treats them with pinpoint accuracy.

4.1.3 Challenges and Solutions

Farming with AI and robots comes with some real-world challenges. There's often not enough data to properly train AI systems, fields can be rough and unpredictable, some farmers may be hesitant to use new tech, and internet access isn't always reliable in rural areas. But there are smart solutions to these problems. AI can now generate realistic fake data to help models learn better. Robots can be built tougher and equipped with advanced navigation to handle tricky environments. Friendly, easy-to-use interfaces like chat bots that explain things in plain language can make the tech more approachable for farmers. And with edge computing, these systems can keep working even when there's no internet, thanks to offline backup modes. The Table 2 shows the challenges and solutions.

Table 2. Challenges and solutions

Challenge	Solution
Data scarcity for AI models	Use generative models to synthesize realistic training data
Harsh/unstructured environments	Reinforced Cobots designs with rugged SLAM systems
Farmer adoption	Human-friendly interfaces, natural language explanations from AI
Connectivity issues	Edge computing with fallback offline modes

4.2 Analyzing the Real-World Applications and Impacts of Agriculture 6.0 through Global Case Studies

The evolution of agriculture from its earliest forms to the present digital era has been marked by transformative technological milestones. The latest phase, often referred to as Agriculture 6.0, integrates cutting-edge technologies such as artificial intelligence (AI), robotics, the Internet of Things (IoT), block-chain, and synthetic biology. Unlike previous revolutions that focused solely on yield increases or mechanization, Agriculture 6.0 seeks to establish sustainable, resilient, and intelligent systems capable of adapting to climate change, managing limited resources, and feeding a growing global population. It represents a fundamental shift in how food systems are designed, monitored, and managed.

This paradigm is underpinned by the concept of ultra-precision agriculture, in which interconnected systems continuously gather, analyze, and respond to data across all stages of the agricultural value chain. Decision-making is driven by real-time analytics and automation, allowing for site-specific responses and minimal human intervention. Through cloud computing, machine learning, and edge computing, Agriculture 6.0 enhances productivity while minimizing waste and environmental impact. This data-centric model aims not only to increase efficiency but also to promote ecological sustainability and socioeconomic resilience.

To investigate the real-world applications and implications of Agriculture 6.0, a comparative case study methodology was adopted. The selection of case studies was guided by diversity in geography, economic development, and technological infrastructure. Both qualitative and quantitative data were drawn from academic literature, government and industry reports, and interviews with key stakeholders. The cases include advanced economies with high-tech agriculture as well as developing countries leveraging mobile-based solutions, offering a comprehensive view of global adoption patterns and contextual limitations.

In the United States, AI and robotics have been integrated into large-scale crop farming. A notable example is John Deere's acquisition of Blue River Technology, which developed machine vision systems capable of identifying and targeting individual weeds with herbicide spray. These "see and spray" systems use machine learning to significantly reduce chemical inputs by as much as 90% without compromising yields. This approach demonstrates how autonomous farming equipment can simultaneously lower environmental impact and production costs, contributing to more sustainable large-scale agriculture.

Japan presents a different approach through its widespread adoption of vertical farming. Companies such as Spread Co. utilize multi-layered indoor farms controlled by AI and IoT systems that precisely regulate light, humidity, temperature, and nutrient flow. These controlled-environment agricultural systems use up to 98% less water and require no pesticides, addressing both land scarcity and food safety concerns. Furthermore, automation helps compensate for Japan's aging agricultural workforce, illustrating how high-tech solutions can address both productivity and demographic challenges.

The Netherlands, known for its agricultural innovation, has embraced precision livestock farming. Lely Industries manufactures robotic milking systems equipped with biometric sensors that monitor each cow's health, productivity, and behavior in real time. The use of AI-driven analytics enables individualized care and early disease detection, improving both animal welfare and operational efficiency. These technologies illustrate the applicability of Agriculture 6.0 principles beyond crop farming, extending their benefits to animal husbandry.

Australia provides a strong example of block-chain implementation for food traceability. Companies like BeefLedger have established block-chain systems that track livestock through every stage of the supply chain. This traceability ensures authenticity, improves food safety, and allows producers to demonstrate the ethical and environmental standards of their products. Such systems are particularly valuable in export markets, where transparency and trust are critical to maintaining competitive advantage.

In Israel, smart irrigation systems integrate IoT sensors, real-time data analytics, and satellite imagery to optimize water use. Companies like Netafim have developed drip irrigation solutions that apply water precisely at the root zone, guided by soil moisture and crop health data. These systems can reduce water usage by up to 50% while increasing yields, proving essential in arid regions. Israel's success illustrates how technology can help reconcile the need for productivity with the realities of water scarcity.

Kenya offers an important perspective on how Agriculture 6.0 can be adapted for smallholder farmers in developing contexts. Initiatives such as Digital Green and IBM's Watson Decision Platform have delivered mobile-based tools that provide localized weather forecasts, pest alerts, and crop management advice. Farmers receive personalized recommendations via SMS or smart-phone apps, enabling informed decision-making despite limited infrastructure. This case underscores the importance of mobile technologies in democratizing access to advanced agricultural solutions.

The socioeconomic implications of Agriculture 6.0 are multifaceted. On one hand, increased productivity, efficiency, and market access offer significant benefits. On the other hand, there are concerns about digital exclusion, particularly among small-scale farmers without access to training, financing, or infrastructure. Inclusive models such as cooperatives, subsidized tech programs, and community-based platforms can mitigate these risks and ensure that the benefits of digital transformation are equitably distributed.

From an environmental standpoint, Agriculture 6.0 aligns with sustainable development objectives. Precision nutrient application, automated irrigation, and controlled environment agriculture all contribute to reduced emissions, lower resource use, and minimal ecological disruption. The reduced dependency on chemical inputs and land use intensity also supports biodiversity conservation and climate change mitigation. These improvements are consistent with global policy frameworks, including the United Nations Sustainable Development Goals.

However, ethical concerns around data governance and corporate control are increasingly prominent. As farmers generate vast amounts of data using digital tools, questions arise about who owns and benefits from this information. Moreover, the consolidation of Agritech markets raises fears of monopolistic practices and increased dependency on a small number of technology providers. Regulatory frameworks that address transparency, data privacy, and equitable participation are essential to ensuring that technological progress does not exacerbate existing inequalities.

Infrastructure gaps remain a significant barrier to widespread adoption, particularly in developing countries. Reliable internet, electricity, and cloud services are prerequisites for many Agriculture 6.0 technologies. In regions where such infrastructure is lacking, phased or modular approaches starting with mobile phones or low-cost sensors can create entry points for broader digital integration. Government and donor-led investments in rural connectivity will be crucial in closing this digital divide.

Integrating indigenous knowledge systems with technological innovation can further enhance the effectiveness of Agriculture 6.0. In India, AI-based pest forecasting tools have been successfully combined with traditional knowledge through farmer cooperatives, resulting in more timely and culturally accepted responses to pest outbreaks. This hybrid approach fosters community engagement, builds trust, and increases the contextual relevance of digital solutions.

Financial investment remains a critical enabler for Agriculture 6.0. Countries with strong innovation ecosystems such as the Netherlands and Israel benefit from a combination of venture capital, public-sector incentives, and university-industry collaboration. In developing regions, blended finance mechanisms involving public, private, and philanthropic partners have shown potential in de-risking innovation and scaling up successful pilots. A supportive policy environment is equally important in incentivizing technology adoption and innovation.

Education and skills development are essential for long-term success. Agriculture 6.0 requires new competencies, including data interpretation, system maintenance, and digital tool management. Programs in Brazil and India that focus on farmer training and digital literacy have demonstrated substantial improvements in adoption and outcomes. Bridging the knowledge gap through tailored education will be key to inclusive agricultural transformation.

Gender dynamics must also be addressed to avoid reinforcing existing inequalities. In many contexts, women farmers face limited access to land, credit, and technology. However, targeted interventions such as mobile platforms designed with women users in mind have shown success in enhancing access to markets, financial services, and extension support. Mainstreaming gender considerations in Agriculture 6.0 initiatives can lead to more equitable and effective outcomes.

One of the strengths of Agriculture 6.0 is its scalability through modular implementation. Technologies can be adopted incrementally, allowing farmers to progress from basic advisory tools to more complex automation systems over time. This phased approach facilitates technology uptake in resource-constrained settings and accommodates a variety of operational scales, from smallholders to commercial enterprises.

Looking ahead, emerging technologies such as quantum computing, synthetic biology, and augmented reality are poised to further transform agriculture. Quantum models may enhance breeding efficiency and climate modeling, while synthetic biology could lead to climate-resilient crops with customized traits. As these innovations mature, they will extend the reach and impact of Agriculture 6.0, reinforcing its potential as a cornerstone of global food security.

In conclusion, Agriculture 6.0 represents a transformative moment in the history of food production. By integrating advanced technologies with sustainable practices, it offers a pathway to more efficient, resilient, and equitable food systems. Case studies from diverse regions highlight both the promise and the challenges of this paradigm. Ensuring that Agriculture 6.0 benefits all stakeholders will require collaborative governance, inclusive innovation, and strategic investment. As the world grapples with climate change, resource constraints, and population growth, this new model of agriculture offers not only productivity but also hope.

4.3 Challenges in Generative AI and Collaborative Robotics in Agriculture 6.0

The advent of Agriculture 6.0 signifies a paradigm shift in global food systems, integrating Generative Artificial Intelligence (GenAI) and Cobots to revolutionize how farming is conceptualized, executed, and optimized. This new wave of agricultural evolution builds on previous technological eras, progressing from traditional manual labor and mechanization to data-driven, AI-augmented precision farming. Agriculture 6.0 introduces an ecosystem where machines are not only autonomous but also capable of learning, adapting, and generating novel insights that optimize production and sustainability. However, despite its promising potential, this transformation raises several critical challenges chief among them being data ownership, equitable access, and job displacement. These issues must be addressed to ensure that the benefits of such technological innovation are shared equitably across different regions, demographics, and economic strata.

A major concern in the deployment of GenAI and Cobots in agriculture is the question of data ownership. Modern agricultural systems generate vast quantities of data through sensors, drones, robotics, and AI models. This data encompasses vital information such as soil composition, crop health, climate variations, and even individual farming practices. As collaborative robots collect this data and GenAI systems learn from it to produce recommendations or take autonomous action, the question arises: who truly owns this data? Often, technology providers and AI platforms retain control over this information via proprietary software and cloud-based licensing models, marginalizing farmers from their own data ecosystems. Furthermore, if AI-generated solutions are customized based on individual farm data, the associated intellectual property rights become contentious. This ambiguity poses serious implications for farmer autonomy and long-term sustainability. Establishing clear legal and ethical frameworks around data governance is thus crucial to prevent exploitation and ensure transparency.

Equitable access to these advanced technologies forms the second major challenge. While Agriculture 6.0 tools such as autonomous drones, robotic harvesters, and AI-driven decision support systems promise to enhance productivity, they are predominantly accessible to large-scale commercial farms due to high capital requirements. Smallholder and marginalized farmers, particularly in developing countries, face significant barriers to entry including financial limitations, digital illiteracy, and inadequate infrastructure. This digital divide threatens to exacerbate existing inequalities in the agricultural sector, potentially leaving millions of farmers behind in an era of unprecedented technological progress. Moreover, generative AI systems often lack localization features, such as support for regional languages and context-specific training data, further alienating rural communities. Addressing this challenge requires multi-level intervention through policy reform, targeted subsidies, cooperative technology-sharing models, and the development of inclusive, open-source AI systems designed for low-resource settings.

Another pressing issue arising from the integration of robotics and AI in agriculture is the risk of job displacement. Cobots are increasingly capable of automating essential farming tasks such as planting, weeding, and harvesting, while generative AI systems can handle managerial and decision-making roles including supply chain forecasting and climate adaptation strategies. This dual-layered automation threatens to disrupt rural labor markets, particularly in regions where agriculture constitutes the primary source of employment. Not only does this affect manual laborers, but it also undermines the livelihoods of agronomists, extension officers, and informal agricultural workers. The socio-economic impact could be profound, resulting in increased unemployment, urban migration, and the erosion of traditional farming knowledge passed down through generations. A proactive strategy is needed to address this challenge, including workforce re-skilling initiatives, the design of human-robot collaboration frameworks, and the implementation of tax policies or social welfare systems that support displaced workers.

Beyond these core challenges, additional ethical and environmental considerations merit attention. The design and deployment of AI models are often biased toward high-yield monoculture systems prevalent in Western contexts, which may be unsuitable or even harmful in diverse agricultural environments. This could lead to ecological imbalances and undermine local food sovereignty. Furthermore, the opacity of AI decision-making raises concerns about accountability and trust. Farmers may find themselves reliant on "black-box" systems whose recommendations they do not understand or cannot contest, further deepening the power asymmetry between technology developers and end-users.

In conclusion, while Agriculture 6.0 holds immense promise for enhancing food security, sustainability, and global resilience, the path forward must be navigated with caution. Challenges surrounding data ownership, equitable access, and job displacement are not merely technical or logistical concerns they are deeply human issues tied to dignity, fairness, and inclusion. To truly realize the "seeds of intelligence" planted by generative AI and collaborative robotics, the agricultural revolution must be guided by principles of ethical design, inclusive policymaking, and a commitment to leaving no farmer behind.

5. CONCLUSIONS

In conclusion, the integration of generative AI and collaborative robots (Cobots) marks a pivotal shift toward Agriculture 6.0, transforming traditional farming into a more adaptive, data-driven, and intelligent system. By examining global case studies, it becomes clear that these technologies are already reshaping agricultural practices, improving precision, sustainability, and productivity. However, as this transformation accelerates, it is equally important to address critical challenges such as data ownership, ensuring fair access to advanced tools, and mitigating potential job displacement. Balancing innovation with inclusivity and ethical considerations will be a key to the successful and responsible adoption of smart farming technologies.

6. FUTURE DIRECTIONS

Move away from agricultural tech towards open-source platforms. Establish global, open data protocols for seamless device and platform collaboration. Foster vendor-neutral ecosystems where diverse Cobots and AI systems interoperate. Prioritize affordable, modular tech solutions for smallholders in low-income regions. Develop solar-powered, offline-capable sensors and decision-support tools. Emphasize linguistic and cultural contextualization of AI training data and interfaces. Leverage quantum computing for advanced crop modeling and supply chain optimization. Utilize synthetic biology for super-resilient, climate-adapted crops. Implement edge-AI for real-time, low-latency decision-making, especially in connectivity-poor areas.

Integrate carbon tracking, biodiversity, and soil health into AI-driven management. Connect to carbon markets, green financing, and certification schemes. Develop generative AI to simulate ecological impacts and promote agro ecological diversity.

Define digital property rights for farmers and smallholders. Establish standards for AI transparency, auditability, and ethical compliance. Promote equitable data governance and prevent monopolistic corporate control.

Design Cobots and AI to augment not replace human labor. Innovate wearable robotics and human-in-the-loop systems. Invest in large-scale digital literacy and re-skilling programs for rural populations.

Develop continuous, gamified, AR/VR-based learning platforms. Tailor extension services to evolving digital agricultural tools. Support farmer communities with participatory technology development. Mobilize international bodies (FAO, World Bank, WTO) to harmonize Agri-AI policies. Foster public-private partnerships and regional innovation hubs. Ensure equitable technology access to developing countries. Link Agriculture 6.0 to One Health, climate science, urban development. Explore circular bio-economy and digital nutrition tracking. Pursue interdisciplinary research to maximize sustainability impact.

Data Sovereignty & Farmer Empowerment: Consider mechanisms like data co-operatives or block-chain-based farmer-controlled data marketplaces to strengthen farmers' negotiating power.

AI Explainability & Trust: Investigate user-friendly explainability interfaces so farmers understand and trust AI recommendations. Resilience beyond Productivity: Encourage research on system-level resilience metrics beyond yield to include social, economic, and ecological dimensions. Sustainability Incentives: Explore innovative incentive models such as pay-for-ecosystem-services using smart contracts. Monitoring & Evaluation: Develop impact assessment frameworks to track the social and environmental outcomes of deploying AI and robotics in agriculture.

REFERENCES

Abayaratne, S., Su, D., & Qiao, Y. (2024). "CropFinder: AI-based Detection and Tracking of Crops for Precision Agriculture," 2024 IEEE 33rd International Symposium on Industrial Electronics (ISIE), Ulsan, Korea, Republic of, pp. 1-6, DOI: 10.1109/ISIE54533.2024.10595716

Dhavale, C., Pawar, T., Singh, A., Pole, S., & Sabat, K. (2024). "Revolutionizing Farming: GAN-Enhanced Imaging, CNN Disease Detection, and LLM Farmer Assistant"," 2024 2nd International Conference on Computer, Communication and Control (IC4), Indore, India,, pp. 1-6, DOI: 10.1109/IC457434.2024.10486501

Du, B. et al., (2024). "YOLO-Based Semantic Communication With Generative AI-Aided Resource Allocation for Digital Twins Construction," in IEEE Internet of Things Journal, vol. 11, no. 5, pp. 7664-7678, 1, .DOI: 10.1109/JIOT.2023.3317629

Hou, G., Niu, R., Chen, H., Wang, Y., Zhu, Z., Zhang, J., Ma, Y., & Li, T. (2025, May 15). A Robotic AI Algorithm for Fusing Generative Large Models in Agriculture Internet of Things. *IEEE Internet of Things Journal, 12*(10), 13511–13524. Advance online publication. DOI: 10.1109/JIOT.2024.3516729

Jockusch, O., Hossain, M. Z., Imteaj, A., & Shahid, A. R. (2024). "Generative AI-based Land Cover Classification via Federated Learning CNNs: Sustainable Insights from UAV Imagery," *2024 IEEE Conference on Technologies for Sustainability (SusTech)*, Portland, OR, USA, , pp. 356-361, DOI: 10.1109/SusTech60925.2024.10553449

Kavipriya, J., & Vadivu, G. (2024). "Exploring Crop Yield Prediction with Remote Sensing Imagery and AI," *2024 International Conference on Advances in Computing, Communication and Applied Informatics (ACCAI)*, Chennai, India , pp. 1-5, DOI: 10.1109/ACCAI61061.2024.10601854

Klair, Y. S., Agrawal, K., & Kumar, A. (2024). "Impact of Generative AI in Diagnosing Diseases in Agriculture," 2024 2nd International Conference on Disruptive Technologies (ICDT), Greater Noida, India, pp. 870-875, DOI: 10.1109/ICDT61202.2024.10489759

Kolhalkar, N. R., Pandit, A. A., Kedar, S. A., & Yedukondalu, G. (2025). "Artificial Intelligence Algorithms for Robotic Harvesting of Agricultural Produce," 2025 1st International Conference on AIML-Applications for Engineering & Technology (ICAET), Pune, India, pp. 1-6, DOI: 10.1109/ICAET63349.2025.10932313

Kishor Kumar Reddy, C., Bhavya, T., Anisha, P. R., & Hanafiah, M. M. (2025). Text-Based Prediction and Classification Model of Stress, Anxiety and Depression Among Indians. In Mishra, B. K., Mallik, S., & Le, D.-N. (Eds.), *Smart Sensors for Industry 4.0.*, DOI: 10.1002/9781394214723.ch9

Kumar Reddy, C. K., Reddy, P. S., Pilly, A., & Doss, S. (2025). Transformative Effects of Smarter Chatbots. 333-350. https://doi.org/DOI: 10.1002/9781394234028.ch18

Liu, J., ... (2023). "Exploring the Integration of Digital Twin and Generative AI in Agriculture," 2023 15th International Conference on Intelligent Human-Machine Systems and Cybernetics (IHMSC), Hangzhou, China, pp. 223-228, DOI: 10.1109/IHMSC58761.2023.00059

Majumder, S., Khandelwal, Y., & S. K, (2024). "Computer vision and Generative AI for yield prediction in Digital Agriculture," 2024 2nd International Conference on Networking and Communications (ICNWC), Chennai, India, pp. 1-6, .DOI: 10.1109/ICNWC60771.2024.10537337

Movahed, A. B., Movahed, A. B., Aliahmadi, B., & Nozari, H. (2024). Green and Sustainable Supply Chain in Agriculture 6.0. In Oskounejad, M., & Nozari, H. (Eds.), *Advanced Businesses in Industry 6.0* (pp. 32–45). IGI Global Scientific Publishing., DOI: 10.4018/979-8-3693-3108-8.ch003

Neves, M., Casagrande, B., Cambaúva, V., de Oliveira Teixeira, G., & Toledo, P. (2023). Agriculture 6.0: A New Proposal for the Future of Agribusiness. *Revista de Gestão Social e Ambiental.*, *17*(9), e04004. DOI: 10.24857/rgsa.v17n9-021

Neves, M. F., Casagrande, B. P., Cambaúva, V., de Oliveira Teixeira, G., & Toledo, P. J. F. (2023). Agriculture 6.0: A new proposal for the future of agribusiness. Rev. De Gestão Soc. *Engenharia Ambiental*, *17*, e04004.

Padmaja, P., Sophia, I. J., Hari, H. S., Kumar, S. S., & Somu, K.. (2021). Distribute the Message over the Network Using another Frequency and Timing Technique to Circumvent the Jammers. *J Nucl Ene Sci Power Generat Techno*, *10*, 9.

Patil, Y. H., Patil, R. Y., & Ozer, O. (2024). Next-Gen Agriculture 6.0: A Smart Solution Pioneering the Future of Farming. In Reddy, C. K. K., Doss, S., Khan, S. B., & Alqhatani, A. (Eds.), *Evolution and Advances in Computing Technologies for Industry 6.0* (pp. 167–187). CRC Press. DOI: 10.1201/9781003503934-7

Pratapagiri, S. et al. (2021). "Early Detection of Plant Leaf Disease Using Convolutional Neural Networks," 2021 3rd Interna-tional Conference on Electronics Representation and Algorithm (ICERA), pp. 77-82, doi: DOI: 9.1109/ICERA53111.2021.9538659

Prema Kumar, M. et al. (2024). "Enhanced Plant Disease Detection in Precision Agriculture Using Autoencoder and Transfer Learning," 2024 Third International Conference on Artificial Intelligence, Computational Electronics and Communication System (AICECS), MANIPAL, India, pp. 1-6, DOI: 10.1109/AICECS63354.2024.10957233

Rani, R., Sahoo, J., Bellamkonda, S., Kumar, S., & Pippal, S. K. (2023). Role of Artificial Intelligence in Agriculture: An Analysis and Advancements With Focus on Plant Diseases. *IEEE Access : Practical Innovations, Open Solutions*, *11*, 137999–138019. DOI: 10.1109/ACCESS.2023.3339375

Rizvi, S. M. H., Naseer, A., Rehman, S. U., Akram, S., & Gruhn, V. (2024). Revolutionizing Agriculture: Machine and Deep Learning Solutions for Enhanced Crop Quality and Weed Control. *IEEE Access : Practical Innovations, Open Solutions*, *12*, 11865–11878. DOI: 10.1109/ACCESS.2024.3355017

Shaikh, F. K., Memon, M. A., Mahoto, N. A., Zeadally, S., & Nebhen, J. (2022). "Artificial Intelligence Best Practices in Smart Agriculture," in IEEE Micro, vol. 42, no. 1, pp. 17-24, 1 Jan.-Feb, .DOI: 10.1109/MM.2021.3121279

Shailaja, K., Srinivasulu, B., Thirupathi, L., Gangula, R., Boya, T. R., & Polem, V. (2023). An Intelligent Deep Feature Based Intrusion Detection System for Network Applications. *Wireless Personal Communications*, *129*(1), 345–370. DOI: 10.1007/s11277-022-10100-w

Sonmez, D., & Cetin, A. (2024). "An End-to-End Deployment Workflow for AI Enabled Agriculture Applications at the Edge," 2024 6th International Conference on Computing and Informatics (ICCI), New Cairo - Cairo, Egypt, pp. 506-511, DOI: 10.1109/ICCI61671.2024.10485167

Thamaraikannan, N., & Manju, S. (2023). "Classification of Satellite Agriculture Imagery to Identify Crop Type Using Deep Learning Techniques - A Review," *2023 International Conference on Computer Communication and Informatics (ICCCI)*, Coimbatore, India, pp. 1-5, DOI: 10.1109/ICCCI56745.2023.10128228

Venkata Reddy, P. S., Nandini Prasad, K. S., & Puttamadappa, C. (2024). "Conversational AI bot based on IoT Knowledgebase for Smart Agriculture," *2024 International Conference on Knowledge Engineering and Communication Systems (ICKECS)*, Chikkaballapur, India, pp. 1-7, DOI: 10.1109/ICKECS61492.2024.10616582

Zhao, K., Nguyen, M., & Yan, W. (2024). "Evaluating Accuracy and Efficiency of Fruit Image Generation Using Generative AI Diffusion Models for Agricultural Robotics," 2024 39th International Conference on Image and Vision Computing New Zealand (IVCNZ), Christchurch, New Zealand, pp. 1-6, DOI: 10.1109/IVCNZ64857.2024.10794462

Chapter 2

A Study on Integrating Generative AI and Collaborative Robots to Provide Smart Agricultural Support

Sowmya Pasham
Botho University, Botswana

Jagadeshwari Puttanapura
https://orcid.org/0009-0008-9607-7438
University of South Alabama, USA

C. Kishor Kumar Reddy
Stanley College of Engineering and Technology for Women, India

Shikha Khullar
https://orcid.org/0009-0000-5598-3145
Poornima University, India

Jothi Paranthaman
Botho University, Botswana

ABSTRACT

Advancements in smart farming are being brought on as a result of integrating Generative AI with collaborative robots or cobots. Generative AI models- GANs-, VAEs, transformers- help cobots to be more independent, flexible, and able to

DOI: 10.4018/979-8-3373-4207-8.ch002

take smart decisions on the field. Armed with data from sensors, drones, and IoT devices, AI assists cobots in predicting outcomes, choosing the right actions, and responding within a small window of time to the changing conditions of weather, pests, soil health. For this, it is essential to design a well-established framework that connects both the hardware and software of AI and robots. This practically could be training AI models, managing data flow, or ensuring real-time performance. Another practical application includes crop health monitoring, harvest estimation, and precise spraying, all in favor of saving time, reducing wastage, and maximizing output. On the other hand: limited compute power right there in the fields; the interpretability of AI decision-making; verification of data representativeness and quality constitute a few issues.

1. INTRODUCTION

1.1 Background and Motivation

To feed a growing population, counter labor shortages, and resort to an eco-friendlier way of farming, the agriculture industry is undergoing a great transformation. Traditional agricultural methods can hardly meet the demands, with climate change threatening productivity; second, resource scarcity; and third, soil degradation (FAO, 2017). Thus, with the situation in hand, the use of such advanced technologies as Artificial Intelligence (AI) and robotics has been cited by many as a primary way of improving agriculture efficiency and sustainability. Generative AI is a term used for models that generate new data or simulations based on an input supplied to them; hence, they were highly developed into image synthesis, scenario simulation, and decision-making (Goodfellow et al., 2014). In the agricultural domain, generative AI works with cobots to simulate some situations and define the best solution for planting, harvesting, or pest control, in a world that is more uncertain and data intensive. The collaborative robots or cobots are created to operate safely alongside humans while adapting to complex and unstructured environments like farms. Unlike traditional industrial robots, cobots can carry out more delicate tasks such as picking fruits, monitoring plant health, or applying fertilizers with great precision (Bechar & Vigneault, 2016). Generative AI, when layered on top of cobots, takes them beyond automation and into the realm of prediction, adaptation, and efficiency in the fields. The motivation behind combining generative AI with cobots arises from the prospect of building intelligent farming systems that minimize resource wastage, maximize yields, and support farmers with real-time insights and interventions (Shamshiri et al., 2018).

Also, the rise of smart agriculture and IoT has been offering loads of agricultural data, hence presenting the best scenario for generative models to learn, adapt, and innovate (Kamilaris et al., 2018). Such interplay sets the stage for a future where farming is data-driven, precise, yet sustainable.

1.2 Scope and Objectives

The merging of Generative Artificial Intelligence (AI) with Collaborative Robots (cobots) is defining a new incarnation of farming, injecting intelligent, adaptive, and autonomous abilities into the traditional farming systems. Agriculture is subjected to all sorts of pressure because of expanding population, climate variations, and labor issues; this technology thus provides an apt and impactful solution. Generative AI models, like GANs and VAEs, inject advanced cognitive capabilities into the cobots, i.e., the cobot can learn from big datasets, simulate multiple agricultural scenarios, and, at the end of the day, make real-time, informed, data-driven decisions (Pallottino, Federico & Violino, et al., 2025). Such a convergence represents a evolution towards developing very sophisticated AI-cobot systems that can undertake a wide variety of activities ranging from monitoring crop health to performing complicated operations in the field, while constantly adjusting to the ever-changing environmental context (Etheredge, 2025).Using data from drones, Internet of Things (IoT) sensors, and satellite images, these smart systems can manage agricultural inputs better than ever before - targeting inputs like water, fertilizers, and pesticides improving productivity and significantly contributing to sustainability of the environment (Li.H, et al., 2021). Precision farming is a crucial area to look for use cases in, as AI-powered cobots apply agricultural chemicals only directly to specific areas of a field, minimizing waste and resource consumption while optimizing crop utilization (Ghumman et al., 2025). In addition, machine vision and machine learning allow cobots to provide early warnings of plant stress, diseases, or pest violations giving farmers invaluable information that make remedial actions timely and economical.

The implementation will also include automating critical agricultural procedures, such as seeding, harvesting, and soil quality measurements, thus reducing labor intensity and costs (Pietrantoni et al., 2024). This research will pursue multiple objectives: to comprehend the revolutionary potential of generative AI for enhancing cobots cognitive and mechanical capabilities; to develop a scalable real-time integration between rogue AI algorithms and robotic hardware and software; to examine sustainability and performance parameters of this integration through case studies and simulations.

1.3 Significance in Smart Agriculture

Through the integration of innovative technologies such as the Internet of Things (IoT), artificial intelligence (AI), and data analytics, smart agriculture, or agriculture 4.0, represents an evolutional way of conducting farming (Majeed Y et al., 2024). Through the solution of immediate problems such as scarcity of resources, environmental degradation, and increased demand for food production, such an integration enhances the productivity, sustainability, and efficiency of contemporary farming practices. Resource optimization is a primary advantage of smart agriculture. By utilizing the data obtained from IoT devices and sensors monitoring environmental data, such as moisture and nutrients, almost in real-time, the farmer has a basis to make rational decisions about pest control, fertilization, and irrigation wisdom (Akkem et al., 2023). As a result, overuse is eliminated, expenses are reduced, and the environmental impact is lessened.

Also, AI analytics enable predictive modelling of crop yields and disease outbreaks and support proactive interventions to protect the crop and preserve food security. For instance, AI can consider enormous datasets and help forecast weather and suggest planting days, to improve farm resilience to climate variation (Muhammed D et al., 2024). Smart agriculture also contributes to increased labor efficiency through automation. Technologies such as autonomous tractors and robotic harvesters reduces reliance on human labor and not only aides in scarcity of available labor but also improved efficiency in operations (Dhanaraju, M. & Chenniappan, P et al., 2022). Smart farming also contributes to sustainability by reducing greenhouse gas emissions and conserving biodiversity. Effective use of added inputs avoids allowing chemicals to runoff and avoiding harm to water bodies and soil (Rane et al., 2024).

Table 1 shows that conventional agriculture is manual and intuitive in its farming practices, while smart agriculture leverages technology to enhance precision and efficiency. With regards to smart agricultural practices, yield predictions, resource utilization, and real-time data collection all augment organic and augmented practices.

Table 1. Comparison of traditional vs smart agriculture practices

Feature	Traditional Agriculture	Smart Agriculture
Labor Dependence	High	Low (Automation)
Resource Use Efficiency	Low	High
Decision Making	Manual	AI-Assisted
Scalability	Limited	Scalable
Technology Use	Minimal	Advanced Sensors, AI, Robotics

Figure 1. Traditional vs smart agriculture workflow

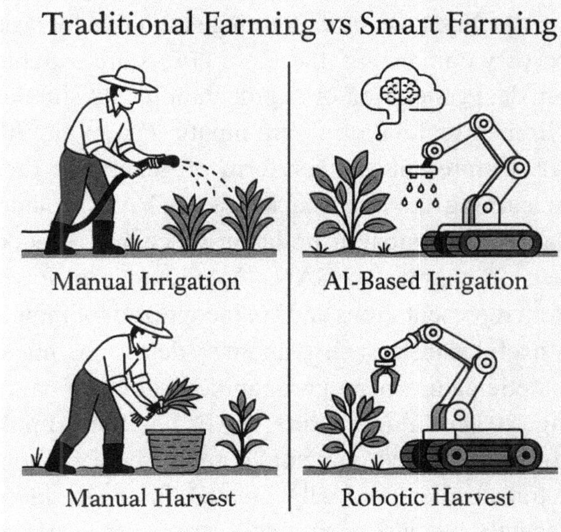

Figure 1 outlines the conventional workflow relies on manual monitoring and standardized procedures, frequently lacking accuracy. Conversely, intelligent agriculture employs sensors, artificial intelligence, and data analysis to automate and enhance farming practices.

2. UNDERSTANDING GENERATIVE AI

2.1 Overview of Generative Models (GANs, VAEs, Transformers)

Generative models are a strong form of machine learning algorithm, used to learn the underlying distribution of input data, and generate new samples of data based on that learned shape. That includes Generative Adversarial Networks (GANs), Variational Autoencoders (VAEs), and transform based models. Each generative model has different features, architectures, strengths and weaknesses for applications.

Generative Adversarial Networks (GANs) are comprised of two neural networks in which a generator and discriminator are trained at the same time in competition to each other. In GANs, the generator creates synthetic data (e.g., images, text) and the discriminator determines if the data was truly real or generated. The goal of the generator is to generate synthetic outputs such that a discriminator cannot distinguish between the synthetic outputs and real outputs, or to confuse (or "fool") the discriminator into thinking that something real is actually synthetic (or vice-versa). In deceiving the user, quality adversarial training is suitable for producing synthetic outputs, and can be very convincing, therefore GANs are suitable for tasks, such as image generation, data generation or augmentation, and simulating agricultural systems that have limited real data training inputs (Pan et al., 2022). Variational Autoencoders (VAEs) represent another form of generative model that merges concepts from deep learning and Bayesian inference. VAEs include an encoder that transforms input data into a probabilistic latent space and a decoder that rebuilds data from this space. In contrast to GANs, VAEs emphasize acquiring smooth, comprehensible latent representations and are recognized for their training stability. They are especially useful in tasks such as anomaly detection, image reconstruction, and generating synthetic data, where a probabilistic grasp of data is advantageous (Kingma & Welling, 2022). In agriculture, VAEs can help simulate crop growth trends, identify anomalies in environmental data, or represent yield distributions. Models based on transformers, initially created for tasks in natural language processing, have recently become adaptable instruments in generative modeling. These models depend on self-attention mechanisms that enable them to grasp long-range dependencies in data, rendering them suitable for generating sequential and structured data. Recent transformer variants such as GPT (Generative Pre-trained Transformer) have been modified for multimodal applications, including the generation of images and time-series data. In smart farming, transformers can be utilized to create predictive weather models, simulate sequential agricultural data, or even produce realistic situations for training autonomous robots (Vaswani et al., 2023). Together, these generative models offer robust resources for producing synthetic data, improving predictive modeling, and facilitating autonomous decision-making in areas such as precision agriculture. Their capability to understand and replicate intricate, high-dimensional data sets renders them essential for contemporary AI-based agricultural uses.

Table 2 talks about how GANs purely generate high-quality synthetic data, VAEs identify efficient configurations for representation learning to reconstruct data inputs, and Transformers are suitable for the sequential modelling of their input. All AI models tested have application differences based on their model formulation.

Table 2. Comparison of GANs, VAEs, and transformers

Model Type	Strengths	Weaknesses	Use in Agriculture
GANs	High-quality generation	Difficult to train	Synthetic data for plant diseases
VAEs	Structured latent space	Lower output quality	Anomaly detection
Transformers	Long-sequence processing	High computation	Predictive analytics

Figure 2. Architecture of a generative AI model

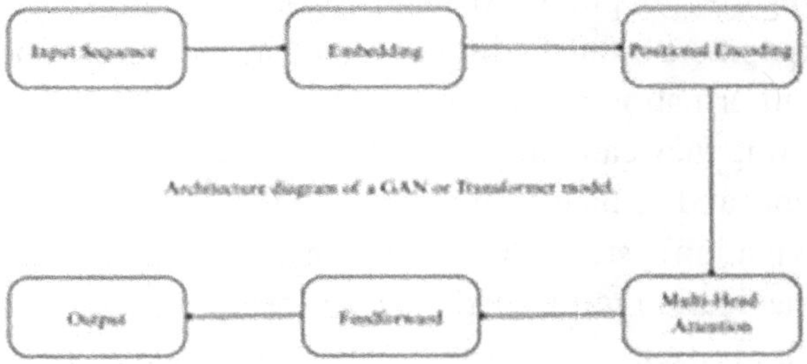

Figure 2 outlines that This architecture generally features an encoder, decoder, or a generator-discriminator combination, based on the type of model. It analyzes input data to recognize patterns and produce new, authentic outputs derived from acquired representations.

2.2 Applications in Automation

Generative models enable automation to lead the next wave of technological change across various sectors, particularly agriculture, manufacturing, and logistics. Based on the work with Generative Adversarial Networks (GAN), Variational Auto-encoders (VAEs), and Transformer-based approaches, machines can not only follow programmed commands but can learn, create, predict, and adapt to complex scenarios intelligently. In agriculture, generative models are transforming smart farming by enabling collaborative robots (cobots) and drones to mimic crop growth in different environmental conditions, forecast pest invasions, and recommend ideal harvesting times, thus automating tasks that once demanded manual monitoring and action. As an example, GANs can create sample images of satellites or fields to prepare a cobot to, at both practical and research scales, recognize crops or difference weeds from crops - even when labelled data is sparse (Lu et al., 2022).In a similar manner, VAEs can be applied to analyze and reconstruct streams of data from sensors to rapidly detect anomalies in irrigation systems or to monitor the operational health of some machinery (Choi et al., 2021).

In manufacturing, generative models can improve automation in planning and design process, quality assurance and production workflows. They can create multiple versions of prototypes, all predefined ranges of parameters, and tell the engineers: "Go ahead and explore these virtual models before we physically fabricate anything, which will save us time in the design process and production costs". Generative models based on transformers, which in these cases are working with sequential data, are also being used for predictive maintenance: predicting break down calls and failure rates of equipment by analyzing continuously measured time-series data from sensors into the information environment (Qi et al, 2023). In logistics and supply chain after modeling they can forecast demand and optimize routes, using realistic scenarios for supply and demands, and they enable dynamic inventory management. In automated navigation systems, by simulating terrain conditions, possible hazards and optimizing navigation routes under scenarios for agricultural vehicles, drones or delivery robots. In general, the use of generative models in automation adds a factor of precision, efficiency and scale, while also providing a degree of cognitive intelligence where machines can work together autonomously in a more agile and sustainable way in constantly evolving surroundings.

3. COLLABORATIVE ROBOTS IN AGRICULTURE

3.1 Definition and Features

Collaborative robots, cobots, exist as robots designed to function alongside humans within their shared workspace. The advanced safety and sensing technology along with control systems enable cobots to operate directly and safely alongside human workers while traditional industrial robotic systems require fixed containment areas. The agricultural sector will experience a transformative change through cobots which enable precision farming techniques while human operators complete time-sensitive and labor-intensive and error-prone tasks. The technical features of agricultural cobots include sensors and cameras along with artificial intelligence vision systems and robotic arm tools that allow the machines to perform planting, pruning, harvesting, soil analysis, and pesticide spraying operations with minimal human guidance (Horabik, J et al., 2016). Another feature of agricultural cobots is adaptive learning through machine learning algorithms. Cobots can observe changes in the environment (e.g., light levels, health of the plant, soil moisture) and subsequently adapt their actions accordingly whereas, traditionally, adaptive actions require a time gap for human intervention. They are lightweight and designed so a farmer can program and operate them without much technical expertise (Duckett et al., 2018). Their modular and mobile design allows them to be deployed on numerous terrains and with various crop types-from greenhouses to open fields. Effectively, cobots serve to augment human labor rather than replace it. They assist in promoting productivity and reduce the work intensity of agricultural workers, making agricultural activities much more efficient, precise, and less immersing labor-intensive, which becomes crucial in areas suffering from agricultural labor shortages. Nonetheless, the technological evolution can further penetrate sustainability by making cobots intelligent, autonomous, and IoT-oriented (Liakos et al., 2018).

Table 3 talks about how Agricultural cobots are intended to reduce human labor in tasks such as harvesting, monitoring and spraying, with high levels of precision and safety. Agricultural cobots have flexibility and adaptability, useful for humans and robots working collaboratively, with an AI integrated to operate optimally.

Table 3. Key features of agricultural cobots

Feature	Description	Benefit
Safety Sensors	Detect human presence	Prevents accidents
Lightweight Arms	Easy to mount	Mobility
Precision Control	Accurate movement	Better task execution
Programmability	Custom tasks	Flexibility

3.2 Role in Agricultural Tasks

Collaborative robots empower the modern agricultural enterprise by automating many vital farming activities with remarkable precision, flexibility, and efficiency. Now, being increasingly used in open-field and controlled-environment agriculture, they solve problems relating to labor shortages, climate changes, and growing demands for productivity. An apt example of a highly mechanized operation requiring cobots would include seed sowing, transplanting, pruning of crops, harvesting, and sorting. Afterward, equipped with advanced sensor technology, computer vision, and AI algorithms, cobots can detect the ripeness of fruits, assess plant health, and perform selective harvesting with minimal damage, and in many cases, they have selected operations in which they excel beyond human capabilities (Baeten et al., 2008). Besides harvesting, cobots assist with crop and soil monitoring. Equipped with multispectral and hyperspectral imaging capacities, they can detect early signs of nutrient deficiency, water stress, pest incidences, and diseases-thus enabling quick remedial actions and reduced input costs (Kamilaris & Prenafeta-Boldú, 2018). These robots could be integrated with GPS and IoT systems to acquire data along with location information so such data could be employed for location-based crop treatment. Within greenhouses, cobots perform automated environmental monitoring activities with adjustments to lighting, humidity, and temperature to sustain optimum plant growing conditions. Cobot-assisted spraying and fertilization are precision applications with the definite goal of greatly averting agrochemical overuse (Sørensen et al., 2010).

The information that cobots can collect and analyze in real time has benefits at the farm management level by providing support to farmers in actionable decision-making in relation to crop growth, determination of performance, and optimization in use of production inputs. The rationalization of operational efficiency also enables the development of predictions on the next onward farming cycle to ultimately inform operations. Cobots do not tire, and work on a continuous basis, which is particularly useful when the need for agricultural practices such as harvesting or planting a crop only occurs in specific periods. This not only means that intelligent automation supports better productivity but also reduces the risk of human error and negative environmental effects as well as operational costs. As technology advances, it is envisaged that the application of cobots will only grow, and one day, they will likely become a considerable partner for sustainable and smart agriculture.

Figure 3. Collaborative workflow between human and cobot in agriculture

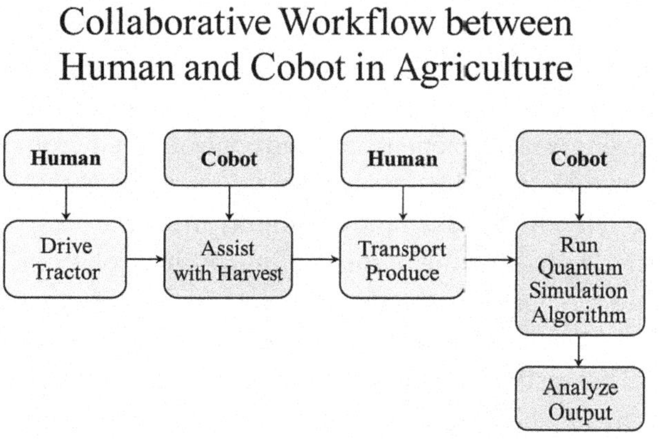

Figure 3 gives a basic idea of how humans handle planning and making decisions, whereas cobots carry out repetitive or precision tasks such as harvesting or spraying. This collaboration boosts productivity, lessens labor burden, and guarantees effective farm administration.

3.3 Human-Robot Collaboration Benefits

Collaborative workflows between humans and robots in agriculture offer a synergistic method that gains the benefits of human brains and robotic accuracy, allow for greater productivity, safety and sustainability in agriculture. An important opportunity with this conceptual partnership is addressing labor shortages through the automation of repetitive, physically demanding and/or time-sensitive tasks, freeing human workers to engage in strategic thinking and monitoring roles. Cobots (collaborative robots) are designed to work safely alongside people, with sensors and smart control systems to prevent collisions and allows for performing tasks collaboratively (Baxter et al, 2018). The immediacy of this interaction encourages flexible workflows, that rely on human talent to increase robotic productivity and more agile and efficient operations. In addition, human-robot coordination is another facet of human-robot collaboration that also improves the adaptability and personalizatio of agriculture spaces. For instance, although cobots scale up discarding or spraying, people are able to respond to complex dynamic environments that require judgement such as with very fragile plants or circumstances the farmer did not expect (Yerebakan, Mustafa & Hu, Boyi 2024). This division of action maximizes effective productivity in agricultural work but also minimizes physical fatigue and risk of injury by limiting exposure to inclement weather, repetitive tasks, or toxic chemicals.

Furthermore, the combination of robotics with human intuition and cognitive data evaluation improves decision-making and learning in the space of agriculture. Through advancements in artificial intelligence and machine learning, cobots can now learn from human feedback and lessons learned from real experiences, which can make their algorithms for assessing crop health, managing pests, and predicting yields more robust. Then paired with farmer knowledge, leads to a co-adaptive context for both humans and robots to share in great learning and valuable sustainable agriculture practices (Bac et al., 2014). Overall, the human-robot collaborative relationship enables a balance between productivity enhancement with automated farming while still allowing for human monitoring and input concerning farm safety and conservation.

4. AGRICULTURAL NEEDS DRIVING AI-COBOT INTEGRATION

4.1 Challenges in Modern Farming

Contemporary agriculture is now at a critical point characterized by a variety of challenges that could undermine food production, escalate climate change, and ultimately affect the economic feasibility of their practices. One of the foremost challenges facing modern agriculture is that of climate change, which involves uncertainty with changing weather, warmer temperatures, and changing climate patterns, for example, droughts vs. floods, which all affect agricultural yields (Lesk et al., 2016).Human responses must deal with climate change, but utilization of technology is disparate at best, especially in developing areas where smallholder farmers may face technological gaps due to infrastructure, knowledge, and lack of fund necessary to adopt new climate-resilient systems, leaving farmers susceptible to potential harvest failures and economic distress. Concurrently, soil degradation from over-farming, chemicals, and erosion is associated with decreasing the quality of arable land which limits future production (Lal, 2011). Continuous cropping and improper nutrient management are associated with poor soils, which lead to decreased soil fertility, declining biodiversity in soils, and lower agricultural outputs .

Additionally, many regions face significant challenges due to a lack of workers, especially as younger generations migrate from agriculture to urban areas in search of employment (Lowder et al., 2021). In addition to reducing the number of workers available for physically demanding agricultural jobs, this demographic shift also reduces the amount of inherited generational and physical knowledge. This is happening in the context of increasing reliance on mechanization and automation which increases efficiencies but requires significant capital and technical skills often not available to smallholder farmers, and risks widening the gap between actors based on discounting the resource-poor farmer thus reinforcing disparities in rural economies. Water scarcities - often exacerbated through irregular climate and insufficient irrigation methods - present one prominent challenge to agricultural systems requiring continuous water supply (Jägermeyr et al., 2015). In many instances, agricultural users assume competing demands for valuable water resources will restrict their access to irrigation at critical confluences. These intersecting problems require comprehensive and innovative solutions that consider the complex balance between productivity, sustainability, and equity in the global food system. For farming to remain viable and resilient in the 21st century, integrated policies, investment in Agri-Tech, and support for climate-smart agricultural practices are critical components.

Table 4 emphasizes about AI-powered cobots can help address issues in agriculture, such as declines in the labor supply, crop diseases, resource wastage, and other challenges. They add value to crop production in a way that lowers human errors, minimizes physical stress and maximizes productivity in terms of resource allocation and labor shortages while enabling the sustainability of farming.

Table 4. Modern farming challenges and AI-Cobot solutions

Challenge	Description	AI-Cobot Solution
Labor Shortage	Aging workforce	Automation via cobots
Climate Uncertainty	Unpredictable weather	Data-driven planning
Pest/Disease Detection	Late detection	Vision-based monitoring
Programmability	Custom tasks	Flexibility

4.2 Need for Intelligent Automation

The increasing sophistication of agricultural systems, labor shortages, climate variability, and increasing societal demands for productivity are driving change to intelligent automation in agriculture. Intelligent automation is a reconnecting of advanced technologies, including artificial intelligence (AI), robotics, machine learning, and the Internet of Things (IoT), to create systems that can perceive, learn, and adapt while operating in real-time. Intelligent automation includes the capability to not only execute repetitive tasks, but to engage in predictive decision making, operate autonomously, and continually optimize numerous agricultural processes (X Jin et al., 2017). For example, we now have tractors that are either autonomous or connected to drones that, through AI-powered vision, analyze crop conditions, optimize pesticide application, detect anomalies, and the like - drastically reducing the workload and environmental impact (Milioto et al., 2018). As the labor pool dwindles and rural areas continue to age behind urban centers, intelligent automation will be critical to filling this labor gap even while protecting precision and productivity (Oliveira et al., 2023). These technologies are scalable and can be capable of being employed on farms of all shapes and sizes - not just those of industrial scales.

In addition to coordinating information flows, intelligent automation can facilitate sustainable farming practices by supporting precision agriculture; to note, precision agriculture utilizes measured inputs (e.g. water, fertilizers, chemicals) only when, and where, required (Chlingaryan, A et al., 2018). We can understand how precision agriculture would decrease waste and cost. It also assists in mitigating the environmental costs of farming. Second, the ability to collect and utilize and analyze data in real-time allows farmers to make better decisions in compliance and improve mitigation plans quickly in the context of pest's outbursts or climate change effects. Lastly, the potential of intelligent automation systems to leverage past data and adapt to changing conditions permits farmers to create a dynamic knowledge framework. The significance of these coordinating capabilities will increase as global use of food expands aggressively in coming decades. In the face of uncertainties, intelligent automation can be a scalable and sustainable pathway to strengthen farm resilience and food security.

5. SYSTEM DESIGN FOR INTEGRATION

5.1 Data Collection and Processing

Data collection and processing are fundamental to modern-day agriculture, making precision farming, smart automation, and strategic decision-making possible.

Thanks to a boom in sensors, drones, satellite imagery, and other IoT devices, farmers can continually (high frequency) collect high-resolution data across a multitude of attributes (soil moisture, temperature, humidity, crop condition, pest activity, etc.) At some update interval, data can be geolocated/time-stamped, allowing farmers to fine-tune field conditions, adding granularity in place of continuous field inspections and subjective judgement (Wolfert et al., 2017). With such a large quantity of new, useful data across many different scales of space and time, targeted remediation and resource management alternatives can be provided. The data is subsequently analyzed using advanced analytics, machine learning models, and cloud computing tasks to extract valuable insights from it. Employing these tools for automating the assessment of extensive datasets not only creates new paths for analysis but also presents unclear relationships that only a human can decipher. For example, data from drone images can be analyzed using deep learning algorithms that accurately detect crop species or diseases enabling farmers to understand and act quickly (Tsouros et al., 2019).

Moreover, the accumulation of data from different sources— such as satellite images, weather forecasts and terrestrial sensors - will enhance the forecasting capabilities of smart agriculture systems. This data fusion method will offer insights into the dynamics of the farm and supports the development of dynamic models critical to proactive farm management. For example, predictive models can forecast pest invasions based on details such as weather patterns, recommend optimal planting dates for better yield, etc. Such information gives farmers the opportunity to make nuanced decisions, alter irrigation schedules, develop and implement fertilizer schedules, opt to manage pest, and reduce resource waste while improving production results (Talaviya et al., 2020). However, there are still existing barriers related to data format standardization, data privacy, and the digital skill training required to help farmers make important sense of predictive analytics (to name a few). These barriers need to close in order to increase the functionality and access of inconsistent advances to all sizes of agriculture. As agriculture will continue to be influenced by data, making the ability to access, manage data rapidly and apply it timely will be increasingly important to being competitively sustainable and resilient within agriculture.

Table 5 shows data relevant to agriculture comes from different sources such as satellites, sensors, cameras, drones, and IoT products and is generally comprised of soil data, climate data, or growth data relating to plant health. AI models collect either structured or unstructured data to generally inform their decisions.

Table 5. Data types and their sources

Data Type	Source	Use
Soil Moisture	Sensors	Irrigation control
Plant Health Images	Cameras	Disease detection
Weather Data	IoT Devices	Scheduling tasks
Crop Yield History	Databases	Yield prediction

5.2 AI Model Deployment on Cobots

The deployment and integration of artificial intelligence (AI) models on collaborative robots (cobots) represents a new development for smart farming, as it allows cobots to work with greater independence, precision, and flexibility. AI allows cobots to make real-time decisions based on the analysis of complex datasets, received from sensors, cameras, or environmental monitoring systems - effectively using AI in the moment and allowing cobots to operate in mode of high autonomy without repetitive human interaction. As an example, deep learning models allow cobots to identify crop classifications, evaluate plant health, recognize weed species, and identify the presence of pests or diseases (Patel et al., 2022) with accuracy and speed. This enables cobots to execute targeted actions with such as selective harvesting, precision spraying, and adaptive pruning behavior enabling increased efficiency and adoption of sustainable practices. AI integrated cobots also provide the benefit of undertaking tasks consistently well and continuously; this is particularly useful during busy times of the growing season, when demands for labor will peak and shift daily.

Edge computing plays a critical role in deploying AI models on cobots by allowing computations to be performed locally on the robot itself, rather than relying entirely on cloud-based systems.This reduces latency, enhances responsiveness, and increases operational reliability—especially in remote agricultural locations that may not have consistent internet (Albanese et al., 2021). Data from the field can be processed on the cobots in real time so they can react to field changes instantly, with changing weather patterns or pest activity, for example. Furthermore, AI models on cobots are normally updated continuously by a machine learning pipeline, where earlier missions frame the feedback to retrain and adjust the AI modelmaking to better decisions over time. The capacity to learn and adapt from earlier experience improves the performance of cobots as they are familiar with more tasks and varieties of conditions. It must also be noted that interfacing AI to cobots relies on the development of effective control architectures and interface layouts that allow farmers to monitor, readjust, and assign robot behavior based upon their judgement of the real-world conditions. Cobots and intuitive interface designs must provide data visualization tools and decision-making insights that allow farmers to keep a hand on automation; experience-based, responsive feedback of a cobot allows active engagement in the subjective interpretation of 'actionable insights'.For example, using reinforcement learning and transfer learning through robots will enable cobots to use knowledge of tasks related to different crop types, farming systems, or agro-ecosystems and environments while minimizing reconfiguration (Li et al., 2023). These adaptive learning methods provisioned with onboard AI, therefore allow for cobots to be adaptive to performing tasks efficiently regardless of layers of new or changing operational realities. This adaptable intelligence, in a scalable way, allows for a signification decrease in human workload, optimization of resource use, and ultimately increase resilience of agricultural systems so that AI-enabled cobots will take on a critical role in the future of farming systems.

Figure 4. Block diagram: AI model, sensors, cobot interface, and decision flow

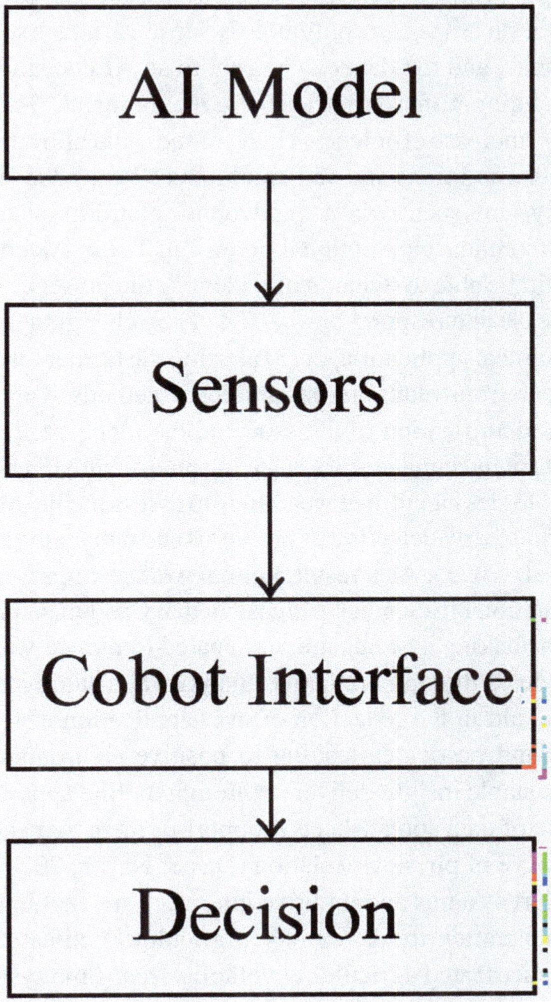

Figure 4 demonstrates how sensors gather real-time information, which the AI model processes to produce actionable insights. These findings are subsequently conveyed via the cobot interface to execute accurate farming operations.

5.3 Real-Time Decision Making

Real-time decision making in agriculture is the collection and immediate evaluation of data so that suitable changes can be made in the moment while executing farming tasks. With the help of the Internet of Things (IoT) devices, cloud computing and artificial intelligence (AI), farmers benefit from continuous knowledge of the fields and can deploy immediate action when necessary. For example, IoT sensors placed in the soil or on crops can continuously sense variables such as temperature, moisture or nutrients, and the data can be sent to an AI decision system to initiate an action like providing water or nutrients "in the moment". This will increase the health of the crops, increase efficiencies in resources, and allow the farmers to adapt to variable weather conditions and soil conditions (Wang, 2024).

Autonomous systems such as drones and robotic platforms also leverage real-time decision-making to enhance operational precision. These systems, enabled by AI-powered vision will be able to dynamically identify the presence of pests, weeds, or other signs of disease and respond on-the-fly: (1) by changing their flight path, (2) changing their chemical application, or (3) alerting the farmer for follow-up actions. These systems depart from traditional observational methods by effectively shortening the time between identification of an issue and taking corrective action, which is vital in minimizing plant damage and improving plant health. Furthermore, livestock monitoring technologies can utilize wearables to assess the health status of animals in real-time, enabling early detection of problems and enhancing the management of the herd (Kant et al., 2023). As a result, animal welfare improves, veterinary costs decline, and production efficiencies increase in dairy and meat production. Finally, real-time decision-making systems are anticipated to reduce waste, limit resource use and improve yields through predictive analytics and automated feedback loops. These systems can aid in the reduction of over-application of resources including water, fertilizers, and pesticides leading to positive environmental implications. The real-time actionable insight delivered through mobile apps and dashboards (as opposed to the use of data spreadsheets) support farmers to make timely informed decisions irrespective of physical isolation (Das & Nayak, 2024). The capabilities of personalized alert systems and intuitive interfaces are enabling timely decisions and action, and adaptation to uncertainty. Agricultural climate change and labor challenges have exacerbated agricultural volatility, real-time systems are a requirement for resilient, responsive, and sustainable agriculture.

6. APPLICATIONS IN SMART FARMING

6.1 Precision Irrigation and Spraying

Precision irrigation and spraying are disruptive shifts with modern farm practices, gaining traction with the adoption of advanced technologies to reduce utilized resources like water and agrochemicals. Specifically, precision-like technologies include; Internet of Things (IoT), artificial intelligence (AI), and remote sensing to facilitate decisions supported by data labeling, increasing yields and conserving resources; which also creates a more efficient use of labor. Various technologies can more accurately support decisions by continuously monitoring the environmental conditions and allowing for some intervention(s) to be made with the exact timing and dose for maximum effectiveness. For example, precision irrigation externally combines data from soil moisture sensors, weather forecasts, and artificial intelligence algorithms to give the right time of application and amount of water to apply; rather than following past farmer experiences, research protocol, or unreasonable assumptions as guided by the best available science. These are all desirable problems that could promote sustainability of future and current producers as the improved efficiencies deal with limited water resources and potential fertilizer runoff occurring within agricultural production. Also, zonal portions of fields can be used to guarantee water utilization matches the unique moisture conditions and subsequent crop type. For instance, AI using a control with built-in access to various forms of real-time data can actuate liquid delivery through a hose upon integration with current soil and weather conditions, increasing water efficiency (Verma & Verma, 2023) and over time, reducing variable costs of production and reliance of manual labor.

Precision spraying takes advantage of drones and autonomous vehicles with machine vision and AI to locate pest infected areas or diseased places within a field. With these systems, growers are accurately applying either pesticides or fertilizers only were needed at the proper application rates. As a result, growers are using less chemistry and ensuring they don't over-apply. Additionally, the systems enable continuous scouting and variable-rate applicators, treating only the affected areas and leaving the healthy areas alone. Recently, for example, researchers have validated the applicability of robotic spot-spraying where the use of herbicide is reduced without loss of weed control (Rahimi Azghadi et al., 2024). Further, this technology directly addresses farmers' concerns in terms of chemical runoff and their footprint on the environment. The farmer incorporates new artificial intelligence systems, with Internet of Things (IoT) technologies, to increase operational efficiencies and sustainable farming practices through lower resource and environmental impact. These systems enable adaptive approaches dealing with changing climate conditions, pest pressures, and resource limitations. As agriculture faces unprecedented challenges, including climate change and resource scarcity, precision irrigation and spraying highlight these approaches to extend resilient and sustainable food production systems (Monchusi, Kgopa, & Mokwana, 2024).Furthermore, cloud-based analytics and mobile platforms are increasingly being used to deliver real-time alerts and prescriptions to farmers, ensuring that the right actions are taken at the right time, even in geographically remote areas.

6.2 Crop Health Monitoring

Crop health monitoring is a fundamental aspect of precision agriculture, allowing farmers to predict and solve nutrient deficiencies, insect issues and diseases. Through the use of new technologies such as the Internet of Things (IoT), artificial intelligence (AI), and remote sensing, farmers can observe real-time changes in crop conditions, which encourages timely actions that relate to better yield and needless waste of inputs. Therefore, in farm management, the ability to observe crop health in real time is important since it might allow management of farm activities to be proactive instead of reactive and it will encourage management before the visible part of crop productivity is negatively impacted. Also, monitoring allows producers to monitor site-specific management, for example, when several interventions occur in one area of a field commercially known as localized interventions. Site-specific management accommodates lower use inputs and lower operational costs. Another example of crop health monitoring is by combining IoT-based sensors on AI algorithms that monitor soil properties including moisture, pH, and nutrient availability. With monitoring systems such as this, farmers can conduct almost real-time assessments of crop health to make automated interventions when an environmental change occurs. An example of this, can be found in a recent study regarding a proposed AI-centered system to help producers monitor crop health through production data obtained from IoT sensors that collected soil properties and processed the IoT through an AI model that identified those discrepancies and directed what to do with (Raj et al., 2025). This sensor- constantly, completely, and continuously increases precision; reduces manual in-field assessments; decreases labor costs; and increases scalability for larger farm numbers. Additionally, the data collected over time may lead to developing long-term soil management practices and crop rotation programs.

Remote sensing technological advances, like the Normalized Difference Vegetation Index (NDVI), in combination with deep-learning models, represent a more practical approach to classifying the overall health of crops. This study explores how NDVI data can be used with Fully Connected Neural Network (FCNN) models to classify crop health conditions, attaining an overall classification accuracy of 97.80%. This method allows for the large-scale monitoring of agricultural fields and drastically reduces the amount of manual effort that needs to happen while providing a scalable model for addressing the global crisis of food security (Judith et al., 2025). The types of precision detections that can be made through remote sensing models for crop and plant monitoring could also be monetized through dissipating these technologies into early-warning systems that could better alert farmers to emerging threats and provide alternative courses of action based on the data in comparison to the emerging issue. These remote-sensing technologies have the capability of decentralizing agricultural producers' and owners' perception of crop and plant health from a moment-in-time measurement of crop health to a linear development of dynamic data that can help mitigate stress conditions, disease outbreaks, or nutrient deficiencies days before the naked-eye can detect the issue. Continuing to research the further integration of powerful technologies like those found through AI and IoT to monitor and maintain crop health will only help with monitoring crop health. Full advancements in these technologies are contributing to a sustainable farm practice and greater crop yields when used effectively (Bais et al ., 2023).These tools not only optimize the use of agricultural inputs like fertilizers and pesticides but also align with climate-smart agriculture initiatives by promoting resource-efficient and environmentally responsible farming. Ultimately, crop health monitoring systems are playing a vital role in improving agricultural resilience, reducing economic losses, and ensuring food security in the face of increasing environmental uncertainty and population growth.

Figure 5. Smart farming dashboard

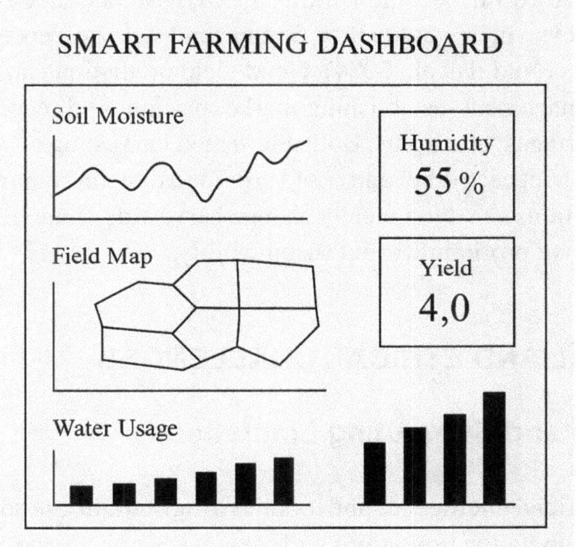

Figure 5 is a dashboard which visually displays information such as irrigation areas, crop health conditions, and yield predictions. It allows farmers to observe and control their operations effectively through real-time analytics and notifications.

6.3 Automated Harvesting and Yield Prediction

Automated harvesting and the immediate ability to predict yield are two of the newer process innovations in agriculture and farming that allow for greater efficiencies and even improvements in productivity. Automated harvesting committed machinery to take over some portion of the work done by human labor, such as harvesting fruits and vegetables. These innovations alleviate the issues with labor shortages in many agricultural markets and, can get the job done faster. Robotic harvesting can also take advantage of computer vision and AI to assist in completing the task while limiting damage and improving quality while assessing what is the crop item, how to move or handle it properly (Mavridou et al., 2019). Where automation is involved in harvesting, the speed can be improved, and the consistency of harvesting can be guaranteed; this is particularly important when considering large farm operations. Yield prediction consists of estimating crop outputs prior to harvest based on data analysis of remote satellite images, weather reports, and field-sensing devices. Powerful machine-learning models use vast datasets to predict yield with high accuracy, which would allow farmers and stakeholders to better plan and allocate resources (Nikhil et al., 2024). Good yield prediction can be beneficial for supply chain management, establishing market pricing, and recognizing potential risks such as shortages, or surplus. Both automated harvesting and yield prediction methods complement each other and could provide a dynamic approach, where real-time yield predictions can dynamically change harvesting strategies and schedules, which may improve profitability and sustainability.

7. TECHNICAL AND ETHICAL CHALLENGES

7.1 Hardware and Computing Limitations

Although artificial intelligence and robotics in agriculture are advancing quickly, hardware and computation limitations still present huge challenges for adoption and successful implementation. Many of the new AI models (especially deep learning-based ones) require significant computation capability, including processing speed, memory, and power consumption. On the other hand, agricultural environments usually will execute AI models on edge devices like autonomous robots, drones, and IoT sensors, which can be limited in processing capability due to size, cost, and energy. Edge devices need to compute decisions and develop actions in real time, and the difference between the processing power needed for their sophisticated

AI algorithms and the computing capabilities of the edge devices is unreasonable (Lloret, 2022).

In these remote rural farming regions, the gap is even wider at the same time because the challenges because of infrastructure. The power supply in these areas is unreliable, or there may not be any reliable power source available, constituting a major barrier to the continuous use of the HPC hardware. This means that the edge devices may be forced to rely on intermittent power sources, such as solar panels or batteries. Unfortunately, this greatly limits the degree of computational intensity and the duration an artificial intelligence (AI) application can run in such settings. Owing to the lack of stable internet connectivity in these remote rural areas, the use of cloud processing solutions is also restricted, which would otherwise offload the many complex computations from the device side. Instead, the edge devices in these scenarios are forced to depend on local processing without having the potential of deeper computing power that, as noted, has equipment limitations. Moreover, the physical environment of agricultural fields is also extremely punishing on electronic hardware. These devices must endure dust, dirt, humidity, heavy precipitation, and extreme ranges of temperatures. These multiples factors can hurt sensor accuracy, fail hardware, and greatly decrease the overall servicing and equipment life cycle, resulting in an unnecessary maintenance expense and lost time (Moshayedi et al., 2024).

7.2 Ethical Concerns in Deployment

The application of artificial intelligence and robotic technologies in farming proportions multiple complex ethical issues that need to be responsible to allow for safe and fair development of technologies. One of the main ethical issues involves data security and privacy. Agricultural Ai and robotics frequently rely on the collection of large and sometimes sensitive data, which may include sensitive information such as farm location, and soil types, crop types, and yield data, proprietary farming practices. This detailed level of data gathering, if mishandled, exposes farmers to threats of data breaches, unauthorized distribution, or exploitation by third parties such as corporations and governments. These violations threaten the privacy of people and could also threaten farmers' competitive advantage and intellectual property (Bazargani et al., 2024). Therefore, strong data governance processes are needed, including clearly defined consent processes, secure data storage, and public usage guidelines, to establish trust with farmers and stakeholders that encourage widespread adoption of AI-assisted agricultural innovations. Beyond data privacy, the ethical implications of using AI in agriculture need to address significant socio-economic issues related to labor and equity. The potential mechanization of jobs previously performed by human laborers—ranging from planting, weeding, and harvesting, to monitoring—could result in the displacement of millions of agricultural workers globally, many of whom depend on farming for their income. This is a serious threat in developing countries, where agriculture is the basis of local economies and a significant proportion of employment relies on manual labor (Amiri-Zarandi et al., 2022). If mitigating measures are not taken, dislocation could exacerbate existing social inequalities, worsen rural poverty, and contribute to social instability. Furthermore, any productivity gains afforded by AI would likely accrue primarily to those large commercial producers who can afford to invest in sophisticated technologies, thus exacerbating the growing divide among prosperous agribusinesses and small farmers.

Table 6 shows that data privacy, job loss, and algorithmic bias are some of the concerns, as well as issues of accuracy in models or reliability and confidence of the hardware used in agriculture. It is proposed that barriers to the uptake of AI will be addressed through policy-making, and strategic approaches taken by a multitude of stakeholder groups and public consultation when designing AI technologies for agriculture.

Table 6. Ethical and technical barriers in AI-Cobot deployment

Category	Challenge	Example
Technical	Hardware limitations	Edge computing constraints
Data	Bias in training data	Favoring certain crop types
Ethical	Job displacement	Reducing human roles
Interpretability	Black-box models	Hard to explain AI decisions

8. FUTURE DIRECTIONS IN AI-COBOT AGRICULTURE

Advancements in Artificial Intelligence (AI), coupled with new collaborative robots, or cobots, are changing modern agriculture by increasing the potential for responsive, smarter, labor-efficient agricultural systems. Growing global pressure for sustainable food production amid climate change, labor issues, and resource scarcity signals the future of AI-Cobot agriculture will follow enhanced routes toward autonomy, precision, and scale. One direction for cobot development suggests building cobots with more intelligence and self-learning capabilities so they can dynamically adjust processes in response to the different requirements to perform a variety of agricultural tasks including seeding, pruning, spraying, and harvesting. Through reinforcement learning and sophisticated computer vision, future cobots will learn to perform not as pre-programmed robots but will observe their environment while learning from their previous actions enhancing their performance as they evolve. Future cobots will not only desegregate crops and identify crop species with maturity levels but will also autonomously determine pest or disease dispositions, and decide what action to take, all independently. This is critical for maintaining large and heterogeneous farm landscapes where conditions change rapidly over time. Swarm robotics is another emerging trend in AI-Cobot agriculture, as physical cobots are modified in AI to perform separately yet collaborate with other AI-enabled cobots to replicate agricultural tasks. These swarm systems offer resilience (as the failure of one unit doesn't disrupt the whole operation), efficiency, and the ability to cover larger areas in a shorter time frame. Combined with IoT-based communication, cobots will be able to share data in real-time, optimizing task distribution, reducing redundancy, and ensuring uniform crop management. Edge AI and energy-efficient hardware are also expected to play a major role. Future cobots will be designed to operate in resource-limited environments with minimal connectivity and power. Onboard edge processors will allow cobots to perform real-time data analysis and decision-making without relying heavily on cloud infrastructure. This advancement is especially beneficial for smallholder farmers in remote or underdeveloped regions. In addition, human-robot collaboration will see further enhancements through intuitive interfaces and natural language processing. Farmers will be able to command cobots using voice or gesture-based controls, making the technology more accessible to users with limited technical expertise. This aligns with the goal of inclusive technology that supports rather than replaces human labor—allowing humans and cobots to work side by side in tasks requiring dexterity, judgment, or empathy (such as animal handling). From a systems perspective, AI-cobot agriculture will become increasingly integrated with other smart farming technologies, such as drone surveillance, satellite imaging, blockchain for supply chain traceability, and predictive analytics for weather and yield. These integrated systems will form a cohesive ecosystem that enhances decision-making, reduces waste, and improves food safety and quality. Lastly, as sustainability becomes central to agricultural innovation, future AI-cobot designs will emphasize eco-friendly materials, low-emission operations, and carbon-smart farming practices. Cobots may even assist in environmental conservation tasks, such as planting cover crops or managing biodiversity zones, contributing to regenerative agriculture and climate resilience. In summary, the future of AI-Cobot agriculture lies in building intelligent, autonomous, and collaborative systems that not only boost productivity but also align with environmental and social sustainability goals. The evolution of these systems will be key to feeding a growing global population in an equitable, efficient, and eco-conscious manner.

CONCLUSION

The integration of Generative AI and collaborative robots (cobots) in agriculture marks a significant leap toward smarter, more resilient, and sustainable farming methods. This study has explored how state-of-the-art AI models—such as Generative Adversarial Networks (GANs), Variational Autoencoders (VAEs), and Transformers—can be effectively applied not only for automation but also for complex, data-driven decision-making and predictive analytics. These models, when embedded in the structure of agricultural cobots, enable them to perform higher value functions like scheduling irrigation, precise spraying, monitoring crop health, and harvesting automatically with great precision - changing traditional agricultural practices into intelligent adaptive systems. Collaborative robots are built to work with humans and augment human labor instead of removing it completely. In agricultural sectors, they provide the capability to inform decision-making and implement real-time sensors on a human operator's side and perform laborious tasks with mechanized precision. The human-robot combination is addressing some of the major challenges within agriculture by alleviating chronic labor shortages, responding to climate impacts that are varied and unpredictable, and managing the pressure of doing more with less inputs. Moreover, the AI integration is a form of learning from data that ultimately leads to the robot working autonomously, based on the needs of the crop status, and constantly learning from new data and dynamics in the field. In high-variability environments like those in agricultural sectors, the robots should provide valuable help, however, there are very significant technical and ethical problems. Hardware limitations in rural, off-grid areas are barriers to deploying high-performance computing to support bots with real-time AI inference. Besides, ethical challenges surrounding data privacy, algorithmic fairness, and socioeconomic consequences, including dislocation of rural workers, must be resolved with inclusive policies, strong data governance, and community involvement. In the future AI-Cobot systems will be developed on the back of advancements in edge AI, reinforcement learning and swarm robotics will lead to giving the AI-Cobots more autonomy, larger scale deployment, and lower costs thus more easily deploy AI-Tech for a wider range of agricultural applications. More research, multidisciplinary frameworks, and coherent policies can move agricultural technologies closer to a transformational smart, data-driven industry that provides a productive balance between productivity and sustainable practices. Other words, the future of AI-Cobot agriculture consists of the creation of co-creational, intelligent ecosystems that new technologies to match citizen-, environmental-, and social-values; machine intelligence connected to agricultural system core processes. This abroad, resilient farming systems will not only be able to provide food for a growing population but also advance towards a more just and ecologically desirable future.

REFERENCES

Akkem, Y., Biswas, S. K., & Varanasi, A. (2023). Smart farming using artificial intelligence: A review. *Engineering Applications of Artificial Intelligence, 120*, 105899. DOI: 10.1016/j.engappai.2023.105899

Albanese, A., Nardello, M., & Brunelli, D. (2021). Automated Pest Detection with DNN on the Edge for Precision Agriculture. *arXiv preprint arXiv:2108.00421*. https://arxiv.org/abs/2108.00421

Amiri-Zarandi, M., Dara, R. A., Duncan, E., & Fraser, E. D. G. (2022). Big Data Privacy in Smart Farming: A Review. *Sustainability (Basel), 14*(15), 9120. DOI: 10.3390/su14159120

Bac, C. W., van Henten, E. J., Hemming, J., & Edan, Y. (2014). Harvesting Robots for High-value Crops: State-of-the-art Review and Challenges Ahead. *Journal of Field Robotics, 31*(6), 888–911. DOI: 10.1002/rob.21525

Baeten, J., Donné, K., Boedrij, S., Beckers, W., & Claesen, E. (2008). Autonomous Fruit Picking Machine: A Robotic Apple Harvester. In Laugier, C., & Siegwart, R. (Eds.), *Field and Service Robotics. Springer Tracts in Advanced Robotics* (Vol. 42). Springer., DOI: 10.1007/978-3-540-75404-6_51

Bais, D., Tiwari, V., Kolhe, S., Jain, S. (2023). IoT and AI Enabled Framework to Monitor Soil and Crop Health for Sustainable Development in Precision Farming-A Study.

Baxter, P., Cielniak, G., Hanheide, M., & From, P. (2018).. . *Safe Human-Robot Interaction in Agriculture., 59-60*, 59–60. Advance online publication. DOI: 10.1145/3173386.3177072

Bazargani, K., & Deemyad, T. (2024). Automation's Impact on Agriculture: Opportunities, Challenges, and Economic Effects. *Robotics (Basel, Switzerland), 13*(2), 33. DOI: 10.3390/robotics13020033

Bechar, A., & Vigneault, C. (2016). Agricultural robots for field operations: Concepts and components. *Biosystems Engineering, 149*, 94–111. DOI: 10.1016/j.biosystemseng.2016.06.014

Chlingaryan, A., Sukkarieh, S., & Whelan, B. (2018). Machine learning approaches for crop yield prediction and nitrogen status estimation in precision agriculture: A review. *Computers and Electronics in Agriculture, 151*, 61–69. DOI: 10.1016/j.compag.2018.05.012

Choi, K., Yi, J., Park, C., & Yoon, S. (2021). Deep learning for anomaly detection in time-series data: Review, analysis, and guidelines. *IEEE Access : Practical Innovations, Open Solutions*, *9*, 120043–120065. DOI: 10.1109/ACCESS.2021.3107975

Das, S. K., & Nayak, P. (2024). Integration of IoT-AI powered local weather forecasting: A Game-Changer for Agriculture. *arXiv preprint arXiv:2501.14754*. https://arxiv.org/abs/2501.14754

Dhanaraju, M., Chenniappan, P., Ramalingam, K., Pazhanivelan, S., & Kaliaperumal, R. (2022). Smart Farming: Internet of Things (IoT)-Based Sustainable Agriculture. *Agriculture*, *12*(10), 1745. DOI: 10.3390/agriculture12101745

Duckett, T., Pearson, S., Blackmore, S., Grieve, B., Chen, W.-H., Cielniak, G., & Yang, G.-Z. (2018). Agricultural Robotics: The Future of Robotic Agriculture. *arXiv [Cs.RO]*. Retrieved from http://arxiv.org/abs/1806.06762 DOI: 10.31256/WP2018.2

Etheredge, L. M. (2025). *AI in agriculture*. ResearchGate. Li, H., Guo, Y., Zhao, H., Wang, Y., & Chow, D. (2021). Towards automated greenhouse: A state of the art review on greenhouse monitoring methods and technologies based on internet of things. *Computers and Electronics in Agriculture*, *191*, 106558.

Ghumman, S., Di Troia, F., Andreopoulos, W., Stamp, M., & Rai, S. (2025). AGRO: An autonomous AI rover for precision agriculture. *arXiv preprint*, arXiv:2505.01200.

Goodfellow, I. J., Pouget-Abadie, J., Mirza, M., Xu, B., Warde-Farley, D., Ozair, S., Bengio, Y. (2014). Generative Adversarial Networks. *arXiv [Stat.ML]*.

Horabik, J., Parafiniuk, P., & Molenda, M. (2016). Experiments and discrete element method simulations of distribution of static load of grain bedding at bottom of shallow model silo. *Biosystems Engineering*, *149*, 60–71. DOI: 10.1016/j.biosystemseng.2016.06.012

Jägermeyr, J., Gerten, D., Heinke, J., Schaphoff, S., Kummu, M., & Lucht, W. (2015). Water savings potentials of irrigation systems: Global simulation of processes and linkages. *Hydrology and Earth System Sciences*, *19*(7), 3073–3091. DOI: 10.5194/hess-19-3073-2015

Jin, X., Liu, S., Baret, F., Hemerlé, M., & Comar, A. (2017). Estimates of plant density of wheat crops at emergence from very low altitude UAV imagery. *Remote Sensing of Environment*, *198*, 105–114. DOI: 10.1016/j.rse.2017.06.007

Judith, J., Tamilselvi, R., Beham, M. P., Lakshmi, S. S. P., Panthakkan, A., Al Mansoori, S., & Al Ahmad, H. (2025). Remote Sensing Based Crop Health Classification Using NDVI and Fully Connected Neural Networks. *arXiv preprint arXiv:2504.10522*. https://arxiv.org/abs/2504.10522

Kamilaris, A., Kartakoullis, A., & Prenafeta-Boldú, F. X. (2018). A review on the practice of big data analysis in agriculture. *Computers and Electronics in Agriculture*, *143*, 23–37. DOI: 10.1016/j.compag.2017.09.037

Kamilaris, A., & Prenafeta-Boldú, F. X. (2018). Deep learning in agriculture: A survey. *Computers and Electronics in Agriculture*, *147*, 70–90. DOI: 10.1016/j.compag.2018.02.016

Kant, J. K., Sripaad, M., Bharadwaj, A., Rajashekhar, V. S., & Sundaram, S. (2023). An Autonomous Hybrid Drone-Rover Vehicle for Weed Removal and Spraying Applications in Agriculture. *arXiv preprint arXiv:2308.04794*. https://arxiv.org/abs/2308.04794 DOI: 10.1109/AGRETA57740.2023.10262416

Kingma, D. P., & Welling, M. (2022). Auto-Encoding Variational Bayes. *arXiv [Stat.ML]*. Retrieved from http://arxiv.org/abs/1312.6114

Lal, R. (2011). Soil health and climate change: an overview. *Soil health and climate change*, 3-24.

Lesk, C., Rowhani, P., & Ramankutty, N. (2016). Influence of extreme weather disasters on global crop production. *Nature*, *529*(7584), 84–87. DOI: 10.1038/nature16467 PMID: 26738594

Li, T., Xie, F., Xiong, Y., & Feng, Q. (2023). Multi-Arm Robot Task Planning for Fruit Harvesting Using Multi-Agent Reinforcement Learning. *arXiv preprint arXiv:2303.00460*. https://arxiv.org/abs/2303.00460 DOI: 10.1109/IROS55552.2023.10341822

Liakos, K. G., Busato, P., Moshou, D., Pearson, S., & Bochtis, D. (2018). Machine Learning in Agriculture: A Review. *Sensors (Basel)*, *18*(8), 2674. DOI: 10.3390/s18082674 PMID: 30110960

Lloret, J. (2022). Edge Computing in Precision agriculture. 1-1. .DOI: 10.1109/FMEC57183.2022.10062622

Lowder, S. K., Sánchez, M. V., & Bertini, R. (2021). Which farms feed the world and has farmland become more concentrated? *World Development*, *142*, 105455. DOI: 10.1016/j.worlddev.2021.105455

Lu, Y., Chen, D., Olaniyi, E., & Huang, Y. (2022). Generative adversarial networks (GANs) for image augmentation in agriculture: A systematic review. *Computers and Electronics in Agriculture*, *200*, 107208. DOI: 10.1016/j.compag.2022.107208

Majeed, Y., Fu, L., & He, L. (2024). Editorial: Artificial intelligence-of-things (AIoT) in precision agriculture. *Frontiers in Plant Science, 15*, 15–2024. DOI: 10.3389/fpls.2024.1369791 PMID: 38344185

Mavridou, E., Vrochidou, E., Papakostas, G., Pachidis, T., & Kaburlasos, V. (2019). Machine Vision Systems in Precision Agriculture for Crop Farming. *Journal of Imaging, 5*(12), 89. DOI: 10.3390/jimaging5120089 PMID: 34460603

Milioto, A., Lottes, P., & Stachniss, C. (2018). Real-time Semantic Segmentation of Crop and Weed for Precision Agriculture Robots Leveraging Background Knowledge in CNNs. *arXiv [Cs.CV]*. Retrieved from http://arxiv.org/abs/1709.06764 DOI: 10.1109/ICRA.2018.8460962

Monchusi, B. B., Kgopa, A., & Mokwana, T. (2024). Integrating IoT and AI for Precision Agriculture: Enhancing Water Management and Crop Monitoring in Small-Scale Farms. *International Conference on Intelligent and Innovative Computing Applications*. 2024. 151-158. DOI: 10.59200/ICONIC.2024.017

Moshayedi, A. J., Khan, A., Yang, Y., Hu, J., & Kolahdooz, A. (2024). *Robots in Agriculture: Revolutionizing Farming Practices*. EAI Endorsed Transactions on AI and Robotics., DOI: 10.4108/airo.5855

Muhammed, D., Ahvar, E., Ahvar, S., Trocan, M., Montpetit, M. J., & Ehsani, R. (2024). Artificial Intelligence of Things (AIoT) for smart agriculture: A review of architectures, technologies and solutions. *Journal of Network and Computer Applications, 228*, 103905. DOI: 10.1016/j.jnca.2024.103905

Nikhil, U. V., Pandiyan, A. M., Raja, S. P., & Stamenkovic, Z. (2024). Machine Learning-Based Crop Yield Prediction in South India: Performance Analysis of Various Models. *Computers, 13*(6), 137. DOI: 10.3390/computers13060137

Oliveira, R., & Silva, R. D. S. (2023). Artificial Intelligence in Agriculture: Benefits, Challenges, and Trends. *Applied Sciences (Basel, Switzerland), 13*(13), 7405. Advance online publication. DOI: 10.3390/app13137405

Pallottino, F., Violino, S., Figorilli, S., Pane, C., Aguzzi, J., Colle, G., Nerio Nemmi, E., Montaghi, A., Chatzievangelou, D., Antonucci, F., Moscovini, L., Mei, A., Costa, C., & Ortenzi, L. (2025). Applications and perspectives of Generative Artificial Intelligence in agriculture. *Computers and Electronics in Agriculture, 230*, 109919. DOI: 10.1016/j.compag.2025.109919

Patel, D., Gandhi, M., Shankaranarayanan, H., & Darji, A. D. (2022). Design of an Autonomous Agriculture Robot for Real Time Weed Detection using CNN. *arXiv preprint arXiv:2211.12077*. https://arxiv.org/abs/2211.12077

Pietrantoni, L., Favilla, M., Fraboni, F., Mazzoni, E., Morandini, S., Benvenuti, M., & De Angelis, M. (2024). Integrating collaborative robots in manufacturing, logistics, and agriculture: Expert perspectives on technical, safety, and human factors. *Frontiers in Robotics and AI, 9*, 1342130. DOI: 10.3389/frobt.2024.1342130 PMID: 39687349

Qi, Z., Du, L., Huo, R., & Huang, T. (2023). Predictive maintenance based on identity resolution and transformers in IIoT. *Sensors (Basel), 16*(9), 310.

Rahimi Azghadi, M., Olsen, A., Wood, J., Saleh, A., Calvert, B., Granshaw, T., Fillols, E., & Philippa, B. (2024). Precise robotic weed spot-spraying for reduced herbicide usage and improved environmental outcomes: A real-world case study. *arXiv preprint arXiv:2401.13931*. https://arxiv.org/abs/2401.13931

Raj, L., Gaurav, S., Kumar, A., Kumar, H., & Rampal, A. (March 06, 2025). AI-Powered Crop Health Monitoring System. Available at *SSRN*: https://ssrn.com/abstract=5175088 or http://dx.doi.org/DOI: 10.2139/ssrn.5175088

Rane, J., & Kaya, Ö., & Mallick, S., & Rane, N. (2024). Smart farming using artificial intelligence, machine learning, deep learning, and ChatGPT: Applications, opportunities, challenges, and future directions. .DOI: 10.70593/978-81-981271-7-4_6

Shamshiri, R. R., Kalantari, F., Ting, K. C., Thorp, K. R., Hameed, I. A., & Weltzien, C. (2018). Advances in greenhouse automation and controlled environment agriculture: A transition to plant factories and urban agriculture. *International Journal of Agricultural and Biological Engineering, 11*(1), 1–22. DOI: 10.25165/j.ijabe.20181101.3210

Talaviya, T., Shah, D., Patel, N., Yagnik, H., & Shah, M. (2020). Implementation of artificial intelligence in agriculture for optimization of irrigation and pesticide spraying. *Artificial Intelligence in Agriculture, 4*, 58–73. DOI: 10.1016/j.aiia.2020.04.002

Tsouros, D. C., Bibi, S., & Sarigiannidis, P. G. (2019). A Review on UAV-Based Applications for Precision Agriculture. *Information (Basel), 10*(11), 349. DOI: 10.3390/info10110349

Vaswani, A., Shazeer, N., Parmar, N., Uszkoreit, J., Jones, L., Gomez, A. N., & Polosukhin, I. (2023). Attention Is All You Need. *arXiv [Cs.CL]*. Retrieved from http://arxiv.org/abs/1706.03762

Verma, S. K., & Verma, A. (2023). Precision irrigation management: An innovative technique for efficient water management. *Indian Farming, 73*(4), 21–24.

Wang, C. (2024). Intelligent Agricultural Greenhouse Control System Based on Internet of Things and Machine Learning. *arXiv preprint arXiv:2402.09488*. https://arxiv.org/abs/2402.09488

Wolfert, S., Ge, L., Verdouw, C., & Bogaardt, M. J. (2017). Big data in smart farming – A review. *Agricultural Systems*, *153*, 69–80. DOI: 10.1016/j.agsy.2017.01.023

Yerebakan, M., & Hu, B. (2024). Human–Robot Collaboration in Modern Agriculture: A Review of the Current Research Landscape. *Advanced Intelligent Systems*, *6*(7), 2300823. Advance online publication. DOI: 10.1002/aisy.202300823

Chapter 3

AI-Driven Collaborative Robotics in Agriculture: Optimizing Soil and Crop Yield

Md. Shoeab Akhter
https://orcid.org/0009-0004-4582-2937
International University of Business Agriculture and Technology, Bangladesh

Israt Jahan
East West University, Bangladesh

Mahima Tasnim
Underwood University, USA

M. Shohidullah Miah
International University of Business, Agriculture, and Technology, Bangladesh

A. J. M. Sirajul Karim
International University of Business, Agriculture, and Technology, Bangladesh

ABSTRACT

Emerging trends in precision agriculture are the consequence of the convergence of AI with collaborative robotics, delivering smart and dynamic soil and crop management systems. This chapter covers the theoretical background and application of AI-driven robots in farm environments, with a specific focus on the analytical optimization of soil health and the assessment of crop yield. It emphasizes data-driven AI techniques, including machine learning, deep learning, and sensor data modeling, to simulate interventions by collaborative robots, such as soil spatial mapping, nutrient deficiency detection, predictive analytics for yield estimation, and decision-making support for farmers. The chapter serves as a prelude to

DOI: 10.4018/979-8-3373-4207-8.ch003

forthcoming AI-robot applications and real-world use case scenarios. By bridging human expertise and AI, this chapter moves toward creating resilient, efficient, and sustainable agricultural ecosystems. Furthermore, this approach aligns with global sustainability goals by enhancing resource efficiency and empowering data-driven agricultural policymaking.

1. INTRODUCTION

Combining Artificial Intelligence (AI) and collaborative robotics can change farming from being very hands-on and traditional to using data and precision for better results (Eze et al., 2025; Onteddu et al., 2025). With a world population estimated to grow up to 9.7 billion by 2050, the challenge to produce more food—some 70% more, according to the FAO—is hardly surging (Lal, 2016; Molotoks et al., 2021). Trying to meet the demand, however, old farming styles don't cut it either, considering these are typically manual labor and generationally poor habits of annual practices that drain resources and do no help for the environment. Agriculture is also one of the world's greatest consumers of fresh water and a major contributor to global warming and pollution; its reliance on chemicals, including synthetic fertilizers and pesticides, has increasingly been linked with soil and water degradation and loss of biodiversity.

Figure 1. AI in agriculture

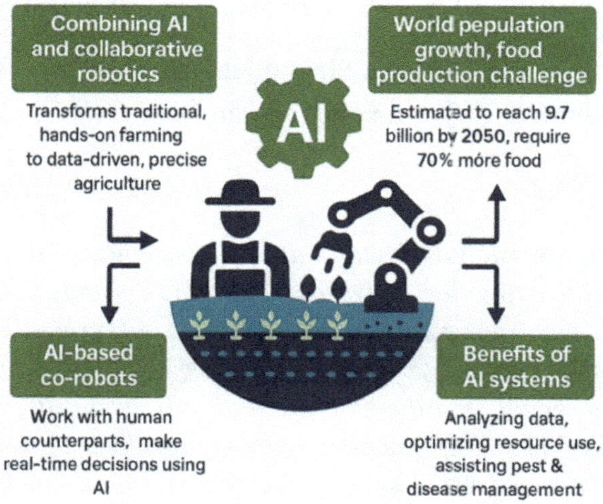

Sources: *Authors' Own*

AI-based Co-robots represent a potential solution. These robots, intended to work with human counterparts, make decisions in real time about soil health, crop yields, and resource allocation using AI. Modern robotic systems work alone because they have limited abilities, but cooperative robots can share a workspace with humans, using human guidance while also having their automated functions. These robots, equipped with sensors, cameras, and machine learning algorithms, vigilantly monitor the soil, climate, and resource utilization. An important feature of AI-based systems is their ability to instantly analyze large volumes of data, providing farmers with actionable outcomes regarding what actions to take and when to take them to achieve optimal results (Olabimpe Banke Akintuyi, 2024). For example, combining soil sensors and weather data with AI models allows accurate predictions for soil moisture, nutrient needs, and crop health. This capability empowers farmers to proactively make data-driven decisions that enhance crop yields and reduce waste. In a similar vein, we have also seen how AI models trained on soil, weather, and satellite data can predict yield much better, thus again assisting in better resource allocation and less overuse of fertilizers and water. Research shows that using AI-based precision irrigation can reduce water use by 23–30% and increase crop yield by 10–20%, and, in some cases, even up to 30%, giving farmers a clear advantage, especially in dry areas (Schwalbert et al., 2020).

Additionally, these AI systems facilitate improved pest control and disease management. By processing images, we allow AI-powered robots to analyze high-resolution images of crops to detect early stages of pest infestation, disease, etc. Robots can use deep learning techniques, such as CNNs, to interpret complex visual data, for which the robot can detect an exact diagnosis and recommendation for intervention. Such an approach would reduce the application of broad-spectrum pesticides and lower the chemical input to a field, hence limiting the input cost and the possible environmental impacts of agriculture. Another important consideration is the incorporation of reinforcement learning (RL) to assist the robot's behavior (Luo et al., 2025). RL allows robots to change and adapt with the environment as they receive feedback and modify their behavior (Bauer et al., 2025). For instance, an RL-pretrained robot can dynamically modify the amount of fertilizers it disperses, adapting to identify the most effective crop-growing strategies (P. Roy et al., 2025). This self-learning capability will drive AI robots to learn continuously as they perform more tasks, improving their effectiveness over time by collecting increasing amounts of data, which will optimize soil health and maximize agricultural yields (Eze et al. 2025). A simulation-based approach can apply these technologies to agriculture by simulating various realistic agricultural settings before deploying the real robots. The data collected from soil sensors, weather stations, and satellite images can be used to simulate different scenarios to see how different variables—like the type of soil, how often to irrigate, and what variety of crop to plant—affect overall yield. This technique could be valuable, especially in resource-poor regions and developing countries where cutting-edge technology is scarce." AI-based farming systems, for instance, have shown tantalizing potential for boosting yields by 50%–60% in sub-Saharan Africa, where smallholder farmers have little access to technology and are faced with extreme climate variability (Chavula et al., 2024). Combining AI with information about local weather and soil types, for instance, may help mitigate climate change by giving farmers tailored advice for their specific circumstances. Second, AI-powered simulations are meant to be scalable, which means the model can be ported to a range of settings and kinds of farms, from big commercial ventures to small, family-run farms. For farmers struggling to make sense of weather-induced chaos or volatile, degenerated soil, AI-powered robots and simulations offer help that is tailored to their needs, helping them to both improve the productivity and the sustainability of their farming operations (Javaid et al., 2023). There is enormous potential in the use of AI-based robotics for agricultural activities; however, several issues are present. Good data is also key to accurately predicting crop yields, and it can be challenging to integrate every kind of data source—soil health, weather, and pest control information. Moreover, the investment in such technology might also represent a barrier to its adoption by most of the growers, especially in developing countries. Crafting approaches to overcome these challenges will require continued development of affordable iot solutions, such as inexpensive sensors, open-source AI models, and partnerships between tech developers and farmers (Cob-Parro et al., 2024).

In conclusion, AI and collaborative robotics in agriculture represent a significant paradigm shift in food production. By better optimizing the management of soil, increasing crop yields, and reducing waste, these new technologies can have a large impact on major global food security challenges. In the future, next-generation AI robots that are used for simulation-based and real-time data analysis are expected to create a better agricultural future (Dai & Ke, 2022). Access and cost may place barriers to entry, yet the pure excitement over the evolution of these technologies promises a better future not only for the farmers but for the field (and field practitioners) overall.

2. THEORETICAL FRAMEWORK: ROBOTICS AND AI FOR COLLABORATIVE ROBOTICS IN AGRICULTURE

Artificial intelligence (AI) has greatly transformed the agricultural sector through promising solutions for augmented crop yields and efficient farming. The fusion of AI with collaborative robotics also overcomes related environmental issues and promotes sustainable farms (Eze et al., 2025). Theoretical constructs reviewed in this chapter should inform future trajectories on how AI robots could synergize with humans for better soil management, decision support, and environmental sustainability. These are the theories that underpin the pragmatic application of AI in agricultural robotics and their disruptive influences on the farming industry (Spagnuolo et al., 2025).

2.1 Precision Agriculture and Multi-Agent Systems

Theoretical framework of precision agriculture Theoretical framework of Precision Agriculture draws attention to the application of emerging technologies, including AI, sensors, and robotics, for improving farm management. Multi-agent system theory in precision agriculture suggests that multiple robots, which can work on their own or together, can team up to solve complicated farming problems (Chevalier et al., 2015).

Figure 2. Theories on AI in collaborative robotics for agriculture

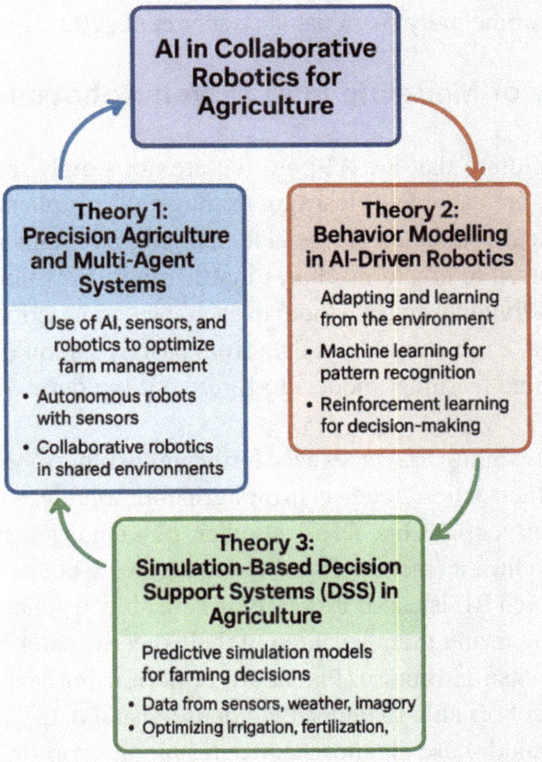

Sources: *Author's Own*

Sensors embedded in these robots are sensors that capture real-time field data on soil health, moisture levels, temperature, and pest presence. The robots analyze this data using AI algorithms to make informed decisions on watering, fertilizing, and pest control that minimize waste and maximize yields. The underlying idea behind this theory is the notion of operating side-by-side with one another and collaborating with human farmers. This partnership means robots can respond to human direction and real-world factors. Robotic cooperation is AI-based, and the robots are capable of learning from the actions of other robots and performing better as a group (Hong, 2001). As one example, robots cooperating could cache irrigation by flowing real-time information from soil sensors to water dissemination to optimize watering, increasing efficiency of a natural resource.

The resultant theory has positive and negative rights as end points, which facilitates the optimization of a proactive robotic agriculture system in which robots are collecting several observations about farms, and they are tuning their fitness and farm parameters to be sustainable and productive. Such multi-agent systems based on AI can represent a breakthrough in the traditional agricultural practices, and automation can be introduced to traditional farms to make them more resource-efficient and environmentally sustainable (Bauer et al., 2025).

3.2 The Behavior Modeling in AI-Driven Robotics

Robots in agriculture that are AI-powered are not simply loaded with pre-set instructions—they are capable of learning as they go. This phenomenon is mainly due to the theoretical principles of machine learning and reinforcement learning. AI has also been used to create machines that learn, rather than requiring human organizations or individuals to train them in each new transaction (Robinson et al., 2012). For instance, a robot could put data from past irrigation events and weather patterns into machine learning models to figure out the optimal amount of water for crops.

Reinforcement learning (RL) goes even further in enabling robots to take decisions based on feedback their actions receive. In the agricultural context, this feedback is the performance of interventions like fertilization or pest management. Afterward, the robot will learn to adjust its means of quality for achieving better decision-making. The concept on which RL is based is that robots are able to get better all the time at performing tasks by trying them, evaluating their success, and changing behavior accordingly. Kokonotsu and hotaru / Pixiv Furthermore, using deep learning, they've managed to make robots able to understand dense visual data. This information is very helpful for plant disease diagnosis and determining crop stress level using image recognition (Praveen et al., 2025). The psychological foundations of modeling behavior for AI in robotics focus on building robots that learn to act adaptively in changing environments, improving their performance over time.

3.3 Simulation-Based Decision Support System (DSS) in Agriculture

The Decision Support System (DSS) is one of the important theoretical constructs that support the use of AI-based agricultural robotics, especially in terms of simulation-based modeling. In conventional agriculture, decisions related to irrigation, fertilization, and anti-pest measures are typically intuitive or based on scarce evidence as well. But with DSS, such decisions are taken on the basis of simulation models taking packages of data from different sources—soil sensors, weather stations, satellite images, etc. It finds its roots in system dynamics theory, which is essential to simulation-based DSS. The theory describes how changes in one part of a system can affect other parts. In the context of farming, such variables might consist of soil parameters, climate, crop maturity, and pest densities. Simulating these exchanges allows AI to forecast the outcomes of different farming methods, letting farmers refine their approaches without incurring potentially expensive blunders (Jana, 2024).

Simulation-oriented environments may simulate virtual agriculture practices to predict crop yields, water use efficiencies, and soil health maintenance. This is especially beneficial in resource-limited regions, as field testing may be unfeasible. Take farmers in the developing world: with the help of AI informed simulations, they can now determine optimal irrigation schedules and compare the effect of various fertilization strategies without having to sink the large sums required for real-world trials. In addition, the theory of sim-based DSS does not only have predictive value in the improvement of agricultural performance; it also provides the farmers with information to help them to make better decisions and adopt more sustainable farming (Qi & Altinakar, 2011).

4. AI IN AGRICULTURE AND CURRENT ROBOTIC APPLICATIONS

Recently, Artificial Intelligence (AI) and automation have made significant changes in farming. Old farming ways cannot survive due to climate change, as food products are needed globally more than ever, and the farming resources are becoming difficult to obtain. Currently, AI is an essential tool for raising farm results, making farming a sustainable business, and improving decision-making. The technology allows farmers to utilize real-time data, predict what the future holds in line, and automate certain tasks that are useful for present farming. Crop forecasting, monitoring soil health, finding pests and diseases, and figuring out irrigation schedules are some important AI applications used in the agriculture industry. These tasks are commonly done with sensors, satellite pictures, computer programs, or even autonomous machines. To name a few, if you train computer vision on crop images, you can predict early disease and take action accordingly at the right time. Weather and soil data can be used to predict yield.

4.1 Principles of Collaborative Robotics and Their Distinction from Autonomous Robotics

Robotics has gone beyond automation and has stepped into intelligent collaboration in the field of evolving smart agriculture. Typically, autonomous robots have been used in mechanical farming tasks, i.e., sowing, spraying, and harvesting, but their functionality generally consists of repetitive and rule-based operations. These systems run in isolation, one being dependent on preprogrammed instructions and GPS guided routines. Collaborative robots ('cobots') are destined to work together with humans as well as other intelligent agents in real time (Demircioglu et al., 2024). They are endowed with real-time sensors, learning algorithms, and communication modules that make them dynamically adapt to complex and unstructured agricultural environments. Continuous monitoring, real-time decision-making, and an adaptive approach to changing agroecological conditions (Qu et al., 2024) are a must while conducting both activities, namely optimizing soil health and crop yield. Current autonomous robots do not have contextual awareness for these tasks, e.g., soil nutrient profiling and plant stress detection. If not pre-trained, every possible scenario that could be done by the traditional weeding robot might not be able to tell the difference between the harmful and beneficial plants. On the opposite side, a collaborative robot with the ability of machine learning and reinforcement learning is able to analyze the multispectral data from the soil sensors and the aerial imagery, detect the areas with nutrient deficiency, and adjust the strategy of the fertilization according to the type and the stage of the growing of the crops (Spagnuolo et al., 2025). Being responsive to day-to-day weather and resource variability is important for sustainable soil management and maximization of yield, especially under weather and resource constraints. The concepts of shared autonomy, safe human-robot interaction, and multi-agent cooperation are the enablers of collaborative robotics. In an AI augmented framework, eventually multiple cobots can work in a micro network together; one will scan/analyze soil moisture and pH levels, the next will scan/apply organic amendments in specified areas, and the third will scan the crop canopy to find its growth estimate. The performance of these agents continuously improves by applying knowledge gained from other agents and humans to optimize field output.

Table 1. Core differences: Collaboration vs. autonomy robot

Feature	Autonomous Robot (Rahmadian & Widyartono, 2020)	Collaborative Robot (Lytridis et al., 2021)
Task Execution	Fixed, preprogrammed	Adaptive, Context-Aware
Human Interaction	Minimal to None	High, Cooperative
Learning Ability	Static Behavior	AI-Driven Learning
Inter-agent Communication	Absent	Multi-Agent Communication
Flexibility	Low	High

Sources: Authors' Own

With the use of reinforcement learning or multi-agent deep learning algorithms, Cobots learn continuously from their environment and adjust their behavior as per the context. The use of this dynamic capability allows them to be more useful in various agroecological settings (Demircioglu et al., 2024). When these cobots are applied in a fleet rather than deployed as an isolated unit, the real benefit becomes apparent. For example, poor nitrogen levels in a soil mapping data set yield an appropriate fertilizer dose from a fertilization cobot, and another robot can evaluate crop response using, for instance, leaf color indices or data from a monitor measuring chlorophyll fluorescence. Performing these actions can be simulated through the echoes of AI models, projecting likely yield outcomes and thereby allowing farmers to act wisely and early (Rashid & Kausik, 2024).

Collaborative robots make for a new approach to precision agriculture by coupling intelligence in acquiring soil health and crop productivity. Integrating with AI, they are used for site-specific, adaptive, and sustainable farming, forming a future when robots become not only companions but also a way to feed a growing population worldwide.

4.2 Data Sources Relevant to Robots in Agriculture

Collaborative robots rely on a multitude of data sources to work effectively in agricultural environments, particularly for optimizing soil and crop yield. In addition to driving robot behavior, these data inputs inform the basis of AI enabled analytics and real-time decision-making as well as autonomous adaptation. As a result, robots can perform context-aware actions that are useful for precision farming by utilizing multimodal data from soil, crops, and the environment. Then study key data sources in this section and categorize them based on functionality and show how they participate in intelligent robotic intervention in agricultural systems (Onteddu et al., 2025).

4.2.1. Soil Sensors and Ground-Based Monitoring Systems

The soil sensors play a crucial role in evaluating the real-time soil health. The critical parameters that these sensors measure include moisture content, pH level, electrical conductivity, nitrogen, phosphorus, potassium (NPK) levels, and organic matter. Sensor modules, or in-field sensors, enable spatially distributed soil data that can be gathered by such robots. This allows them to create soil fertility maps and to detect nutrient imbalances to take corrective actions such as micro fertilizer application. For example, a robotic unit may detect patches of dry soil and send out the signal to turn on localized irrigation in other areas, or it can send out a signal to the missed zones and add organic compost or what have you, because it has lost too much nitrogen. Based on ground input, these micro decisions directly influence the quality of the soil and, in turn, promote maximum crop yield.

4.2.2. Remote Sensing and Aerial Imagery

Field-wide crop and soil assessment is possible through the use of drones and satellite-based remote sensing, which offer images at high resolutions. Such vegetation indices as NDVI (Normalized Difference Vegetation Index), SAVI (Soil Adjusted Vegetation Index), or thermal maps assist robots in understanding at first glance plant health, water stress, or degradation of soil. With aerial imagery, robots can cross-correlate with on-ground sensor measurements to find inconsistencies and hence also be able to make judicious decisions. An example of this type of dynamic behavior could be in the case of a plot of the field with lower NDVI values, which may require the inspection of root zone anomalies in moisture or nutrient uptake by a robotic unit (Milella et al., 2019).

4.2.3. Weather Stations and Climate Data

A similar type of robot also acts on real-time data, including rainfall, temperature, humidity, and wind speed, to time their activities. Weather data is used for scheduling seeding, irrigation, and fertilization events as a part of yield optimization. Take robots that handle possibly heavy rain, for they might avoid irrigating if not necessary so as to conserve water and lessen the nutrient loss that comes with runoff (Singh et al., 2024). These robots learn aggressive signs of climate change by running predictive AI models on past climate patterns and can forecast the best time to plant or harvest.

4.2.4. Farm Management Systems and IoT Integration

The robot's decision loop (and therefore, control loop and state estimation loop) can be fed with data obtained from cloud-based farm management platforms such as FieldVibe, GPS devices, and IoT systems in the farm. Robots prioritize tasks based on the history of crop rotation, the yield records, and the field boundaries. For example, in a case where a historically low yield for the field has been a problem, a collaborative robot can analyze the soil within the field section by focusing efforts in more densely populated areas where compaction or salinity may occur and then provide suggestions on how to rectify the problem.

Essentially, data is the fuel for collaborative intelligence in agriculture. AI-powered robots can take raw data from sensor networks, remote imagery, climate feeds, and digital farm records to provide actionable insights. This intelligent fusion of data sources allows them to make intelligent decisions about the condition of the soil as well as how they can increase crop yield with precision and sustainably (Getahun et al., 2024).

5. ALGORITHMIC FRAMEWORKS FOR ADAPTIVE ROBOT BEHAVIOR IN AGRICULTURE

Collaborative robots in agriculture are powered by artificial intelligence, which allows them to interpret complex data, make real-time decisions, and perform tasks independently. Robots should be able to dynamically interact with field conditions using the intelligent application of different AI methods to support intelligent systems that will result in high productivity and sustainable farming in the soil health and crop yield optimization domain. As stated above, in this section, we summarize the most effective algorithmic models in agriculture as supervised learning, unsupervised learning, deep learning, reinforcement learning, and hybrid approaches (Sibona et al., 2025).

5.1. Supervised Learning for Soil and Crop Prediction

Supervised learning is one of the most common AI techniques being used in agricultural robotics. Training models with this also means training models based on labeled data sets, such that inputs can be compared to outputs, e.g., soil pH, soil moisture, and soil nutrient levels against yield (crop health status). Some of the algorithms that are used are Decision Trees, Support Vector Machines (SVM), and Random Forests (Savitha & Talari, 2025). In the practical applications in agriculture, robots predict key outcomes from supervised models. As an example, a robot can explore historical soil data and predict with confidence if there is a nitrogen deficiency in a certain zone. Then it will be able to target precision fertilization just on those areas, lessening the waste and enhancing plant health. Robots can predict yield from the crop on the basis of the type of seed, the level of planting density, the level of irrigation, and weather conditions. The forecasts are vital for planning farms to maximize yield. Besides, supervised learning is used for predicting the early stage of disease and the exact date of the pest outbreak and harvesting time, which in turn directly results in an increase in productivity and overall soil vitality in one way or another (Bhaba Krishna Kuli et al., 2025).

Figure 3. AI-driven robot collaboration for managing soil health and maximizing crop yields

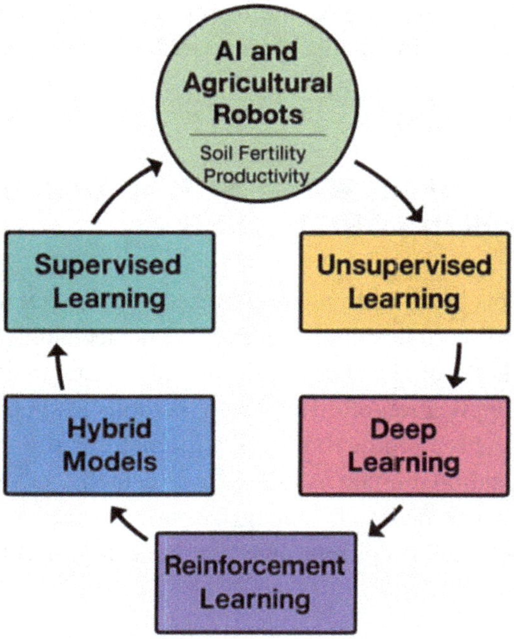

Sources: *Authors' Own*

5.2. Unsupervised Learning for Soil and Crop Pattern Detection

Data that is not labeled is unsupervised learning. It cannot make predictions, but rather it finds patterns, groups, and concealed patterns within the data. Some of the agricultural robots commonly use algorithms like KMeans Clustering, Hierarchical clustering, and Principal Component Analysis (PCA) (Adamu & Ahmad, 2024).

For example, robots can use clustering techniques to divide fields into soil fertility zones that are associated with organic matter content, salinity, and soil water. According to this zoning, site-specific farming can be provided to each zone separately. In a parallel way, the physiological response of crops to stress can be used to cluster crops to identify microclimates in a farm that require specific treatment. Another advantage of this method is that it is also able to detect anomalies, like an abnormal change in chlorophyll content or a change in the structure of the canopy, which are early indications of a disease, water stress, or lack of nutrients. Based on these insights, robots can autonomously perform soil testing, begin targeted spraying, and inform the farmers to prevent losses and to securely keep the production (Bauer et al., 2025).

5.3. Deep Learning for Image and Sensor Data Interpretation

Robots that can process complex, high-dimensional data like images, videos, or time-series sensor readings (e.g., pose and velocity traces) can be constructed with the aid of deep learning, which covers Convolutional Neural Networks (CNNs) and Recurrent Neural Networks (RNNs), among others. This capability is essential for agriculture, where it is needed to interpret inputs from drone imagery, hyperspectral cameras, LiDAR, and ground-based sensors (Farhangmehr et al., 2025).

For example, CNNs are able to work on aerial imagery to spot diseased or pest-infested plants, identify crop maturity stages, or calculate biomass. Remote sensing and CNN-generated soil maps help in soil classification, moisture mapping, and also to detect erosion (Senathipathi et al., 2025). This information helps robots tailor when to turn off watering cycles, when to add nutrients, and when to avoid 'messing up' machinery compaction. Based on deep learning, the robots, when deployed in greenhouses or open fields, can visually inspect crops, observe the development of plants, and, in real time, measure the soil parameters such as temperature, porosity, and salinity. The level of granularity of the data is such that timely intervention may be adopted, which helps in soil conservation and higher crop yield.

5.4. Reinforcement Learning for Adaptive Decision-Making

Reinforcement Learning (RL), or trial and error learning, is the learning approach in which robots learn optimal actions by operating and interacting with the environment and getting feedback. As time goes by, they will fine-tune their strategy to maximize a cumulative reward (e.g., higher yield, better soil moisture balance). In farming, the power of RL for dynamic decision-making (Bauer et al., 2025) is very particular. Robots can autonomously learn approaches to irrigating based on ever-changing weather patterns or soil saturation levels. In other cases, for example, an RL agent can be trained to decide when and how much water to apply so as to maintain optimal root zone moisture without over irrigation—saving water and at the same time maximizing yield (Li et al., 2025).

Further, navigation and path planning in unstructured or obstacle-rich terrain is done by robots using RL. In fertilizer scheduling, weed control, and pesticide optimization, RL can also be used by telling the robot to treat only those areas that need it, thereby reducing the amount of chemicals used as well as having a healthy soil microbiota.

5.5. Hybrid Models and Ensemble Learning

Hybrid models fuse several AI techniques and take the best from all of them. In other words, if a robot has a CNN to detect crop disease visually and reinforcement learning, it can decide on how to treat the disease (ignoring what it is yet) by spraying only those plants without considering forecasted weather (Senathipathi et al., 2025). Bagging and boosting promote ensemble learning, i.e., combinations of multiple weak learners that produce a strong predictor. For yield estimation, soil quality forecasting, and environmental risk assessments in agriculture, the ensemble models are more reliable. Such hybrid approaches have been shown to derive more stable, accurate, and adaptive performance, particularly in diverse and uncertain situations encountered in real fields in agriculture.

Therefore, AI driven algorithmic frameworks are instrumental for making collaborative robots that facilitate smarter and more data-oriented agricultural practices. By utilizing supervised predictions and unsupervised pattern recognition in deep sensory analysis and adaptive decision-making, these robots can change how farmers take care of soil health and optimize results of crops, thus establishing the framework for a sustainable and productive agricultural future (Daraghmah & Younis, 2025).

6. FUNCTIONAL FRAMEWORK OF COLLABORATIVE AI ROBOTS FOR FIELD-LEVEL AGRICULTURAL TASKS

Artificial intelligence in agriculture can only be successfully applied given a functional combination of different types of robots that complete different, autonomous tasks to improve soil quality and to optimize crop yield. In this section, rather than being centered on architecture schematics or hardware integrations, a practical and agriculture-oriented framework is provided on how collaborative AI robots operate in real-time agriculture (Spagnuolo et al., 2025). This appeals to show how different robotic agents, such as ground-based and aerial, interact with each other for intelligent data sharing and intelligent decision-making to address critical problems in soil and crop management.

In this functional framework, personal aerial drones serve as the scouts of the field. They use high-resolution cameras and multispectral sensors to fly over crop fields, thereby monitoring canopy health, leaf coloration, moisture stress, and thermal patterns. Based on these observations, early signs of crop disease, pest infestation, or nutrient deficiency can be identified. If certain problems are identified by air robots, ground robots with soil probes and environmental sensors are dispatched for the root-zone level targeted investigations (Rai et al., 2025). Ground agents record such indicators as soil moisture, pH, nitrogen content, and compaction. With this two-layer system of monitoring in place, there exists a macro and a micro view on the field's conditions. All the intelligence of the system is in how the data is passed between the agents and is analyzed by the embedded or cloud-based AI models. For example, if a certain drone sees that in a particular zone, there is pale leaf coloration, which signals to the drone that it might be a zone with a nitrogen deficiency. In-soil nutrient sensors are then confirmed by a ground robot. Based on this validation, we train an AI model using supervised learning that can determine the level of nitrogen deficiency and decide on the best action (S. Roy et al., 2025). A robot then performs the corrective action, such as localized fertilizer spraying, with the precision needed. Reinforcement learning will, over time, let the system learn how its response strategies improve yield or soil health and, thereby, fine-tune the responses accordingly.

Table 2. AI technologies in collaborative robotics for sustainable agriculture in enhancing soil health and crop yield

Focus Area	AI Technology Used	Robotic Application	Contribution to Soil Health	Sustainable Impact	Contribution to Crop Yield	Sources
Soil Health	Supervised Learning (e.g., Random Forests, SVM)	Soil nutrient prediction and deficiency mapping	Enables precise soil analysis and targeted interventions, improving soil structure and microbial balance	Enables precise, need-based fertilizer application, reducing chemical overuse	Informs optimal planting and irrigation timing, boosting yield consistency	(Botero-Valencia et al., 2025)
	Unsupervised Learning (e.g., K-Means, PCA)	Soil texture, compaction, and fertility zone classification	Discovers soil fertility zones to guide variable-rate application, reducing chemical load	Facilitates variable-rate tillage and site-specific soil amendments	Helps customize planting strategies across zones, improving productivity	(Javidan et al., 2023)
	Deep Learning (CNNs)	Soil erosion detection from drone/ aerial imagery	Identifies signs of soil erosion, compaction, or poor drainage using multispectral imaging.	Helps early intervention in degraded zones, preserving topsoil	Detects early plant stress, enabling timely interventions to prevent yield loss.	(Soori et al., 2023)
	Reinforcement Learning (RL)	Adaptive soil sampling strategies, autonomous irrigation, and fertilization bots	Continuously adapts soil treatment strategies based on feedback, reducing over-fertilization and soil degradation	Optimizes robot routes and sampling frequency, minimizing fuel and soil disturbance	Maximizes yield by optimizing timing and dosage of resources like water and nutrients.	(Bauer et al., 2025; Li et al., 2025)
	Multi-Modal Sensor Fusion with AI	Real-time soil moisture, pH, salinity monitoring	Cross-validates soil data (e.g., moisture vs. rainfall) to guide sustainable land use.	Promotes efficient irrigation, reducing water logging and salinity buildup	Synchronizes input application with weather forecasts and crop stages for peak productivity	(Korthals et al., 2018)
	Predictive Analytics + Edge AI	On-field diagnostics using ground robots	Facilitates immediate response to soil health data without cloud latency, preserving soil response cycles	In-field decision-making with reduced latency, improving response time to soil issues	Enables timely planting, pruning, and spraying—directly improving crop output	(Fountas et al., 2024)

Dividing the work in a coordinative manner results in an efficient use of resources and minimizes unnecessary interventions. Instead of applying the field uniformly, AI collaboration allows for zone-specific care that saves resources, builds sustainability, and preserves long-term soil quality. Robots can specialize, also; in fact, one robot may do just sample while another does weeding or irrigation. By synchronizing these functions using data, AI insights, and shared objectives, agricultural activities are timely and accurately executed. Using the functional framework, the transformative potential of AI integrated robot collaboration is capable of providing practical farming outcomes to precision agriculture. It adds particular value in areas with an imperative to effectively use inputs and to conserve soil health to achieve food security and climate resilience.

7. USE CASE SIMULATIONS FOR SOIL AND CROP YIELD OPTIMIZATION

With AI integration, collaborative robots are rapidly emerging as the latest tool in real-world agricultural scenarios, offering data-driven, smart solutions to the problems of soil health and optimum crop yield (Gadekallu et al., 2025). Three uses intended for robotic systems—precision irrigation, nutrient management, and disease detection—are shown in this section on how robotic systems can act in concert to change traditional workflows with automation, real-time analytics, and adaptive decision-making (Redhu et al., 2022).

7.1 Precision Irrigation: Watering the Right Way

Irrigation precision is certainly one of the most important tasks of water-sensitive agriculture and plays a great role in terms of yield optimization and soil conservation. With AI-enabled ground robots and aerial drones, they work collaboratively to monitor soil moisture, evapotranspiration, leaf surface temperature, and weather forecasts. To obtain this spatial variability in water requirements for a field, these agents rely on data derived from in-situ moisture sensors, thermal imaging, and remote sensing (Mohammed & Munir, 2025). For example, a ground robot can walk across a plot collecting fine-grained data about the soil, while a drone will incorporate an aerial thermal map that identifies regions of water stress. Data is then processed by AI algorithms through the use of predictive models and reinforcement learning to produce an optimal irrigation schedule. The robots aren't deployed everywhere; they're only used where water is needed so that the robots, although capable of overwatering, won't overwater or underwater at all, leaching nutrients and compacting soil while at the same time stressing crops and reducing their yields. This irrigation by means of sprinklers and miniature sprinklers protects against water waste, conserves the soil structure, and ensures the proper moisture content to improve the development of the plant (Lakhiar et al., 2024).

7.2 Smart Nutrient Management: Feeding the Soil Intelligently

Nutrient delivery is crucial because bringing one more ton of nutrients to the plant reduces our environmental impact and, at the same time, maximizes yield and keeps the soil fertile in the long term. In real time, soil chemical sensors and spectrometers mounted on collaborative robots analyze nutrient concentrations (N, P, K, pH, etc.) and provide a corresponding grade for each plant. Using historical soil crop productivity data, AI models made site-specific fertilizer mixes as well as timing of fertilizer application recommendations (S et al., 2025; Gopi et al., 2025). Micro-dosing is not possible with conventional blanket fertilization, whereas ground robots can do so by putting the right amount of nutrients in deficient zones. At the same time, aerial drones can cover the whole field, seeing if there are any discoloration or stunted growth symptoms of nutrient stress. By fusing these multi-source inputs, the central AI system continuously refines the nutrient strategies to reduce the possibility of overfertilization as well as nutrient runoff (Huang et al., 2025). It offers this precision in the use of nutrients, it protects the microbial soil health, and it sustains high yields with lower input costs (Xu et al., 2025).

7.3 Early Disease Detection: Safeguarding Soil and Crops

Pathogen buildup as well as side effects of chemical treatment affect not only crops but also soil ecosystems. Early detection and containment of plant diseases is made possible using AI-integrated robotic platforms, especially drone robot teams. Ground robots inspect specific leaves, stems, or soil patches, and drones fly over crops and record high-resolution images (Chen et al., 2025). Visual patterns are classified by the deep learning models, especially the CNNs, to diagnose diseases like rust, blight, or mildew. The robots will then immediately begin spot treatments using biopesticides or UV light systems once confirmed. At the same time, the central AI system also updates disease maps to prevent cross-contamination between zones. Collaborative surveillance of this nature minimizes yield loss, reduces chemical use, and protects beneficial soil organisms. By gathering longitudinal data, it can even forecast future outbreaks, which will increase the resilience of farms against biotic stress (Asrehegn et al., 2025).

The real-world simulations of the feasibility of AI-driven collaborative robotic implementation use cases show how AI-powered collaborative robotics are a smart, responsive, and sustainable solution to the challenges found in agriculture (Urrea &

Kern, 2025). Modern, climate-resilient farming (Bhaba Krishna Kuli et al., 2025) relies primarily on preserving the soil and the yield, which such systems directly contribute to by refining the irrigation, fertilization, and protection of the crops.

8. STRATEGIC CHALLENGES AND POLICY PERSPECTIVES IN DEPLOYING AI-DRIVEN AGRICULTURAL ROBOTICS

Collaborative robotics deployment in agriculture for soil and crop yield optimization is itself the intersection of many facets of strategic, technical, economic, and ethical concerns. Overcoming barriers to innovation calls for a comprehensive policy framework that combines provisions for innovation support, socioeconomic inclusion, appropriate regulatory guidelines, and long-term sustainability goals. Furthermore, this section discusses the major roadblocks of this system and presents policy actions based on local and global contexts (Onteddu et al., 2025).

8.1. Technical and Infrastructure Challenges

Agricultural robots that use AI face a significant problem: they require steady access to good-quality data, internet connectivity, GPS accuracy, and a consistent electricity supply, all of which are scarce in rural and developing regions. In rural areas in countries such as Bangladesh, India, and many African nations, the internet penetration is under 40%, impeding the real-time decision-making for edge- or cloud-based robotic systems (Steele & Lo, 2013). Lack of digital infrastructure further prevents the spread of sensor networks and drone-based data collection. Therefore, without investment in rural digitization, collaborative robotics will be little more than a theoretical promise and not a practical solution.

Governments must encourage rural ICT (Information and Communication Technology) infrastructure development through public and private partnerships to develop 5G, LoRaWAN, rural cloud platforms, and others. Local incentives for localized solar or hybrid power systems will make sure that robotic operations keep running (van Hilten & Wolfert, 2022).

Figure 4. AI-Driven robotic system workflow for agriculture

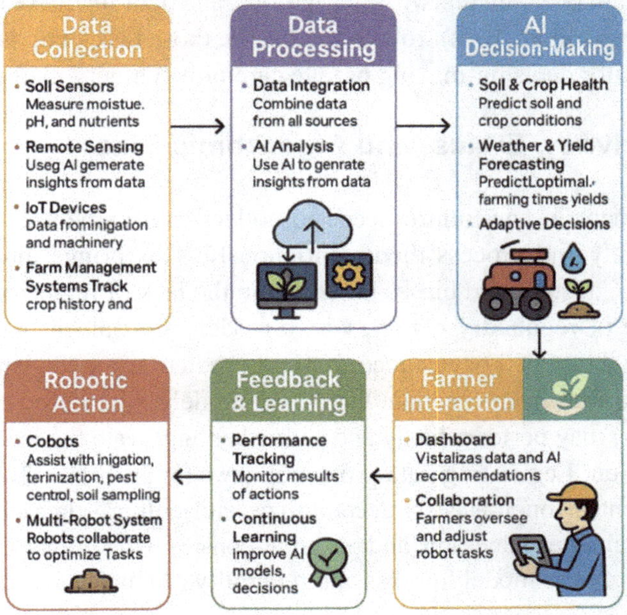

Sources: *Authors' Own*

8.2. Financial Constraints and Economic Viability

Agricultural robots tend to be viewed as expensive, capital-intensive solutions, unattainable by smallholder farmers, who account for 70–80% of the agricultural workforce across the globe (Kim et al., 2020). The limited access to agricultural credit, along with the high upfront costs associated with training AI models and maintaining robotic hardware, widens the adoption gap. For instance, less than 10 percent of farmers in sub-Saharan Africa have access to formal financial institutions that are able to finance such innovations (Kim et al. 2020). Recommendation: Introduce targeted subsidy schemes, cooperative leasing models, and soft loans for the adoption of agri-tech. Create 'Robotic Hubs' or shared service centers for farmers to rent AI-equipped robots for a season to lower entry barriers.

8.3. Workforce Displacement vs. Human-Robot Synergy

One of the main concerns with regard to robotics is labor displacement, particularly when agriculture is the largest employment sector in some countries. However, strategically placed collaborative robots can supplement human labor, not necessarily replace it. Certain tasks are repetitive, hazardous, and data-intensive (e.g., pesticide spraying or deep soil analysis). robots can handle those tasks, and human workers can either monitor decision-making or take care of machinery maintenance.

8.4. Data Privacy, Ethics, and Algorithmic Bias

AI deployment in agriculture needs to collect environmental, crop, and soil data permanently and process them continuously. This brings into question the data ownership, misuse, and biases in AI due to the lack of regulatory clarity and, in the absence of regulatory clarity, where models are trained based on limited or non-diverse datasets. For instance, if yield prediction models are trained only with high-yield data from excellent farms, the models, when used on marginal or degraded lands, may perform badly and so further aggravate the inequality. Policy Recommendation: Legislating data protection laws for agri-data. To help promote the development of open-access, decentralized agricultural datasets that include samples from climate-vulnerable and resource-constrained regions. Create AI audit mechanisms to assess model fairness and reliability (Alzubi et al., 2025).

8.5. Cultural Resistance and Awareness Gaps

Many farms in farming communities, particularly when involving older generations, are skeptical of AI technologies because of low digital literacy or suspicion of machine-based decision-making. Despite the technological feasibility, cultural barriers have the potential to significantly hinder adoption. Recommendation for Policy: Agricultural extension officers have planned to launch community-driven demonstration projects and an awareness campaign. The study recommends creating localized, language-sensitive educational material that demonstrates the practical advantages of robotic farming, such as increased income, reduced drudgery, and sustainable yield growth (Rakshit et al., 2025).

10. DISCUSSION AND RECOMMENDATIONS

10.1 Discussion

Combining AI and robots is a big achievement for sustainable farming practices. AI-powered robots are proven in the chapter to effectively track soil health with little use of resources, as a result of which farmers can produce more crops. They are better than previous tools since they keep track of security at all times. They manage issues one by one and always monitor the field to make choices based on combined information. The combination of supervised learning, non-supervised fertility zoning, and deep learning that detects diseases from images in an image gives a smart system that affects how farmers work on their land. For example, reinforcement learning enables robots to strengthen their feelings based on comments as time goes by and improve irrigation, use of fertilizers, and methods for dealing with pests on their own (Bauer et al. 2025, Li et al. 2025). Dealing with alterations in the climate, soil decline, and crop health, the adaptive mechanism assumes great importance. Examples such as steady watering, correct handling of nutrients, and looking for diseases prove that these systems are successful and also bring benefits to the environment and wallets. Because water is saved, fewer chemicals are used, and the soil structure improves with increased yield per hectare, this method is worth using on a large scale (Elavarasan and Vincent 2020). The chapter pointed out a number of issues and possible obstacles. In particular, the initial expenditure for robotic equipment, most farmers not being familiar with technology, issues with privacy and data safety, and infrastructure, mainly in far-off areas without resources (Huang et al. 2025), are considered major concerns. As a result, different places require different approaches for AI models to perform well due to issues with learning data cooperation between researchers and decision-makers is recommended.

10.2 Recommendations

Share open-access datasets on agriculture, especially those that include many areas with different climates and are clearly labeled to increase the AI model's ability to work well for crop and soil improvements.

Subsidize AI robots for smallholders: The government and development agencies must give aid and operate AI robotics alongside these small farmers.

Investment in Rural Digital Infrastructure: Rural locations require good internet, IoT services, and cloud-edge computer connections to cooperatively use AI in farming. Public-private partnerships should be put to use to solve the gap in infrastructure.

Capacity Building and Training for Farmers: Training on digital topics and robotics through vocational and university courses for farmers provided by agricultural extension centers will make it possible for them to use these technologies at their farms.

Create clear and culturally sensitive AI Systems: Make AI Systems both clear and based on the local knowledge of farming. Such an approach supports both trust and easy use of the tools by all users. Consequently, farmers' responses and area-related know-how are used to develop the design.

11. CONCLUSION

The combination of artificial intelligence (AI) and collaborative robotics has heralded a new age for modern agriculture, specifically in soil health management and crop yield. As evidenced throughout the chapter, AI robotic systems are not just automations; they are intelligent agents that can observe, analyze, and take action on dynamic environmental data while acting autonomously with precision. They change the way we think about farming—moving away from input-based and reactive agriculture toward proactive, efficient, and informed agricultural practices. With algorithmic intelligence (e.g., supervised and unsupervised learning, deep learning, and reinforcement learning) embedded in both ground and aerial robotic platforms, agriculture is able to perform site-specific decision-making at scale. This chapter optimizes irrigation systems, customizes nutrient applications, and monitors disease outbreaks while responding to them often before serious outbreaks occur; it ultimately maximizes productivity and promotes sustainable soil health over time. This chapter also considers real-life use case simulations and examples related to precision irrigation, nutrient zone mapping, and robotic health monitoring, which illustrate the potential of this technology to revolutionize agriculture. Furthermore, the chapter explores the potential of this technology to address challenges in developing nations and areas susceptible to climate change. Low-cost modularity and collaborative service models create opportunities for small-scale farmers to access robotic solutions equitably. With sufficient policy support, infrastructure, and capacity building, AI collaborative robotics can be part of an equitable innovation that supports climate-resilient, sustainable, and productive agriculture in the Global South. After all, while the rise in technology presents vast potential, the context of policy, education, infrastructure, and research will dictate future trajectories of collective change. There has to be collective action of stakeholders to create ethical data governance, transparent AI decision-making, and adaptability to local socio-economic conditions. In addition, interdisciplinary research will always need to continue to build more context-aware, energy-efficient, and understandable AI models, with smallholder farming in different agro-ecological systems in mind. AI-powered collaborative robotics is at the edge of innovation in agriculture, with the potential to not only improve soil and crop productivity but also provide new definitions of sustainability and resilience for food systems. As these smart machines become more capable and available, they will act as a channel for feeding an increasing world population, managing climate risks, and consequently empowering rural communities through the next generation of precision and regenerative agriculture.

12. FUTURE STUDY

As artificial intelligence (AI) as a field continues to grow and mature, it is important that future research achieve better contextual adaptability, scalability, and sustainability in agricultural ecosystems. Such an approach allows for swarm intelligence, whereby robots are able to create and alter knowledge in real time and would greatly enhance task efficiency in swarms of robots, soil management practices, and crop identification and monitoring in larger or fragmented fields. The combination of blockchain and AI-robotics provides an opportunity to look into how data could be openly and securely shared, which is important for developing trust and compliance following prescribed rules in agri-food supply chains. Similarly, edge and 5G technologies may be able to reduce latency for better real-time decision-making in remote parts of the agri-food sectors. In conclusion, future research should also take an interdisciplinary approach, bringing together AI perspectives on robotics, agriculture, economics, and ethics. Future smart agricultural systems should not only be intelligent but also inclusive, equitable, and environmentally sustainable (Muhammed, 2025).

REFERENCES

Adamu, A. Y., & Ahmad, A. (2024). Machine learning: Algorithm, real-world influence, and path to innovation. ISAR Journal of Multidisciplinary Research and Studies, ISAR International Journal of Mathematics and Computing Techniques, 70-77.

Al-Zu'bi, M., Mabhaudhi, T., Daher, B., & Brouziyne, Y. (2025). Inclusive policy development from the ground up: Insights from the household water-energy-food nexus. *Environmental Science & Policy*, *169*, 104084. DOI: 10.1016/j.envsci.2025.104084

Bauer, B., Schulz, M., Hufnagel, T., Schramm, D., & Wittenberg, C. (2025). Reinforcement Learning for Robust Control of Individual Wheel Drive Mobile Robots with Passive Articulated Steering for Reverse Maneuvering. In Degen, H., & Ntoa, S. (Eds.), *Artificial Intelligence in HCI* (pp. 123–137). Springer Nature Switzerland., DOI: 10.1007/978-3-031-93429-2_8

Botero-Valencia, J., García-Pineda, V., Valencia-Arias, A., Valencia, J., Reyes-Vera, E., Mejia-Herrera, M., & Hernández-García, R. (2025). Machine Learning in Sustainable Agriculture: Systematic Review and Research Perspectives. *Agriculture*, *15*(4), 4. Advance online publication. DOI: 10.3390/agriculture15040377

Chavula, P., Kayusi, F., & Juma, L. (2024). Leveraging Artificial Intelligence for Enhancing Wheat Yield Resilience Amidst Climate Change in Sub-Saharan Africa. *LatIA*, *2*, 88. DOI: 10.62486/latia202588

Chen, Z., Li, X., Zhao, T., Wei, X., Song, Z., Wang, J., & Jian, J. (2025). PhenoBee: Drone-based robot for advanced field in vivo contact-based phenotyping in agriculture. *Computers and Electronics in Agriculture*, *236*, 110483. DOI: 10.1016/j.compag.2025.110483

Chevalier, A., Copot, C., De Keyser, R., Hernandez, A., & Ionescu, C. (2015). A Multi-Agent System for Precision Agriculture. In Busoniu, L., & Tamás, L. (Eds.), *Handling Uncertainty and Networked Structure in Robot Control* (pp. 361–386). Springer International Publishing., DOI: 10.1007/978-3-319-26327-4_15

Cob-Parro, A. C., Lalangui, Y., & Lazcano, R. (2024). Fostering Agricultural Transformation through AI: An Open-Source AI Architecture Exploiting the MLOps Paradigm. *Agronomy (Basel)*, *14*(2), 2. Advance online publication. DOI: 10.3390/agronomy14020259

Dai, C.-P., & Ke, F. (2022). Educational applications of artificial intelligence in simulation-based learning: A systematic mapping review. *Computers and Education: Artificial Intelligence*, *3*, 100087. DOI: 10.1016/j.caeai.2022.100087

Daraghmah, R. H., & Younis, N. R. (2025). A Comprehensive Review of Machine Learning Algorithms in Autonomous Robotics: Challenges and Future Prospects. *Estonian Journal of Engineering*, *1*(1), 1. Advance online publication. DOI: 10.63496/eje.Vol1.Iss1.35

Demircioglu, P., Bogrekci, I., Durakbasa, M. N., & Bauer, J. (2024). Autonomation, Automation, AI, and Industry-Agriculture 5.0 in Sustainable Agro-Ecological Food Production. In Durakbasa, N. M., & Gençyılmaz, M. G. (Eds.), *Industrial Engineering in the Industry 4.0 Era* (pp. 545–556). Springer Nature Switzerland., DOI: 10.1007/978-3-031-53991-6_42

Elavarasan, D., & Vincent, P. M. D. (2020). Crop Yield Prediction Using Deep Reinforcement Learning Model for Sustainable Agrarian Applications. *IEEE Access : Practical Innovations, Open Solutions*, *8*, 86886–86901. DOI: 10.1109/ACCESS.2020.2992480

Farhangmehr, V., Imanian, H., Mohammadian, A., Cobo, J. H., Shirkhani, H., & Payeur, P. (2025). A spatiotemporal CNN-LSTM deep learning model for predicting soil temperature in diverse large-scale regional climates. *The Science of the Total Environment*, *968*, 178901. DOI: 10.1016/j.scitotenv.2025.178901 PMID: 39987832

Fountas, S., Espejo-García, B., Kasimati, A., Gemtou, M., Panoutsopoulos, H., & Anastasiou, E. (2024). Agriculture 5.0: Cutting-Edge Technologies, Trends, and Challenges. *IT Professional*, *26*(1), 40–47. DOI: 10.1109/MITP.2024.3358972

Gadekallu, T. R., Kumar Reddy Maddikunta, P., Boopathy, P., Deepa, N., Chengoden, R., Victor, N., Wang, W., Wang, W., Zhu, Y., & Dev, K. (2025). XAI for Industry 5.0—Concepts, Opportunities, Challenges, and Future Directions. *IEEE Open Journal of the Communications Society*, *6*, 2706–2729. DOI: 10.1109/OJCOMS.2024.3473891

Getahun, S., Kefale, H., & Gelaye, Y. (2024). Application of Precision Agriculture Technologies for Sustainable Crop Production and Environmental Sustainability: A Systematic Review. *TheScientificWorldJournal*, *2024*(1), 2126734. DOI: 10.1155/2024/2126734 PMID: 39421732

Gopi, A., L.R, S., & S., I. T. J. (. (2025). Enhancing soil data classification for cassava crop productivity using swarm robotics optimization (SRO) and convolutional neural networks. *International Journal of System Assurance Engineering and Management*, *16*(4), 1409–1423. DOI: 10.1007/s13198-025-02727-2

Hong, W. (2001). *Modeling, Estimation, and Control of Robot-Soil Interactions*. https://dspace.mit.edu/handle/1721.1/7075

Huang, J., Xu, Y., Wang, Q., Wang, Q., Liang, X., Wang, F., Zhang, Z., Wei, W., Zhang, B., Huang, L., Chang, J., Ma, L., Ma, T., Liang, Y., Zhang, J., Guo, J., Jiang, X., Fan, X., An, Z., & Fei, A. (2025). Foundation models and intelligent decision-making: Progress, challenges, and perspectives. *Innovation (Cambridge (Mass.))*, *6*(6), 100948. Advance online publication. DOI: 10.1016/j.xinn.2025.100948 PMID: 40528892

Jana, P. (2024). AI-powered IoT solutions for sustainable water management in cities. *Uncertainty Discourse and Applications*, *1*(2), 2. Advance online publication. DOI: 10.48313/uda.v1i2.37

Javaid, M., Haleem, A., Khan, I. H., & Suman, R. (2023). Understanding the potential applications of Artificial Intelligence in Agriculture Sector. *Advanced Agrochem*, *2*(1), 15–30. DOI: 10.1016/j.aac.2022.10.001

Javidan, S. M., Banakar, A., Vakilian, K. A., & Ampatzidis, Y. (2023). Diagnosis of grape leaf diseases using automatic *K*-means clustering and machine learning. *Smart Agricultural Technology*, *3*, 100081. DOI: 10.1016/j.atech.2022.100081

Kim, J., Shah, P., Gaskell, J. C., & Prasann, A. (2020). *Scaling Up Disruptive Agricultural Technologies in Africa*. World Bank Publications. DOI: 10.1596/978-1-4648-1522-5

Korthals, T., Kragh, M., Christiansen, P., Karstoft, H., Jørgensen, R. N., & Rückert, U. (2018). Multi-Modal Detection and Mapping of Static and Dynamic Obstacles in Agriculture for Process Evaluation. *Frontiers in Robotics and AI*, *5*, 28. Advance online publication. DOI: 10.3389/frobt.2018.00028 PMID: 33500915

Kuli, B. K., Debnath, J., Sheikh, A., Das, S., & Balai, P. S. (2025). Smart Farming Revolution: AI, IoT, and Robotics in Precision Agriculture and Soil Conservation. *International Journal of Scientific Research in Science, Engineering and Technology*, *12*(2), 688–706. DOI: 10.32628/IJSRSET25122193

Lakhiar, I. A., Yan, H., Zhang, C., Wang, G., He, B., Hao, B., Han, Y., Wang, B., Bao, R., Syed, T. N., Chauhdary, J. N., & Rakibuzzaman, M. (2024). A Review of Precision Irrigation Water-Saving Technology under Changing Climate for Enhancing Water Use Efficiency, Crop Yield, and Environmental Footprints. *Agriculture*, *14*(7), 1141. DOI: 10.3390/agriculture14071141

Lal, R. (2016). Feeding 11 billion on 0.5 billion hectare of area under cereal crops. *Food and Energy Security*, *5*(4), 239–251. DOI: 10.1002/fes3.99

Li, S., Liu, J., Yang, Y., Shen, R., & Jiang, J. (2025). Thinking as human: Self-reflective reinforcement learning framework for fertilization decision-making. *Smart Agricultural Technology*, *10*, 100841. DOI: 10.1016/j.atech.2025.100841

Luo, L., Liu, B., Chen, M., Wang, J., Wei, H., Lu, Q., & Luo, S. (2025). DRL-enhanced 3D detection of occluded stems for robotic grape harvesting. *Computers and Electronics in Agriculture*, *229*, 109736. DOI: 10.1016/j.compag.2024.109736

Lytridis, C., Kaburlasos, V. G., Pachidis, T., Manios, M., Vrochidou, E., Kalampokas, T., & Chatzistamatis, S. (2021). An Overview of Cooperative Robotics in Agriculture. *Agronomy (Basel)*, *11*(9), 1818. DOI: 10.3390/agronomy11091818

Milella, A., Reina, G., & Nielsen, M. (2019). A multi-sensor robotic platform for ground mapping and estimation beyond the visible spectrum. *Precision Agriculture*, *20*(2), 423–444. DOI: 10.1007/s11119-018-9605-2

Molotoks, A., Smith, P., & Dawson, T. P. (2021). Impacts of land use, population, and climate change on global food security. *Food and Energy Security*, *10*(1), e261. DOI: 10.1002/fes3.261

Muhammed, D. (2025). *Artificial Intelligence of Things (AIoT) for Smart Agriculture: Architectures and Solutions* [Phdthesis, Sorbonne Université, Faculté des Sciences et Ingénierie]. https://hal.science/tel-05046536

Onteddu, A. R., Kundavaram, R. R., Kamisetty, A., Gummadi, J. C. S., & Manikyala, A. (2025). Enhancing Agricultural Efficiency with Robotics and AI-Powered Autonomous Farming Systems. *Malaysian Journal of Medical and Biological Research*, *12*(1), 1.

Praveen, R., Mittal, M., Parida, P., Begum, R., Kumar, Y., & Nijhawan, G. (2025). Deep Learning Applications for Detecting Crop Diseases from Image Data. *2025 International Conference on Computational, Communication and Information Technology (ICCCIT)*, 566–571. DOI: 10.1109/ICCCIT62592.2025.10928061

Qi, H., & Altinakar, M. S. (2011). Simulation-based decision support system for flood damage assessment under uncertainty using remote sensing and census block information. *Natural Hazards*, *59*(2), 1125–1143. DOI: 10.1007/s11069-011-9822-8

Qu, J., Cui, G., Li, Z., Fang, S., Zhang, X., Liu, A., Han, M., Liu, H., Wang, X., & Wang, X. (2024). Advanced Flexible Sensing Technologies for Soft Robots. *Advanced Functional Materials*, *34*(29), 2401311. DOI: 10.1002/adfm.202401311

Rahmadian, R., & Widyartono, M. (2020). Autonomous Robotic in Agriculture: A Review. *2020 Third International Conference on Vocational Education and Electrical Engineering (ICVEE)*, 1–6. DOI: 10.1109/ICVEE50212.2020.9243253

Rakshit, S., Gorai, S. K., Das, B., Sahu, S., Gills, R., Ray, P., & Mukherjee, A. (2025). Empowerment of Rural Women and Youth Retention as Drivers of the Agricultural Economy. In Sharma, N. K., Rai, P. K., & Rai, D. C. (Eds.), *Indian Agriculture: Challenges, Priorities and Solutions* (pp. 149–181). Springer Nature., DOI: 10.1007/978-981-96-5273-0_8

Rashid, A. B., & Kausik, M. A. K. (2024). AI revolutionizing industries worldwide: A comprehensive overview of its diverse applications. *Hybrid Advances*, *7*, 100277. DOI: 10.1016/j.hybadv.2024.100277

Redhu, N. S., Thakur, Z., Yashveer, S., & Mor, P. (2022). Artificial intelligence: A way forward for agricultural sciences. In Sharma, P., Yadav, D., & Gaur, R. K. (Eds.), *Bioinformatics in Agriculture* (pp. 641–668). Academic Press., DOI: 10.1016/B978-0-323-89778-5.00007-6

Roy, P., Padhiary, M., Hoque, A., Sahu, B., Roy, D., & Kumar, K. (2025). Machine Learning for Precision Agriculture and Crop Yield Optimization: Techniques and Applications. In *The Role of Artificial Intelligence in Advancing Applied Life Sciences* (pp. 189–234). IGI Global Scientific Publishing. DOI: 10.4018/979-8-3693-9208-9.ch007

Roy, S., Ansal, D. K., & Kumar, M. (2025). Soil Texture Prediction: Advances in Remote Sensing, Image Analysis, and Machine Learning. *2025 Fourth International Conference on Smart Technologies, Communication and Robotics (STCR)*, 1–4. DOI: 10.1109/STCR62650.2025.11020387

Savitha, C., & Talari, R. (2025). Evaluating the performance of random forest, support vector machine, gradient tree boost, and CART for improved crop-type monitoring using greenest pixel composite in Google Earth Engine. *Environmental Monitoring and Assessment*, *197*(4), 437. DOI: 10.1007/s10661-025-13880-3 PMID: 40108031

Schwalbert, R. A., Amado, T., Corassa, G., Pott, L. P., Prasad, P. V. V., & Ciampitti, I. A. (2020). Satellite-based soybean yield forecast: Integrating machine learning and weather data for improving crop yield prediction in southern Brazil. *Agricultural and Forest Meteorology*, *284*, 107886. DOI: 10.1016/j.agrformet.2019.107886

Senathipathi, S., Abirami, N., Narayanasamy, P., Mathankumar, S., Lotus, A. A., & Muthulekshmi, M. (2025). Smart Robotic Crop Replanting and Recovery Using CNNs and Edge Computing. *2025 3rd International Conference on Disruptive Technologies (ICDT)*, 1349–1354. DOI: 10.1109/ICDT63985.2025.10986524

Sibona, F., Harrison, N., Wallace, N., Stevens, C., Sukkarieh, S., Raubenheimer, D., Simpson, S. J., & González, L. A. (2025). A Framework to Enable Eco-Cyber-Physical Systems for Robotics-Focused Digital Twins in Smart Farming. *Journal of Field Robotics*, rob.22562. DOI: 10.1002/rob.22562

Singh, T., Kundroo, M., & Kim, T. (2024). WSN-Driven Advances in Soil Moisture Estimation: A Machine Learning Approach. *Electronics (Basel)*, *13*(8), 8. Advance online publication. DOI: 10.3390/electronics13081590

Spagnuolo, M., Todde, G., Caria, M., Furnitto, N., Schillaci, G., & Failla, S. (2025). Agricultural Robotics: A Technical Review Addressing Challenges in Sustainable Crop Production. *Robotics (Basel, Switzerland)*, *14*(2), 9. DOI: 10.3390/robotics14020009

Steele, R., & Lo, A. (2013). Telehealth and ubiquitous computing for bandwidth-constrained rural and remote areas. *Personal and Ubiquitous Computing*, *17*(3), 533–543. DOI: 10.1007/s00779-012-0506-5

Urrea, C., & Kern, J. (2025). Recent Advances and Challenges in Industrial Robotics: A Systematic Review of Technological Trends and Emerging Applications. *Processes (Basel, Switzerland)*, *13*(3), 3. Advance online publication. DOI: 10.3390/pr13030832

van Hilten, M., & Wolfert, S. (2022). 5G in agri-food—A review on current status, opportunities and challenges. *Computers and Electronics in Agriculture*, *201*, 107291. DOI: 10.1016/j.compag.2022.107291

Xu, J., Ren, C., Zhang, X., Wang, C., Wang, S., Ma, B., He, Y., Hu, L., Liu, X., Zhang, F., Lu, L., Li, S., Zhang, J., Zhu, Y.-G., Vitousek, P., & Gu, B. (2025). Soil health contributes to variations in crop production and nitrogen use efficiency. *Nature Food*, *6*(6), 1–13. DOI: 10.1038/s43016-025-01155-6 PMID: 40234681

Chapter 4
Exploring the Technical Foundations and Integration Strategies of Generative AI Within Collaborative Robotics Systems

Laya Billinty Varra
https://orcid.org/0009-0001-1205-3789
Stanley College of Engineering and Technology for Women, India

Jagadeshwari Puttanapura
https://orcid.org/0009-0008-9607-7438
University of South Alabama, USA

C. Kishor Kumar Reddy
Stanley College of Engineering and Technology for Women, India

Shikha Khullar
https://orcid.org/0009-0000-5598-3145
Poornima University, India

Ghita Lazrek
https://orcid.org/0009-0004-7306-7369
Lab LIASSE, ENSA, Université Sidi Mohamed Ben Abdellah, Fez, Morocco

Jothi Paranthaman
Botho University, Botswana

ABSTRACT

The fusion of Generative AI with collaborative robots (cobots) represents a new frontier in Agriculture 6.0, by allowing intelligent decisions, adaptive learning, and increased automation. This chapter provides a thorough technical overview of the architectures, methodologies, and implementation approaches of integrating generative AI models (GANs, Transformers, and diffusion models) with cobots for

DOI: 10.4018/979-8-3373-4207-8.ch004

agricultural applications. The focus of this chapter is on real time adaptability, data efficient learning, and multi-modal sensor fusion for cobots and AI generation techniques. The chapter discusses hardware-software co-design, model deployment on edge processing hardware, interoperability issues, and use cases in the field. The information presented is designed to benefit researchers, technologists, and practitioners working to build sustainable, intelligent, and collaborative agricultural environments.

1. INTRODUCTION TO GENERATIVE AI IN AGRICULTURE

Overview of Generative AI: The way agriculture has been incorporated and integrated with artificial intelligence (AI) has modernized traditional agriculture into an efficient, data-driven industry. Emerging technologies available in AI including Generative AI (Gen AI). Generative AI is a class of models that learn from data and produce new, realistic data that resemble original data consisting of consistent patterns. Generative models, such as Generative Adversarial Networks (GANs), Variational Autoencoders (VAEs) and Transformer based architectures have learned to generate images of crops, simulate environmental conditions, and make artificial sensor data. This ability is particularly useful in agriculture where observable high-quality labeled datasets are often limited due to the costs, variability and complexity of the field. (Sai, S., et al., 2025)

Historical perspective of cobot integration in smart farming: The development of collaborative robots (cobots) overlay the evolution of AI. Cobots, designed to accompany humans in the workplace were introduced into agriculture to eliminate labor intensive tasks. The early iterations of agricultural cobots were traditional robots that used rule based systems, had a limited range of autonomy and performed repetitive tasks such as seeding, spraying, and harvesting. Due to advances in robotics and AI over the past 10 years, agricultural robotics has advanced to the point where agri-cobots can navigate their surroundings in a context aware fashion. Today's agri-cobots employ visual perception, sense their environment and utilize intelligent control systems to gather and analyse their surroundings before undertaking autonomous actions. Recent advances in generative AI, will enable agri-cobots to now simulate a range of field conditions, improve perceptual fidelity and dynamically adjust their behaviours establishing a new chapter in deep AI and robotics used in smart farming. (Batra, I., et al., 2024)

Significance of AI-generated data and models in agriculture: A major catalyst for this synergy will be the use of AI-generated data and models. In an environment as dynamic as agriculture, preparing a training dataset for cobots is the labour intensity and effort of preparing such a dataset can be substantial. Generative AI eliminates that labour hurdle via synthetic datasets that can help train models with demonstrably similar tasks such as disease detection, yield prediction or weed identification. For example, a GAN can create photorealistic images of crops with diseased variances to support the robustness of models, even without large collections of real images. This method expedites the formation of dependable AI systems with skin in the game and with smaller dependence on data experiences. Using generative models directly on cobots via edge AI provide the ability to learn and adapt while using a real-time environment, or field context; which it processes as decentralized intelligence for a reduced response time and adoption of its own adapt preferences without detract to the cloud - further allowing crops to operate with minimum human influence using adaptive responses. (Sapkota, R., et al., 2025)

Table 1 summarizes the main benefits of integrating Artificial Intelligence (AI) and collaborative robots (cobots) in modern agricultural contexts. Each benefit category also specifies a performance, sustainability, or decision-making area that is enhanced by AI-cobots. AI-cobots benefit modern agriculture by offering reduced labor dependence, improved accuracy, improved data-driven actions, and synthetic training data from artificial intelligence (AI). These benefits highlight potential for transformative change in AI-cobot integration in various agricultural environments. Each benefit is contextualized with real-world examples that demonstrate deployment in the field.

Table 1. Real-World benefits of integrating AI with agricultural cobots

Benefit Category	Description	Examples
Labor Efficiency	Reduces dependency on manual labor through automation	Robotic harvesting, autonomous spraying
Data-Driven Decision Making	Enables cobots to adapt actions based on real-time AI analysis	Dynamic irrigation adjustments, disease detection
Operational Accuracy	Minimizes human error and ensures precise task execution	Row-specific seeding, targeted pesticide application
Environmental Sustainability	Reduces resource usage through intelligent planning	Lower water and chemical usage via AI-optimized routines
Economic Productivity	Enhances yield quality and throughput by ensuring timely interventions	Predictive maintenance of crops, early stress detection
Scalability	Adaptable for different crop types and field scales	Configurable platforms for vineyards, orchards, and greenhouse operations
Data Augmentation	Uses generative models to synthesize training data in low-data environments	Simulating pest-infested vs. healthy crop images for AI training

2. ARCHITECTURAL OVERVIEW OF COBOTS

Core components of agricultural cobots: Collaborative robots in agriculture are designed for autonomous operation in complex farming environments with limited human direction and interface. Agricultural robots are made up of mechanical, electronic, and computational subsystems or components that can interact to perform numerous agricultural tasks accurately. Agricultural cobots typically consist of a mobile chassis or base to traverse varied terrain, a manipulator or end-effector to perform an operation, embedded computing with hardware for control and processing, and a power supply or energy source (such as rechargeable batteries or solar-powered systems). Their chassis tends to be rugged, and often, modular, allowing tools for the specific agricultural operations such as weeding, harvesting, snipping, or crop monitoring to be attached. (Taesi, C., et al., 2023)

Figure 1. Functional block diagram of an agricultural cobot

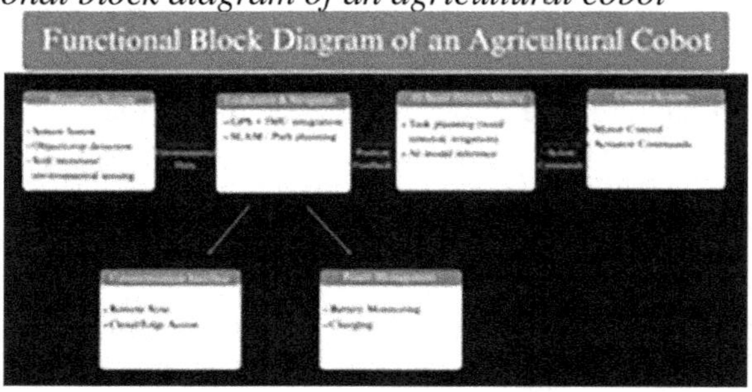

Figure 1 shows the internal functional architecture of a collaborative agricultural robot (cobot). The diagram shows a logical flow of operations from sensing the environment to physically implementing the task to be performed. The representation begins with the Perception System where sensors will implement sensor fusion, object/crop detection, and environmental sensing. It then takes the perceived information and sends it to the Localization and Navigation module where, using GPS and IMU and path planning, the location and trajectory of the cobot will be determined. The spatial and sensory information is inputted into the AI-Based Decision Making block where it will be planned for a specific task (e.g. weed removal or irrigation) in real time employing AI inference. The decisions will then pass to the Control System where motor control and actuator are generated to implement the task. The Control System works simultaneously allowing for controlled motor function while the Communication Interface modules allow for communication and synchronization of data with any remote or cloud-based systems. The Power Management module controls and monitors the energy consumption of the cobot. The labeled arrows between modules (e.g. "Environmental Data", "Position Feedback" and "Action Commands") clearly define the type of data and whether it is feedback, or control flow to the respective modules, adding visibility and traceability to the operation of the system.

Sensor systems, actuator control, and edge computing: Sensor integration, actuator control, and edge computing comprise the building blocks of intelligent cobot behaviors. Sensor arrays provide real-time perception of the environment and include ambient visual sensing (RGB cameras, multispectral cameras), spatial awareness (LiDAR), geolocation (GPS), inertial sensing (motion), and environmental monitoring (temperature, humidity, soil moisture). Sensors generate data streams that are analyzed by edge computing (an embedded platform within the cobot) to allow for real-time response to changes in the environment without relying on remote data centers - that is, while relying on data latency below the processing threshold of the sensor network with respect to the environment. Actuator control systems will interpret high-level intent commands and control the mechanical elements of the cobot (for example, arm movements and attachments, navigation speeds of autonomous vehicles, and tool engagement), allowing the cobot to precisely execute commands such as arm motions, tool harvesting, and guidance through its space with respect to the personalized task at hand. The integration of the sensor network, potential actuators, and processing controls allows the cobots to function efficiently and adaptively in real-time and dynamic agricultural settings. (Ali, Z., et al., 2025)

Figure 2. Anatomy of a typical agricultural cobot

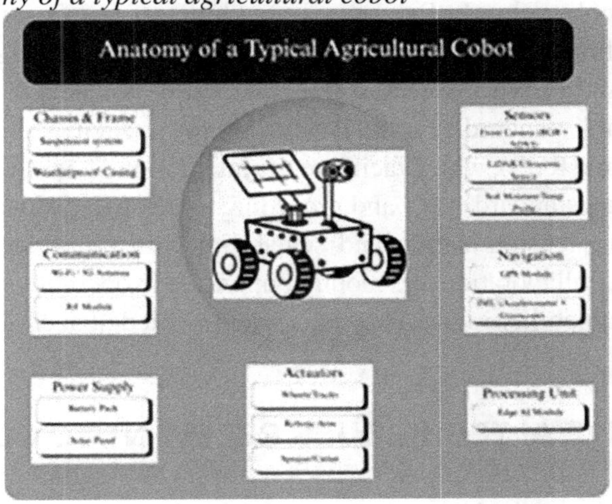

Figure 2 depicts the physical architecture for a typical agricultural collaborative robot (cobot) in a side-profile view showing the necessary hardware components .

The physical embodiment of the recommended cobot is a ground-based rover which is typical of modern precision farming production systems. Various sensor systems (RGB, NDVI cameras, LiDAR, ultrasonic sensors, soil moisture or temperature probes, etc.) are benchmarked on the frame where they obtain the best perception of the surrounding environment. Cobot benchmarks include active navigation (e.g., GPS module and IMU) as well as passive navigation. The processing unit typically utilizes an edge AI module (NVIDIA Jetson, Raspberry Pi, etc...) to achieve real-time status determination and functionality. Cobot actuation systems may include wheels or treads to achieve movement as well as robotic arms or a sprayer for physical tasks. The cobot obtains its power from an onboard battery pack, and may contain solar panels for extended on-field operation. Communications systems (e.g. Wi-Fi, 5G antennas, RF transmitters, etc.) permit remote access monitoring and cloud connection. The physical chassis and frame provides a structural base, weather proofing, and a suspension system to travel off road. These components describe the anatomical structure of a cobot, creating the potential for complete autonomous or semi autonomous behavior in differing environmental conditions for agricultural tasks.

Communication frameworks and interoperability: Communication infrastructure and interoperability are also essential characteristics of the cobot architecture. In modern digital agriculture, cobots are expected to communicate with other machines, human operators, and centralized farm management systems. The messages exchanged between cobots and their environment will be communicated through wireless, machine-to-machine protocols such as Wi-Fi, Bluetooth, Zigbee, and long-distance, low-power protocols such as LoRa, that will enable data transmissions and reception. These systems allow for coordinated coordination, and monitoring throughout the farm in real time. Likewise, software interoperability is linked to hardware communication through standardized middleware platforms such as autonomous Robot Operating Systems (ROS), which facilitate the integration of hardware and software modules through a standard stage and procedure. Interoperability of hardware and software promotes customization and scalability with multiple layers of smart farming systems, collaboration and modular upgrades to operational efficiency, and longer-term sustainability in precision agriculture. (Riedelbauch, D., et al., 2023)

3. GENERATIVE AI TECHNIQUES RELEVANT TO COBOTS

Generative Adversarial Networks (GANs): Generative AI has started to change the way collaborative robots (cobots) function in farming. One of the best methods is Generative Adversarial Networks (GANs), which, during training, capitalize on how both a generator and discriminator neural network learn against each other. In agricultural applications, GANs are especially useful when generating synthetic images of plants under different conditions of a plant stressor such as disease, drought, or pest. Synthetic datasets can help alleviate the shortage of labeled training data and can ultimately teach cobots to identify and respond to numerous field conditions. Furthermore, GANs can regenerate crop patterns, signify future crop growth, or model the weed population distribution in the fields, enabling cobots to execute a better plan of action. (Rahman, Z. U., et al., 2024)

Variational Autoencoders (VAEs): Variational Autoencoders (VAEs) also play a crucial role in enabling cobots to interpret and reconstruct complex agricultural environments. VAEs learn to encode input data—such as images of plants or soil profiles—into a compressed latent representation from which new data can be generated. This ability to generate variations of input data proves useful in anomaly detection and adaptive learning, where cobots must distinguish between healthy and unhealthy crops or recognize unfamiliar environmental patterns. Unlike GANs, VAEs are more stable to train and provide a probabilistic understanding of data distribution, making them suitable for real-time field applications where uncertainty must be factored into the robot's decision-making processes (Akkem, Y., et al., 2024) (Isinkaye, F. O., et al., 2025).

Large Language Models (LLMs) for decision-making: Large Language Models (LLMs), such as GPT-style transformers, have recently found relevance in robotics for their ability to interpret complex instructions and support high-level decision-making. In the context of agricultural cobots, LLMs can be integrated as planning agents that interpret multi-modal inputs—such as sensor data, historical weather records, and human commands—and generate coherent action plans or dialogues. For example, a cobot equipped with an LLM can receive voice or text input from a farmer, interpret the task (e.g., "check for leaf blight in plot B"), and execute an appropriate sequence of actions. This language-driven control paradigm enhances the usability and flexibility of cobots, especially in heterogeneous farming environments where adaptability is essential. (Vizniuk, A., et al., 2025)

Diffusion models for image synthesis and field mapping: Diffusion models, a more recent advancement in generative AI, offer powerful tools for high-fidelity image synthesis and field mapping. These models work by gradually transforming noise into coherent data samples through a diffusion process, enabling the generation of ultra-realistic visual content. In agriculture, diffusion models can be employed to create precise topographic maps, simulate crop stress patterns, or fill in missing sensor data during mapping operations. When integrated with a cobot application, these high fidelity models only improve the ability of the autonomous system to navigate, identify and understand obstacles, and specify the mapping conditions. Their relative ability to provide sight in high resolution data is one of the most interesting prospects as it allows for very fine granularity of visual analysis of situational awareness in a complex, 2-D or 3-D environment. (Heschl, A., et al., 2024)

Table 2 summarizes the use of popular generative AI model types (e.g., Generative Adversarial Networks (GANs), Variational Autoencoders (VAEs), Transformers, and Diffusion Models) in agriculture. It shows the application of each of the models to address specific applications of generative AI (e.g., dealing with lack of data or anomalous data, sequential decision making, and simulating complex environments). The table also demonstrates how these models will improve the capabilities of the agricultural cobots, acting more autonomously, more accurately, and more flexibly than in the real world.

Table 2. Use of GANs, VAEs, transformers, and diffusion models in agricultural context

Model Type	Application in Agriculture	Cobot Relevance
GANs	Generating synthetic crop images under various disease and lighting conditions	Improves vision systems for cobots in low-data scenarios
VAEs	Learning latent representations of soil conditions and crop health	Enables unsupervised monitoring and detection of anomalies
Transformers	Planning crop care tasks based on sequential weather and sensor data	Assists cobots in task scheduling and adaptive behavior
Diffusion Models	Simulating realistic crop fields and terrain for training navigation models	Enhances virtual training environments for cobot path planning

4. SYSTEM INTEGRATION AND PIPELINE DESIGN

Data acquisition and preprocessing from sensors: The workflow for integrating generative AI with cobot platforms in the agriculture sector begins with acceptable system architecture, such that data can flow seamlessly from sensors to the decision-making units. The workflow begins with data acquisition and preprocessing. Once the data has been collected, there are four known different types of onboard sensors that cobots can obtain multi-modal information from. These include RGB cameras, multispectral cameras, GPS sensor for the position of the cobot, spatial mapping, LiDAR for navigating, environmental humidifiers, temperature sensors, soil sensing (gravimetric density), and many more onboard sensors. Raw data in the field is also often very noisy, sometimes inconsistent, or completely incomplete based on local environment variability. At this stage, preprocessing (aka data wrangling) is a necessary step in which normalization, filtering, augmenting, and extracting features occur. Overall, these operations prepared the data in a way that is less redundant, in a way that highly relevant features became fixed and highlighted. As a result, the AI models down-stream are able to receive and analyze better cleaned, non-redundant, and more meaningful data that contributes to making informed and confident inferences. (Lee, S., 2019)

Real-time AI inference on cobot platforms: Data is prepared, and then it is ingested in the inference engine, to run real-time AI on the cobot platform. This stage uses generative AI models such as GANs, VAEs, or diffusion models for predictive tasks, or data synthesis, or anomaly detection. For example, cobot could apply a trained generative model to expect crop disease based on the current visual input as well as infer where the reduced risk of diseased areas on the field is to generate missing imagery from occluded parts of a state map of the field, for better situational awareness. Real-time inference is especially relevant in agricultural contexts where delayed decision may decrease operational efficiency or possibly affect the crop. It is also relevant, as cobot platforms are enabled with additional hardware, such as a dedicated GPU, TPU, or low-power AI accelerator, so that the trained model can work seamlessly with respect to time and energy necessary for executing the model directly on the device.

Edge AI vs Cloud AI deployment models: When deploying the different AI models on cobots, practitioners have to make a determination between Edge AI and Cloud AI, both of which have unique advantages and disadvantages. Edge AI implies that inference and computation happen locally on the cobot and not on remote servers. The benefits of this approach are lower latency, greater reliability in rural areas with poor connectivity, and data privacy. Edge AI is best for real-time tasks such as obstacle avoidance, object detection, and adaptive spraying. Cloud AI is based on a model where centralized servers are used to conduct heavier computations or to train complex generative models that are outside the scope of the cobot's onboard processing. While cloud deployment is inherently scalable and more powerful computing masses larger datasets, these come with dependency on the availability of a network and synchronization delays when transmitting the data via the Internet. In practice, most smart farming systems take a hybrid architecture and keep the critical real-time tasks on the Edge and then the heavier tasks like model updates and historical analysis in the cloud. This loosely coupled pipeline design enables an equitable distribution of compute load in a way that enables the cobots' practical usability without devoting all the computing capacity in an asynchronous state that has the full generative AI support behind it. (Kum, S., et al., 2024)

Table 3 provides an overview of widely used tools and frameworks aligned with each stage of integrating generative AI models into agricultural cobot systems. From initial data acquisition to real-time monitoring and control, these tools enable a seamless workflow that bridges AI development with robotic execution. The listed technologies support sensor integration, data processing, model training and deployment, robotic motion planning, and feedback collection—ensuring a robust and scalable AI-cobot pipeline tailored for precision agriculture applications.

Table 3. Tools and frameworks for each stage of integration

Integration Stage	Common Tools/Frameworks	Purpose
Data Acquisition	OpenCV, ROS Sensors, LIDAR SDKs	Capturing and streaming sensor data
Data Preprocessing	Pandas, NumPy, LabelImg, Albumentations	Cleaning and preparing datasets
Model Training	PyTorch, TensorFlow, Keras, FastAI	Building and training generative AI models
Model Deployment	TensorRT, ONNX, OpenVINO, NVIDIA JetPack	Optimizing and deploying models to edge devices
Execution & Control	Robot Operating System (ROS), ROS2, MoveIt, Gazebo	Integrating AI outputs with robotic motion and behavior control
Monitoring & Feedback	MLflow, TensorBoard, Prometheus, ROS logs	Tracking model performance and gathering operational feedback

5. APPLICATIONS IN PRECISION AGRICULTURE

Crop monitoring with generative models: The collaboration between generative AI and collaborative robots is creating new possibilities in precision agriculture through enhanced monitoring, decision-making, and actuation. One of the most promising applications is in crop monitoring using generative models, where artificial intelligence algorithms can simulate crop health over time and quantify the AI predictions compared to standards. Generative models like GANs and diffusion models can generate realistic images of plants with varying amounts of stress (nutrient deficiency, disease, pests), which allows cobots to recognize subtle signs that may go unnoticed. Generative models also generate predictive visualizations of crop development based on environmental data input to help farmers and cobots preemptively address an issue before it becomes a crisis. AI cobots can classify actual field-based data and learn generative features when trained on both actual video data and synthetic images, resulting in cobots being more accurate and robust in identifying abnormalities, yield estimations and suggesting when to take actions. (Kerdegari, H., et al., 2019)

Soil condition simulation and treatment planning: Generative AI has also improved on soil condemnation simulation and treatment options. Soil health is the basis for agriculture to develop sustainably, and at times monitoring soil in "real-time" across expansive areas can be difficult. Cobots with environmental sensors accumulate data needed to assess factors like pH, moisture, and nutrients, which generative models can simulate the future behavior of the soil undergoing various treatment strategies. A VAE-based after all could develop a model predicting how the soil will behave using fertilizer variations or rotation patterns for irrigation and develop virtual predictions such that an agronomist can read-off the most effective and environmentally optimal treatment option. By utilizing simulation generative models, cobots can reduce the physical trial-and-error learning models of human and animal intelligence and use "minimal" enforcement (or maximum chances of success) while also accelerating resource use efficiencies and crop-productive results. (Shaheb, M. R., et al., 2022)

AI-generated visual guidance for robotic arms: Generative AI is also relevant in the area of visual guidance of robotic arms preferable to mainly harvesting, pruning and precision spraying tasks. Traditional robotic manipulation systems have largely implemented deterministic vision algorithms, which struggle with unstructured outdoor environments given the variability in lighting, the geometry of plants, occlusion and other factors. The use of AI-generated visual maps and depth predictions presented a more holistic and nuanced understanding of the cobot's surrounding workspace. For example, diffusion models can reproduce hidden areas of the plant or 3D visual guides that help robotic arms approach fruit without damage to the fruit itself or surrounding branches (with this all adding a dynamic visual guidance aspect that improves operational efficiency, dexterity, and procedural mechanical error). Thus, depending on how generative models improve in both computational speed and material realism, and can be incorporated into cobot vision and control systems, it will only continue to expand the types of tasks that these systems are able to complete autonomously and reliably.

Table 4 indicates how generative AI improves the key actions undertaken by agricultural cobots, namely planting, weeding, spraying, harvesting, irrigation, and monitoring. In each case, the AI-enabled capabilities, the explicit role of generative models (e.g., generating adaptive plans, generating weed maps, generating natural language reports), and the benefits - such as increasing efficiency, reducing inputs and improving crop performance - are indicated. There are real systems and robots in practice that demonstrate the generative capabilities in precision agriculture.

Table 4. Cobot actions enabled by generative AI

Cobot Action	AI-Enabled Capabilities	Role of Generative AI	Benefits	Example Systems/ Robots
Planting	Generate precise planting patterns and seed placement based on field variability data.	Generates adaptive planting plans optimized for soil and weather conditions.	Maximizes germination rates and resource efficiency.	Autonomous seed planters with AI guidance
Weeding	Identify weed locations and generate targeted	Synthesizes weed maps and optimal removal paths	Reduces herbicide use and crop damage.	Robotic weeders like Blue River's See & Spray
Spraying	Generate spraying schedules and dosage based on crop health and pest data.	Creates adaptive spraying plans minimizing chemical use.	Environmentally friendly and cost-effective.	Precision sprayers with AI control
Harvesting	Generate optimal harvesting sequences and timing predictions based on crop maturity	Predicts best harvest windows and navigational routes.	Reduces crop loss and labor costs.	Autonomous harvesters like John Deere's See & Spray
Irrigation	Generate irrigation schedules customized by microclimate and soil moisture data.	Synthesizes sensor data to optimize water delivery.	Saves water and enhances crop growth	Smart irrigation cobots
Monitoring & Reporting	Generate real-time reports and alerts from multisource sensor data	Automates data fusion and natural language summary generation	Improves farm management decisions.	AI-powered drones and ground robots

6. MODEL OPTIMIZATION AND DEPLOYMENT

Model compression and quantization for cobots: Using generative AI models in agricultural cobots involves more than just correctly generating models; it involves models that are lightweight, efficient, and can run on the limited computational capabilities of embedded hardware. Many of the fundamental methods to achieve this is through model compression and quantization, which involve reducing the size and complexity of neural networks without excessively degrading predictive performance. Methods in compression, like pruning, remove unnecessary weights and connection from trained models, which help reduce inference time and the amount of memory required. Quantization reduces the number of bits/bytes required to store a weight (from high precision floating point to low precision integer representations like 8-bit integers). Quantization halves the memory needed for each weight, and compressing in this way can yield substantial speed improvements when applicable.

These types of optimizations are useful for all models, but particularly valuable for cobots, which often leverage low-power processors or need to operate in real-time and energy-constrained environments. (Narvekar, C., et al., 2023)

Federated learning and on-device training: Another important technique that can help to improve model adaptability while preserving data privacy is federated learning and on-device training. In traditional AI workflows, data from devices is gathered and sent back to centralized servers for training. This results in large privacy and bandwidth concerns. Federated learning allows cobots to train on their own data while sharing the updated parameters with a centralized server instead of the data. This allows for a decentralized method of continual learning from multiple/ different data sources, allowing the model to improve continuously while avoiding compromising sensitive farm data. In addition to federated learning, some cobots can perform lightweight on-device training, where small updates can be made to pre-trained models directly in the field, allowing the system to better adapt to changing environments (e.g., crop types, lighting, soil conditions, etc.) with limited human supervision. (Llisterri Giménez, N., et al., 2022)

Memory and power-efficient deployment strategies: To properly ensure sustainable and scalable deployment of generative AI on cobots, memory and power efficient practices need to be included in the design process. Advanced AI models have an unfortunate tendency to require significant amounts of memory bandwidth and require enormous amounts of power, which is impractical when dealing with cobots capable of field deployment, particularly those relying on battery or solar power. Strategies such as neural architecture search (NAS), efficient model distillation, and specialized inference engines such as TensorRT, OpenVINO, and ONNX Runtime are all critical aspects of safely accessing resources without sacrificing performance-quality. In addition to these strategies, cobot designs are increasingly including edge AI accelerators, which are like custom hardware, meant for AI inference, such as Google's Edge TPU or NVIDIA Jetson modules, to allow cobots to perform autonomously where spending more time in a performance only is being cognizant of the user's energy requirements. All of these strategies are critical when considering deployment and sustainability of AI-powered cobots. By following these strategies, AI-powered cobots can be deployed in environments that lack the infrastructure otherwise needed for long-term operation. Because they work autonomously, AI-powered cobots effectively balance knowledge income, and can be deployed in sustainable, scalable ways, which will create opportunities for gentrification in precision farming and sustainable technology to be developed and used properly. (Arif, M., et al., 2024)

7. CHALLENGES IN INTEGRATION

Figure 3. Venn diagram of challenges: Hardware, software, data, and environment

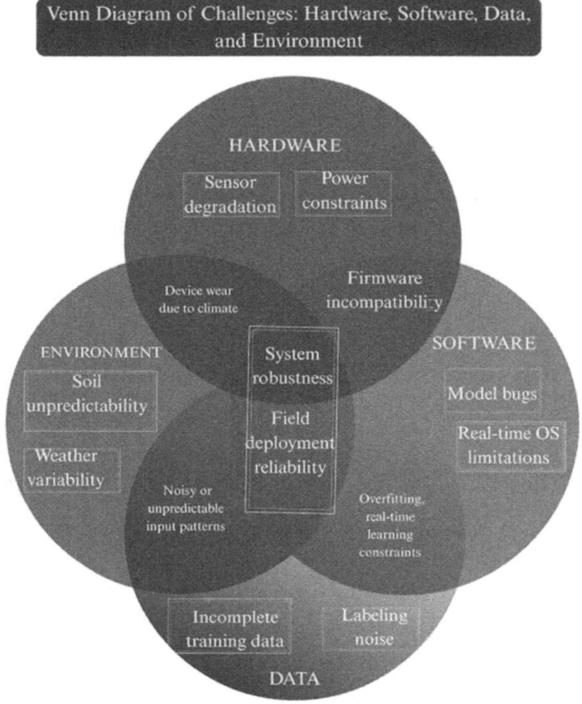

Figure 3 depicts the multi-dimensional and layered issues faced while trying to deploy AI-powered cobots in agriculture. We have organized the major challenges into four areas: hardware, software, data and environment. The hardware area includes problems such as sensor degradation and power limitations, which suggest the nature and limitations of these robotic systems. The software area includes things such as bugs in the model and limitations to how to implement real-time operating systems, which will inhibit system responsiveness and decision-making. In the data area, the problems we show include things like incomplete training data and noise in labeling, that will settle into the accuracy and generalization of any AI models. Finally, in the environment area, the variability in soil and the weather creates uncertainties during field processing. The areas of overlap of the domains present more complicated, cross-sectional challenges. The area between the hardware and environmental domains indicates device wear due to climatic conditions, while the intersection of the data and software domains indicates difficulties inherited when overfitting, or real-time learning is not possible. All of the domains converge toward the center of the diagram at specific issues of focus around likely concerns about the robustness of the system, and reliability of the deployment in the field. These central challenges underscore how the convergence of technical, data-related, and environmental issues must be carefully managed to ensure dependable, scalable AI-cobot operations in agriculture.

Latency and real-time response: Although the convergence of generative AI and collaborative robots (cobots) in agriculture presents transformational possibilities, it also creates a number of technical and operational hurdles. One primary concern is latency and real time response - both very important for field based applications where split second decisions can affect productivity or safety. Generative models, particularly generative models for decision making or image synthesis, require considerable computation time. Likewise, even small amounts of latency have the potential to degrade the altruistic productivity of cobots operating in unpredictable dynamic environments. Even small amounts of inference latency could lead to cobots performing non-ideal actions such as failure to detect an object of interest, or the delayed actuation of an actuator. The need to ensure that generative AI for cobots can be employed in fields such as precision spraying, harvesting and advanced obstacle avoidance must be carried out with highly optimized models and overly efficient runtime environments so the cobots can maintain real time responsiveness. (Makondo, N., et al., 2024)

Interoperability between AI models and robotic OS: A further concern is interoperability between AI models and robotic operating systems. Cobots deployed in agriculture usually work on Robot Operating Systems (ROS), which are modular, indeed allowing the integration of a variety of sensors, actuators, and control algorithms. Contrarily, generative AI models are frequently developed in separate machine learning frameworks such as PyTorch or TensorFlow, and assembling them into ROS capability may involve significant engineering effort. The various AI inference outputs have to be synchronized with sensor streams, robotic control loops, and even with safety protocols, necessitating a carefully designed middleware layer. Likewise, minimizing the variation in communication between edge devices, AI components, and cloud-based services among the field of agricultural technology involves adherence to protocols and APIs, which are being developed at this time. (Tsolakis, N., et al., 2019)

Data limitations and domain adaptation: Data limitations and domain adaptation are another major impediment to better integration. Training generative AI models requires vast amounts of quality, annotated data, but in agriculture, larger amounts of this would be needed because of diverse conditions of crops, environmental conditions, soil differences, and regional tendencies. It is true that even if a dataset is available, the models that are trained in an environment do not generalize to the other domains and thus this issue is called a domain shift. For example, models trained on greenhouse tomato crops will hardly perform to an acceptable level in open field pepper crops in different environments with different soil types and illuminating conditions. Similar to above, domain adaptation approaches such as transfer learning or generating synthetic data can be useful, but are not always successful. In order to permit AI models using a robot or AI device more stable, updates would be done continuously or federated, and subsequently a rigorous evaluation protocol would be in order, allowing researchers to follow along and ensure that the domain adaptation worked in each agricultural domain contexts. Overcoming these inadequacies allows for trustworthy and scalable AI-cobot systems to be learned and also deployed. (López, I. D., et al., 2023)

8. CASE STUDIES AND PILOT DEPLOYMENTS

Examples of AI-cobot integration in greenhouse and field environments: Real-world case studies and pilot deployments provide the reader insights into the practical application and performance of AI-equipped cobots in agricultural settings. Some greenhouse deployments of cobots with generative AI models have included monitoring plant health, detecting early disease, and doing precision automation tasks like pruning and spraying. Particular pilot deployments have included cobots with vision-based generative models for harvesting hydroponic lettuce from farms. These cobots leverage generative adversarial networks (GANs), which create synthetic image data to enhance performing leaf segmentation and supporting growth assessments under multiple light conditions. In field settings, a number of AI-cobot deployments have included autonomous fruit harvesting in orchards and weed detection in row crop environments. Generative models in these deployments support simulating plant structures and occlusion combinations which allow the cobots to navigate dense plant structures while enabling more precision for the harvesting operation.

ROI analysis and scalability assessments: Understanding the return on investment (ROI) and scalability of AI-cobot deployments is critical for the widespread adoption of this technology, beyond technical feasibility. Field trials in a variety of agricultural domains have shown a lot of promise, particularly around reducing labor demand and increasing productivity. For example, in strawberry farms, AI-cobots have been trialed for a variety of applications, including autonomous picking and disease detection, indicating great expansion potential to improve labor efficiency while supporting consistent quality assurance. While several studies have evaluated the feasibility of AI-acobots and similar technologies, scalability evaluations have identified several key issues that will restrict the commercialization of these technologies; these ideals include high capital costs, high degree of customization needed based on crop type and end use, and limits around the availability of training datasets that capture the variability of real-world land and climate conditions that impact crop production. Fortunately, advances in modular robotic hardware, standardized software-as-a-service platforms, and cloud-supported AI models that enable continuous updating by the workforce will support even higher levels of flexibility and affordability of new deployment models. Agritech startups and research institutes are actively working on building AI-cobots so they are adaptable that they could (with some reconfiguration) be used for any farm context from small-holder farms to commercial scale farms. If proof of concept and field evaluation continue, these companies can help find a way to develop successful AI-cobot approaches that are foundational to commercially scalable data-enabled agriculture.

9. FUTURE PROSPECTS

Adaptive generative AI models for autonomous farming: The future of AI-cobot integration in agriculture promises to be heavily influenced by continuing advances in adaptive generative AI models that are crucial to achieving fully autonomous farming. Generative models are maturing from the more traditional notion of data generation in a static format for decisions-making to an active model that learns and makes decisions continuously. In the not-too-distant future, we will see cobots equipped with generative models that can adapt to new and desired crop types, as well as changing environmental conditions and unusual scenarios in the field and would not require extensive reprogramming. We are already seeing the exploration of algorithms, such as diffusion models and transformers, with the capability of modelling complex agronomic situations for evaluation by cobots to allow the cobots to have the chance to virtually test different intervention strategies before executing a given action, In agriculture this would likely mean that these models may eventually allow cobots to very aggressively adapt their strategies to maximize crop yields, conserve resources, and adaptively intervene with undesirable biotic and abiotic stressors in the field, thus allowing the cobot to be a self-optimizing system able to adapt its behaviour to achieve desired goals or limits of actions. (Yin, S., et al., 2025) (Dusadeerungsikul, P. O., et al., 2024)

Role of synthetic data for training cobot AI: An important driver of this autonomy is the increased use of synthetic data to train cobot AI. Using real-world data to train AI in agriculture is problematic due to the inherent barriers in collecting data across a seasonal schedule, labor intensive activities or location-specific variability. Synthetic data produced either from GANs or simulated environments can fill such gaps through the generation of large datasets that are diverse and annotated. Such datasets can be extremely helpful in the training of deep learning models focusing on visual perception, navigation, and decision-making. Furthermore, as synthetic datasets become more realistic with context, it reduces reliance on costly, manually-labeled datasets, and speeds up AI model development that generalizes between agricultural contexts. This capability is especially crucial for underrepresented regions relative to others (e.g. countries), or crops where real world datasets do not exist, or are very limited. (Jabrane, K., et al., 2021)

Regulatory, ethical, and safety considerations: As AI-cobot systems develop, the areas of regulation, ethics, and safety must also evolve. There's an increasing demand for frameworks to ensure the safe use of autonomous cobots in shared spatial environments, especially with farmers, for whom labor deploying a cobot may be part of the normal farm economic landscape. Safety frameworks will need to address not just physical interactions, but also the transparency and trust associated with AI decisions. There are ethical considerations about the use of synthetic data, such as what is visible in model behavior, the provenance of data, and how AI could be biased in its behavior. Regulators will need to set out a standard from the perspectives of validating training data, auditing algorithms, and the interoperability of the systems to ensure equitable and secure AI. Issues of data ownership, farmer consent, the sustainability of the practice, and transparency, to garner trust, and ensure access to these quickly developing technologies for all, must emerge. As this ecosystem matures, it will be through collaborative action by technologists, regulators, and agricultural communities that we will build a future where generative AI and cobots interface more naturally with environments and societies. (van der Burg, S., et al., 2024).

Figure 4. Risk mitigation framework for AI in agricultural robotics

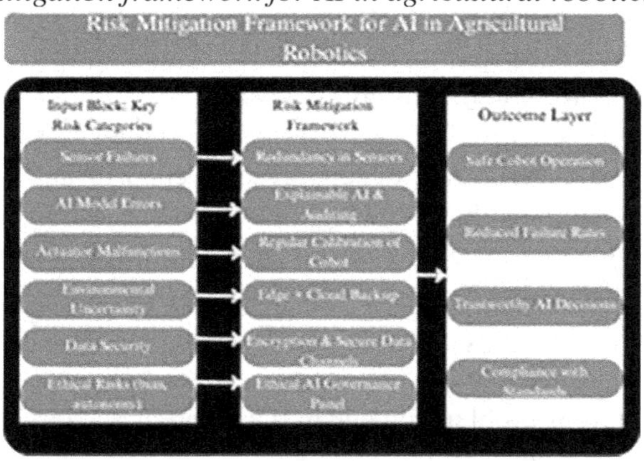

Figure 4 illustrates how systematic responses can comparatively deal with major risks in agricultural cobots powered by AI. The risks are categorized in 6 key areas, which include risks to a sensor system that can affect data collection (e.g., sampling perception through sensors), AI model edge cases that can produce poor or incorrect decisions, actuator failure that can affect mechanical performance of the robot, uncertainties in the physical environment can affect how a system will adapt, data security risks, and moral risks (e.g., bias and lack of transparency). Each risk is matched with a mitigation strategy, such as redundant sensors can improve perception and reduce data collection bias, the regulatory bodies are advocating for audit trails for using Explainable AI models to establish a system of liability, calibration regularly, and coding for anonymity, encryption and backups with cloud migration. Also, an ethical governance framework should be established to oversee the responsible use of AI systems in agriculture. All of these strategies align to promote the reliability, safety, and dependence on cobot operations, while being mindful of the ethical and regulatory requirements, and contributing to the development of sustainable and smart agriculture.

10. CONCLUSION

The use of generative AI with collaborative robotics (cobots) represents a critical advancement in smart agriculture. This technical overview has explained how generative models (e.g., GANs, VAEs, transformers, and diffusion models) are transforming cobots into smarter, more adaptive and autonomous systems. Generative models allow cobots to create synthetic training data, simulate complicated environments, and make decisions based on context—even under constrained environments. Through discussions of system architecture, deployment on the edge and real-world applications, this chapter has demonstrated that AI-assisted cobots can improve labor efficiency, environmentally-sustainable practices, and precision in agricultural activities. Model optimization (e.g., quantization, pruning) and advances in federate learning and on-device learning, are facilitating the implementation of cobots in scalable and energy-efficient ways in the agricultural field. There nevertheless remain barriers that impede our journey—latency, interoperability, domain adaptation, and data governance. Resolving these factors requires bridging disciplines and solutions across hardware, software and legal topographies. The case studies and pilot deployments discussed in this chapter may provide a glimpse into the future of autonomous farming with AI-cobot capabilities that will enhance yield, lower input costs, and allow for environmentally-sustainable practices in a range of agricultural operations. As generative AI continues to evolve, its role in shaping the next generation of cobots will be central—not only in augmenting robotic capabilities but also in enabling more resilient, data-driven, and intelligent food production systems globally.

REFERENCES

Akkem, Y., Biswas, S. K., & Varanasi, A. (2024). A comprehensive review of synthetic data generation in smart farming by using variational autoencoder and generative adversarial network. *Engineering Applications of Artificial Intelligence, 131*, 107881. DOI: 10.1016/j.engappai.2024.107881

Ali, Z., Muhammad, A., Lee, N., Waqar, M., & Lee, S. W. (2025). Artificial intelligence for sustainable agriculture: A comprehensive review of AI-driven technologies in crop production. *Sustainability (Basel), 17*(5), 2281. DOI: 10.3390/su17052281

Arif, M., Khan, M. A., Rehman, A. U., Jan, M. A., Almogren, A., & Kumar, R. (2024). Resource-efficient ubiquitous sensor networks for smart agriculture: A survey. *IEEE Access : Practical Innovations, Open Solutions, 12*, 193332–193364. DOI: 10.1109/ACCESS.2024.3516814

Batra, I., Sharma, C., Malik, A., Sharma, S., Kaswan, M. S., & Garza-Reyes, J. A. (2024). Industrial revolution and smart farming: A critical analysis of research components in Industry 4.0. *The TQM Journal.* Advance online publication. DOI: 10.1108/TQM-10-2023-0317

Dusadeerungsikul, P. O., & Nof, S. Y. (2024). Precision agriculture with AI-based responsive monitoring algorithm. *International Journal of Production Economics, 271*, 109204. DOI: 10.1016/j.ijpe.2024.109204

Heschl, A., Murillo, M., Najafian, K., & Maleki, F. (2024). SynthSet: Generative diffusion model for semantic segmentation in precision agriculture. *arXiv preprint arXiv:2411.03505*. https://arxiv.org/abs/2411.03505

Isinkaye, F. O., Olusanya, M. O., & Akinyelu, A. A. (2025). A multi-class hybrid variational autoencoder and vision transformer model for enhanced plant disease identification. *Intelligent Systems with Applications, 26*, 200490. DOI: 10.1016/j.iswa.2025.200490

Jabrane, K., & Bousmah, M. (2021). A new approach for training cobots from small amount of data in Industry 5.0. *International Journal of Advanced Computer Science and Applications, 12*(10). Advance online publication. DOI: 10.14569/IJACSA.2021.0121070

Kerdegari, H., Razaak, M., Argyriou, V., & Remagnino, P. (2019). Smart monitoring of crops using generative adversarial networks. In Vento, M., & Percannella, G. (Eds.), Lecture notes in computer science: Vol. 11678. *Computer analysis of images and patterns. CAIP 2019* (pp. 517–528). Springer., DOI: 10.1007/978-3-030-29888-3_45

Kum, S., Oh, S., & Moon, J. (2024). Edge AI framework for large scale smart agriculture. In *2024 27th Conference on Innovation in Clouds, Internet and Networks (ICIN)* (pp. 143–147). IEEE. DOI: 10.1109/ICIN60470.2024.10494451

Lee, S. (2019). A novel design and implementation of data acquisition and preprocessing system for precision agriculture. In *Agriculture and environment perspectives in intelligent systems* (Vol. 24, pp. 52–71). IOS Press., DOI: 10.3233/AISE190004

Llisterri Giménez, N., Monfort Grau, M., Pueyo Centelles, R., & Freitag, F. (2022). On-device training of machine learning models on microcontrollers with federated learning. *Electronics (Basel)*, *11*(4), 573. DOI: 10.3390/electronics11040573

López, I. D., Grass, J. F., Figueroa, A., & Corrales, J. C. (2023). A proposal for a multi-domain data fusion strategy in a climate-smart agriculture context. *International Transactions in Operational Research*, *30*(4), 2049–2070. DOI: 10.1111/itor.12899

Makondo, N., Kobo, H. I., Mathonsi, T. E., & du Plessis, D. P. D. (2024). Implementing an efficient architecture for latency optimisation in smart farming. *IEEE Access : Practical Innovations, Open Solutions*, *12*, 140502–140526. DOI: 10.1109/ACCESS.2024.3466994

Narvekar, C., Shinde, S., Mishra, V., & Kshirsagar, D. (2023). Productivity improvement with generative AI framework for data enrichment in agriculture. *International Journal on Recent and Innovation Trends in Computing and Communication*, *11*(9), 679–684. DOI: 10.17762/ijritcc.v11i9.8859

Rahman, Z. U., Asaari, M. S. M., Ibrahim, H., Abidin, I. S. Z., & Ishak, M. K. (2024). Generative adversarial networks (GANs) for image augmentation in farming: A review. *IEEE Access : Practical Innovations, Open Solutions*, *12*, 179912–179943. DOI: 10.1109/ACCESS.2024.3505989

Riedelbauch, D., Höllerich, N., & Henrich, D. (2023). Benchmarking teamwork of humans and cobots—An overview of metrics, strategies, and tasks. *IEEE Access : Practical Innovations, Open Solutions*, *11*, 43648–43674. DOI: 10.1109/ACCESS.2023.3271602

Sai, S., Kumar, S., Gaur, A., Goyal, S., Chamola, V., & Hussain, A. (2025). Unleashing the power of generative AI in Agriculture 4.0 for smart and sustainable farming. *Cognitive Computation*, *17*(1), 1–18. DOI: 10.1007/s12559-025-10420-6

Sapkota, R., & Karkee, M. (2025). Generative AI in agriculture: Creating image datasets using DALL·E's advanced large language model capabilities. *arXiv*. https://arxiv.org/abs/2307.08789

Shaheb, M. R., Sarker, A., & Shearer, S. A. (2022). *Precision agriculture for sustainable soil and crop management*. IntechOpen., DOI: 10.5772/intechopen.101759

Taesi, C., Aggogeri, F., & Pellegrini, N. (2023). COBOT applications—Recent advances and challenges. *Robotics (Basel, Switzerland)*, *12*(3), 79. DOI: 10.3390/robotics12030079

Tsolakis, N., Bechtsis, D., & Bochtis, D. (2019). AgROS: A Robot Operating System based emulation tool for agricultural robotics. *Agronomy (Basel)*, *9*(7), 403. DOI: 10.3390/agronomy9070403

van der Burg, S., Giesbers, E., Bogaardt, M. J., van der Veen, M., & van Est, R. (2024). Ethical aspects of AI robots for agri-food: A relational approach based on four case studies. *AI & Society*, *39*(2), 541–555. DOI: 10.1007/s00146-022-01429-8

Vizniuk, A., Diachenko, G., Laktionov, I., Siwocha, A., Xiao, M., & Smoląg, J. (2025). A comprehensive survey of retrieval-augmented large language models for decision making in agriculture: Unsolved problems and research opportunities. *Journal of Artificial Intelligence and Soft Computing Research*, *15*(2), 115–146. DOI: 10.2478/jaiscr-2025-0007

Yin, S., Xi, Y., Zhang, X., Sun, C., & Mao, Q. (2025). Foundation models in agriculture: A comprehensive review. *Agriculture*, *15*(8), 847. DOI: 10.3390/agriculture15080847

Chapter 5
Overcoming Modern Farming Challenges With Automation Technologies

Rupayan Roy
https://orcid.org/0000-0002-8118-431X
National Institute of Fashion Technology, Kannur, India

Nambiar Arya Rajesh
National Institute of Fashion Technology, Kannur, India

Ayona Padhi
National Institute of Fashion Technology, Kannur, India

Mohana Priya R.
National Institute of Fashion Technology, Kannur, India

ABSTRACT

Modern agriculture faces workforce shortages, climate variability, soil deterioration, pest infestations, and resource inefficiencies. These issues affect global food security, rural livelihoods, and agricultural production. Meanwhile, tremendous advances in automation, AI, the IoT, and robots provide unprecedented prospects to change agriculture. This chapter discusses modern agriculture's major issues and how automation can solve them. The chapter shows how automation might increase farming system efficiency, resource use, and resilience by evaluating worldwide case studies, upcoming technologies, and adoption constraints. It discusses automation's socioeconomic effects and recommends inclusive and sustainable technological integration in agriculture.

DOI: 10.4018/979-8-3373-4207-8.ch005

1. INTRODUCTION

Agriculture is fundamental to human civilization and essential for global food security. Over 60% of the global population relies directly or indirectly on agriculture, which is essential for sustaining economies, providing livelihoods, and assuring nutrition for billions (Ayim et al., 2020) . It plays a crucial role in the GDP of developing countries and is essential for attaining the United Nations' Sustainable Development Goals (SDGs), especially those concerning zero hunger, poverty alleviation, and climate action Global food demand is anticipated to rise by 60% by 2050, propelled by population expansion, urbanization, and evolving dietary habits. This highlights the pressing necessity to rethink agricultural practices and management (Van Dijk et al., 2021).

Agriculture has historically experienced several modifications, ranging from crude hand tools and animal-drawn plows to the mechanical revolution of the 19th century and subsequently, the Green Revolution of the 20th century(Binswanger, 1986).The Green Revolution, characterized by the implementation of high-yield varieties (HYVs), synthetic fertilizers, insecticides, and sophisticated irrigation methods, significantly augmented food output in nations such as India, Mexico, and the Philippines (Evenson & Gollin, 2003). Nonetheless, although these advancements enhanced yields, they also introduced new difficulties including soil degradation, excessive water usage, biodiversity loss, and reliance on chemical inputs (Lavelle et al., 2006). Furthermore, the advantages of the Green Revolution were inequitably allocated, and small-scale farmers frequently encountered difficulties in accessing its technological innovations (Pingali, 2012).As we progress farther into the 21st century, agriculture confronts unparalleled problems. Climate change is causing erratic weather patterns, resulting in a rise in the occurrence of droughts, floods, and heatwaves that impact agricultural productivity (Change, 2013). Labor shortages, particularly in elderly agricultural communities, are intensifying in both developed and emerging nations. Concurrently, the expenses associated with agricultural inputs are increasing, whereas market prices for products exhibit volatility, hence constricting farmers' incomes (Tong et al., 2024). Moreover, consumer expectations about food quality, traceability, and sustainability are swiftly advancing, exerting additional pressure on conventional farming systems. The necessity for a transition in agriculture is, consequently, imperative. This transition must address productivity while ensuring ecological balance, social equality, and economic viability. The fourth agricultural revolution, known as "Agriculture 4.0," anticipates a sophisticated, networked, and sustainable farming ecosystem driven by advanced technologies including artificial intelligence (AI), the Internet of Things (IoT), drones, robotics, and big data analytics. These technologies are poised to improve accuracy in agricultural practices, minimize resource depletion, streamline labor, and deliver prompt insights for decision-making (Fasciolo et al., 2024). Automation, a fundamental component of Agriculture 4.0, provides solutions to numerous persistent and emerging difficulties in agriculture(Bazargani & Deemyad, 2024) . Automated systems can assess crop health, soil conditions, and meteorological data in real time, facilitating more efficient and adaptive management approaches. Autonomous tractors and harvesters execute repetitive and labour-intensive tasks with exceptional precision, diminishing reliance on physical labor (Etezadi & Eshkabilov, 2024) .Intelligent irrigation systems and sensor-driven fertilization methods can substantially reduce water and chemical consumption, hence enhancing sustainability (Lakhiar et al., 2024) .

This chapter aims to analyse the primary concerns facing contemporary agriculture and investigate the development and implementation of automation technologies to mitigate these issues. This chapter will examine future potential, obstacles to widespread adoption, and the socio-economic consequences of automating agricultural activities. By comprehending the constraints of current systems and the possibilities of automation, we can delineate a strategy for a robust, efficient, and sustainable agricultural future.

2. MAJOR CHALLENGES IN MODERN AGRICULTURE

Contemporary agriculture faces a pivotal moment, as traditional methods are increasingly challenged by a multifaceted array of socio-economic, environmental, and technological influences. Agriculture is essential for nourishing a burgeoning population, yet it faces a multitude of interconnected issues that jeopardize its sustainability and production. This section identifies six significant concerns that require immediate attention and creativity.

2.1 Labor Shortages and Aging Farmer Population

The availability of labor in agriculture has markedly diminished in recent decades. A primary factor is rural-to-urban migration, especially among the youth who are increasingly pursuing enhanced possibilities in urban areas. Consequently, agricultural communities are undergoing a demographic transformation marked by an increasing elderly farmer demographic (Ngadi et al., 2023). In countries such as Japan, the average age of farmers exceeds 67 years, but in the United States, it is roughly 58 years. The aging population raises significant issues regarding the sustainability of agricultural techniques, the transmission of information, and the physical capacity to perform strenuous jobs(Fuhrmann-Aoyagi et al., 2024).

Figure 1. Modern agriculture at a crossroads

Mechanization has reduced certain responsibilities; yet, numerous small and marginal farmers in emerging nations continue to depend significantly on manual labor. In the absence of prompt action and personnel replenishment, a significant risk of labor bottlenecks and subsequent reduction in food productivity is imminent. Figure 1, depicts a split landscape in a rural farming area such as Rajasthan, India.

2.2 Climate Change and Unpredictable Weather Patterns

Climate change is undoubtedly the most significant and widespread concern confronting contemporary agriculture. Increasing global temperatures, unpredictable precipitation, droughts, floods, and the escalation of extreme weather phenomena significantly disturb planting and harvesting schedules. Crop yields face persistent threats due to climate variability, which affects soil moisture, growth seasons, and insect dynamics(Yuan et al., 2024) .

Furthermore, numerous climate-sensitive crops, including wheat, rice, and maize, are experiencing yield declines in susceptible areas (Zhao et al., 2017). Climate modelling studies indicate that if global temperatures persist in rising unchecked, agricultural production in low-latitude nations could decline by as much as 25% by 2050 (Heikonen et al., 2025) . Consequently, modifying agricultural systems to endure climate shocks is imperative.

2.3 Decreasing Arable Land and Soil Health

Urban growth, deforestation, industrialization, and inappropriate land-use practices have cumulatively resulted in a substantial decrease in arable land per capita (Seto et al., 2012). In numerous regions of Asia and Africa, arable land is being rapidly repurposed for non-agricultural uses. Simultaneously, intense agricultural practices, excessive application of synthetic fertilizers and pesticides, and monoculture farming are exhausting soil nutrients and deteriorating soil structure.

Global reports indicate a rising incidence of soil biodiversity loss, compaction, acidification, and salinization (Smith et al., 2024). Degraded soils diminish crop resilience, water retention, and yield potential. Without the restoration of soil health by sustainable practices, agricultural productivity will persist in its fall.

2.4 Pest and Disease Outbreaks

Outbreaks of pests and diseases are inherent occurrences in agriculture; however, their frequency and intensity have intensified in recent years as a result of globalization and climate change. Increased temperatures and altered precipitation patterns enable numerous pests to extend their geographical distribution and proliferate swiftly (Skendžić et al., 2021). Invasive species, such as the fall armyworm in Africa and desert locusts in South Asia, have caused significant destruction to crops worth billions of dollars. Furthermore, traditional pest management techniques are becoming progressively ineffectual due to the development of pesticide resistance and the elimination of toxic agents from the market. Crop losses due to pests and diseases are expected to account for up to 40% of annual global food production losses, highlighting the urgent need for more intelligent and sustainable protection strategies.

2.5 Water Scarcity and Inefficient Irrigation

Water is essential to agriculture, representing over 70% of global freshwater withdrawals. Nonetheless, freshwater resources are being used more rapidly than they can be restored, owing to over-extraction, pollution, and climate-induced alterations in hydrological cycles(Biswas et al., 2025). Areas such as the Middle East, certain regions of India, and North Africa are currently experiencing severe water scarcity.

Conventional irrigation methods, such flood irrigation, exhibit significant inefficiency, resulting in water loss of up to 60% from evaporation and seepage (Alexander et al., 2023). Poor water management frequently results in waterlogging or salinity, hence diminishing land yield. The use of precision irrigation technology, including drip and sprinkler systems, is constrained in numerous developing countries due to financial barriers and insufficient understanding.

2.6 Rising Input Costs and Low Productivity

The prices of agricultural inputs namely seed, fertilizers, agrochemicals, and energy have consistently escalated, substantially raising cultivation costs. Simultaneously, numerous places are experiencing stagnant or decreasing food production attributable to issues such as soil deterioration, water scarcity, and pest infestations(Khan et al., 2024) . The gap between escalating input costs and low production is especially severe for smallholder farmers, who function with constrained resources. In the absence of sufficient access to capital, training, or technology, many farmers remain ensnared in a cycle of minimal returns and indebtedness (Hasan et al., 2022) . This economic disparity not only undermines rural livelihoods but also deters the younger generation from seeing farming as a viable profession.

The difficulties facing contemporary agriculture are complex and intricately linked. Labor shortages, climate variability, depleted natural resources, pest outbreaks, water scarcity, and economic limitations collectively provide a grim scenario. Nevertheless, comprehending these issues establishes the basis for formulating effective answers. Automation, precision agriculture, and digital technology present viable solutions to address these challenges, as examined in the next sections of this chapter.

3. EMERGENCE OF AUTOMATION IN AGRICULTURE

The incorporation of automation into agriculture signifies a pivotal stage in the advancement of farming. The agricultural sector is progressively adopting technology advances to improve productivity, efficiency, and sustainability in response to difficult concerns such as labor shortages and climate change. This section examines the definition, breadth, categories, and facilitating technologies that define agricultural automation.

3.1 Definition and Scope of Agricultural Automation

Agricultural automation denotes the utilization of sophisticated technologies to execute farming tasks with minimal human involvement. These technologies aim to enhance many activities including planting, irrigation, fertilization, pest management, harvesting, and post-harvest processing (Wolfert et al., 2017) . The realm of automation encompasses the complete agricultural value chain, ranging from seed selection and soil preparation to production monitoring and distribution.

Historically, agricultural labor was labour-intensive, time-consuming, and reliant on human expertise. Nonetheless, the emergence of intelligent machinery and digital platforms has enabled farmers to administer their fields with more precision and efficiency. Automation is essential for mitigating critical challenges such as labor shortages, resource optimization, and immediate decision-making. It enhances operational efficiency while promoting ecologically sustainable practices by reducing resource waste. A high-tech farm scene featuring IoT sensors embedded in the soil, drones flying overhead, and digital overlays displaying real-time data on soil moisture, nutrient levels, and weather conditions is demonstrated in Figure 2.

Figure 2. Harnessing smart technology in farming

3.2 Types of Automation: Mechanical, Digital, and Biological

Agricultural automation can be classified into mechanical, digital, and biological categories, each encompassing distinct innovations and uses(Duckett et al., 2018). Mechanical automation entails the utilization of machinery and technology to execute manual activities. Examples comprise automated tractors, seeders, harvesters, and milking robots. These machines diminish labor requirements while enhancing operational velocity and precision. Mechanization is particularly crucial in extensive agriculture, where human labor is inadequate for prompt field management (Gupta & Kumar Pal, 2025) . Digital Automation denotes the incorporation of digital technology, including artificial intelligence (AI), machine learning, and big data analytics, into agricultural management systems. These tools facilitate precision agriculture by assessing soil health, meteorological conditions, crop efficacy, and insect risks. Decision Support Systems (DSS) and intelligent farming applications facilitate real-time, data-informed decision-making, thereby enhancing results and mitigating risks (Benke & Tomkins, 2017) .

Biological Automation, a nascent notion, encompasses breakthroughs such as automated breeding, CRISPR-based genetic alteration, and synthetic biology to improve the traits of crops and livestock. Controlled Environment Agriculture (CEA) systems, such as vertical farms and hydroponics, automate plant growth by utilizing optimum environmental controls(Soussi et al., 2024).

These categories are frequently interconnected, with hybrid systems integrating mechanical hardware and digital intelligence to get superior performance results. A robotic harvester may employ sensors and artificial intelligence algorithms to accurately locate and harvest ripe fruits.

3.3 Overview of Technological Enablers

The swift advancement of agricultural automation is propelled by a collection of potent enabling technologies. Among these, the most significant are:

The Internet of Things (IoT): IOT is essential to smart agriculture. Soil sensors, meteorological stations, and machinery gather extensive data regarding environmental conditions, plant vitality, and equipment efficiency. This data is relayed in real-time to centralized systems for monitoring and analysis to facilitate decision-making (Sharma et al., 2025) .

Artificial Intelligence (AI) and Machine Learning (ML): AI improves automation through predictive modelling and adaptive systems. AI algorithms can forecast agricultural illnesses, improve irrigation plans, and evaluate production predictions. Machine learning allows systems to enhance precision progressively by assimilating data inputs and feedback mechanisms.

Robotics and Autonomous Machines: Robots and autonomous vehicles are extensively utilized for planting, weeding, spraying, and harvesting chores. Drones are utilized for airborne photography and the application of pesticides. These devices substantially diminish labor and guarantee uniform performance, particularly in repeated or perilous activities (Abbes et al., 2024).

Drones and Unmanned Aerial Vehicles (UAVs) outfitted with multispectral and thermal cameras offer essential information about crop health, moisture content, and insect infestations. They may rapidly encompass extensive areas, providing farmers with a high-resolution perspective of their fields unattainable to the unaided eye (Demestichas et al., 2020) .

Satellite Imaging and Remote Sensing: Satellite data is essential for extensive monitoring of agricultural conditions, soil moisture levels, vegetation indices, and alterations in land use. The incorporation of satellite imagery into Geographic Information Systems (GIS) facilitates geographical analysis and targeted interventions (Zhang, 2019).

Blockchain and Traceability Mechanisms: Although not universally categorized as automation, blockchain technology enables secure and transparent data sharing throughout the supply chain. It facilitates the monitoring of agricultural products from cultivation to consumption, improving food safety, authenticity, and adherence to regulations(Soussi et al., 2024) .

Collectively, these technologies establish a networked ecosystem that facilitates precision agriculture—a methodology that customizes inputs and interventions according to particular crop requirements and environmental circumstances. The combined application of several technologies aids in lowering input expenses, diminishing waste, enhancing crop quality, and augmenting profitability.

The advent of technology in agriculture signifies a fundamental transformation in food production and management. Modern farms are increasingly efficient, data-driven, and adaptable due to advancements in mechanical systems, digital platforms, and biological discoveries. Automation not only resolves significant challenges such as labor shortages and climate variability but also creates new opportunities for sustainable and scalable agriculture. The forthcoming part will examine how the integration of automation with artificial intelligence and robotics has the potential to further transform agricultural methods.

4. OPPORTUNITIES OFFERED BY AUTOMATION

Agricultural automation has arisen not only as a solution to current difficulties but also as a pivotal catalyst for change in contemporary farming methods. The integration of sophisticated technologies, including robots, artificial intelligence, the Internet of Things, and remote sensing, presents numerous opportunities for enhancing output, sustainability, and profitability. This section delineates significant potential presented by automation, highlighting its increasing impact on the transformation of agriculture worldwide.

4.1 Precision Farming and Smart Irrigation Systems

Precision agriculture represents one of the most significant results of agricultural automation. The process utilizes GPS, IoT sensors, AI, and data analytics to customize agricultural techniques according to particular field variability. Precision farming enables the use of inputs like water, fertilizers, and pesticides in optimal amounts, minimizing waste and environmental harm (Lakhiar et al., 2024) .

Smart irrigation systems illustrate this methodology by utilizing soil moisture sensors and climatic data to automate and manage water distribution. These systems are both resource-efficient and capable of providing continuous crop hydration, hence enhancing production quality. For example, IoT-controlled drip irrigation can modulate water discharge according to real-time moisture metrics and meteorological predictions, guaranteeing that water is utilized exclusively when and where necessary. Research indicates that these devices can decrease water usage by 30–50% relative to conventional approaches (Qu et al., 2024).

4.2 Autonomous Tractors, Drones, and Robotic Harvesters

The advancement of autonomous machinery has substantially enhanced mechanization in agriculture. Tractors outfitted with GPS, LiDAR, and artificial intelligence may function autonomously, doing ploughing, sowing, and harvesting with exceptional accuracy. These machines reduce overlaps, conserve fuel, and maintain steady operation even during prolonged hours or adverse weather conditions (Cho et al., 2024). Figure 3 autonomous tractors and robotic harvesters operating seamlessly across vast farmland.

Figure 3. Autonomous farm machinery in action

Drones are utilized for airborne surveillance, cartography, and pesticide spraying. Their capacity to navigate challenging terrains and swiftly evaluate field conditions provides farmers with an instantaneous aerial perspective of their crops. Robotic harvesters signify a significant advancement, particularly in labour-intensive crops such as strawberries and tomatoes, where manual harvesting is both time-consuming and costly. Furnished with machine vision and soft grippers, these robots can discern ripe fruits and harvest them gently, minimizing harm and enhancing productivity. Automation in machinery diminishes reliance on labor, while simultaneously augmenting production and decreasing operational expenses over time.

4.3 AI for Crop Monitoring and Disease Detection

Artificial Intelligence (AI) is crucial in facilitating proactive agricultural management. AI-driven platforms can assess satellite photos, drone footage, and sensor data to identify early indicators of crop stress, pest infestations, and nutritional deficits (Mohanty et al., 2016). Machine learning algorithms consistently enhance their precision, rendering them indispensable instruments for forecasting outbreaks and averting agricultural losses.

AI-driven models developed for picture identification can detect plant illnesses with an accuracy surpassing 90%, delivering immediate notifications to farmers through smartphone applications. These systems can additionally propose appropriate treatments or modifications in crop management informed by historical and real-time data(Lee et al., 2024). AI-driven decision support diminishes the necessity for regular manual inspections and enables farmers to make informed decisions promptly, thus reducing crop damage and input expenses.

4.4 Smart Greenhouses and Controlled Environments

Automation is transforming greenhouse agriculture and other Controlled Environment Agriculture (CEA) systems. Intelligent greenhouses utilize sensors, actuators, and artificial intelligence to autonomously regulate temperature, humidity, illumination, carbon dioxide concentrations, and fertilizer distribution. This precise management optimizes plant growth and facilitates year-round production irrespective of external climatic conditions (Garzón et al., 2023).

Automated hydroponic and aeroponic systems provide crops with adequate nutrients without soil, hence minimizing land usage and enhancing productivity per square meter. Vertical farming, typically situated in climate-controlled environments, use LED lights and sophisticated software to replicate natural conditions and meticulously regulate plant growth(Rajwar et al., 2024).

Smart greenhouses are particularly advantageous for high-value crops and in areas with extreme or erratic weather, offering stability, scalability, and enhanced profitability.

4.5 Data-Driven Decision Making for Resource Optimization

Data constitutes the foundation of contemporary agricultural automation. Automated systems gather extensive data on soil health, meteorological conditions, crop development, and machinery efficiency from seeding to harvesting. When analysed using analytics and machine learning algorithms, this data yields actionable insights for optimizing resource allocation and enhancing agricultural productivity (Fountas et al., 2015).

Farm Management Information Systems (FMIS) facilitate the consolidation of diverse data sources into a cohesive display, empowering farmers to monitor input utilization, evaluate yields, schedule maintenance, and analyse financial results.

Predictive analytics can anticipate market demand, pest infestations, or unfavourable weather conditions, providing farmers with a competitive advantage(Hoque & Padhiary, 2024) .

Furthermore, automation enhances precision logistics by forecasting the ideal harvesting time and coordinating transportation to minimize waste. By synchronizing operational decisions with real-time data, farms may minimize input waste, enhance production, and elevate overall sustainability(Getahun et al., 2024) .

Automation in agriculture is not a distant aspiration; it is an accelerating reality that provides concrete advantages throughout the farming ecosystem. Automation facilitates a transformative era of productive, sustainable, and intelligent agriculture by enhancing water utilization, crop protection, and enabling data-driven decision-making. The collaboration between human expertise and computer intelligence will be essential for tackling food security, environmental limitations, and economic sustainability.

5. CASE STUDIES AND GLOBAL BEST PRACTICES

The implementation of automation in agriculture is no longer hypothetical—it is currently revolutionizing farming systems globally. Numerous nations have implemented customized technical techniques informed by regional difficulties, agricultural varieties, and resource limitations. Case studies from the Netherlands, Japan, India, and the USA demonstrate how automation is tackling essential agricultural challenges through innovation, providing examples for adaptation by other countries.

5.1 Netherlands: Automated Greenhouses

The Netherlands is internationally acknowledged as a frontrunner in high-tech horticulture, especially because of its sophisticated automated greenhouse systems. Notwithstanding its limited size and high population density, the Netherlands ranks as the second-largest exporter of agricultural products worldwide, primarily attributable to technological innovation (Ariesen-Verschuur et al., 2022) .

Dutch greenhouses widely employ automation for temperature adjustment, lighting, humidity control, and fertilizer delivery. AI and IoT systems gather real-time data on plant health, autonomously modify microclimates, and ascertain ideal harvest timings. The utilization of robotic arms for planting and harvesting in greenhouses diminishes reliance on manual labor and enhances operational efficiency (Hemming et al., 2019). Figure 4, a visually detailed image of a high-tech greenhouse with glass panels allowing controlled climate conditions.

Figure 4. Smart greenhouses: Optimizing crop production

The "Autonomous Greenhouse Challenge," launched by Wageningen University, fosters innovation by urging international teams to create entirely automated cultivation systems utilizing AI and robotics.

These solutions have allowed Dutch farmers to decrease water consumption by as much as 90% relative to conventional farming techniques, while attaining markedly greater yields (49). Their accomplishment illustrates how automation, coupled with data-driven decision-making, can surmount spatial and resource constraints.

5.2 Japan: Robotic Rice Planting and Smart Farming

Japan, confronted with an aging agricultural workforce and a deficit of labourers, has adopted automation as an imperative. A significant innovation is the implementation of robotic rice planting systems that utilize GPS-guided autonomous tractors, which can spread seeds with precision and regularity.

Companies such as Yanmar and Kubota have launched autonomous tractors and drones integrated with AI-driven analytics to enhance seeding, irrigation, and fertilizer application. Furthermore, Japan's “Smart Agriculture Demonstration Project,” initiated in 2019, has expedited the implementation of robot-assisted farming in various prefectures (Stanghellini, 2013) . These robots are engineered for planting, weeding, harvesting, and data collection crucial for improving productivity in rice paddies and vegetable farms.

Smartphone applications and cloud platforms enable farmers to remotely oversee agricultural operations, strategize crop rotations, and obtain weather forecasts. This digitalization enables smallholders to sustain viability despite labor limitations (Li et al., 2023) . Japan's focus on interoperability, wherein diverse automation systems function cohesively, establishes a standard for the integration of disparate agri-tech solutions.

5.3 India: Precision Agriculture in Water-Stressed Regions

The agricultural sector in India, employing over fifty percent of the workforce, confronts significant issues including water scarcity, unpredictable rainfall, and diminishing soil fertility. Precision agriculture and automation are progressively implemented to mitigate these issues, especially in drought-prone regions such as Maharashtra, Karnataka, and Rajasthan.

IoT-based soil moisture sensors activate smart drip irrigation systems, resulting in a reduction of water consumption by as much as 40% (Zhang et al., 2022). Startups such as Fasal, CropIn, and AgNext offer sensor-based crop monitoring, disease prediction, and advising services to farmers through mobile applications.

The Central Institute of Agricultural Engineering (CIAE) has developed semi-automated planters, robotic weeders, and solar-powered irrigation controllers to decrease labor reliance and enhance production per acre . The Indian government's Digital Agriculture Mission (2021–2025) advocates for the implementation of AI, drone-assisted crop surveillance, and blockchain technology into agricultural value chains.

Although the extent of deployment remains constrained relative to developed countries, India's expanding ecosystem of agri-tech firms and pilot automation programs exhibit considerable promise, particularly for smallholder farmers operating under water-scarce or fragmented conditions.

5.4 USA: John Deere's AI-Powered Machinery

The United States has historically been a leader in mechanized and precision agriculture, although current advancements in AI-integrated technology signify a new epoch. Firms such as John Deere, Trimble, and AGCO are advancing next-generation tractors and harvesters integrated with computer vision, machine learning, and real-time analytics functionalities (Kaur et al., 2024) .

In 2022, John Deere introduced the first completely autonomous tractor globally, using a 360-degree video system and deep learning algorithms to traverse fields, circumvent obstacles, and execute agronomic decisions autonomously. Agriculturalists can remotely oversee and manage equipment through smartphone applications, receiving notifications and field data instantaneously(Mehta et al., 2023). The *Global Best Practices in Agricultural Automation*, showcases a visually rich composite Figure 5, highlighting innovative agricultural techniques across multiple regions

Figure 5. Global best practices in agricultural automation

Additional breakthroughs encompass See & Spray™ technology, which employs AI to detect weeds and administer herbicides solely where required, thereby diminishing chemical usage by more than 70%. This method reduces input expenses and promotes sustainable agricultural practices by lessening environmental effect (Liakos et al., 2018).

In the Midwest, extensive corn and soybean cultivators have implemented drone swarms for crop health assessment, in conjunction with satellite data and predictive analytics to enhance optimization.

6. CHALLENGES IN ADOPTING AUTOMATION

Despite the considerable promise of agricultural automation, its extensive implementation is hindered by various substantial obstacles. These obstacles are most pronounced in emerging economies; but, even advanced nations encounter issues related to policy, ethics, and the socio-economic ramifications of automation. Comprehending and mitigating these obstacles is essential for the inclusive, scalable, and sustainable implementation of agricultural technologies.

6.1 High Initial Costs and Investment Barriers

A primary obstacle to the implementation of automation in agriculture is the substantial initial investment needed. Technologies like autonomous tractors, robotic harvesters, AI-driven drones, and IoT systems require significant investment frequently prohibitive for small and marginal farmers, who constitute the majority of the worldwide agricultural demographic (McLandsborough, 2023) .

Although long-term rewards may encompass diminished input costs, heightened yield, and enhanced resource efficiency, numerous farmers lack the financial robustness to absorb initial expenditures. Moreover, the apprehension of investment in unfamiliar technology may hinder its adoption. Research reveals that over 70% of smallholder farmers in nations such as India and Kenya see cost as the principal factor for eschewing digital instruments (Rai et al., 2023) .

Public-private partnerships, subsidies, and equipment-as-a-service (EaaS) models have the potential to reduce economic obstacles; nevertheless, these initiatives remain in their infancy and are unevenly distributed.

6.2 Lack of Awareness and Training Among Farmers

Technological literacy constitutes another significant obstacle. Numerous farmers are either oblivious to the capabilities of automation or deficient in the training required to utilize and sustain these technologies proficiently. Even with the availability of technologies, incorrect utilization or misinterpretation can result in diminished effectiveness or total system failure. (Kendall et al., 2022) In low-income areas, training programs are frequently constrained in scope, irregular, or concentrated on fundamental mechanization rather than advanced technologies such as artificial intelligence or drone operation. Furthermore, generational disparities influence adaptability; younger farmers tend to embrace change more readily, whereas older farmers frequently exhibit resistance to it (Hebsale Mallappa & Pathak, 2023) .

In the absence of organized capacity-building programs, including practical demonstrations, field trials, and hotline assistance, automation is likely to be limited to technologically proficient or large-scale agricultural enterprises.

6.3 Infrastructure and Connectivity Issues in Rural Areas

Efficient automation relies heavily on strong infrastructure and connection. Real-time data transmission from sensors, operation of autonomous machinery, and remote monitoring systems necessitate reliable energy, mobile networks, GPS signals, and internet access—resources frequently deficient in rural and remote agricultural regions .

IoT-enabled smart irrigation systems require continuous electricity and cloud integration for optimal functionality. Likewise, satellite images and drone analytics rely on 4G/5G connectivity to provide timely actionable information. In nations such as Bangladesh and certain regions of Sub-Saharan Africa, more than 60% of rural areas are devoid of dependable internet infrastructure (Abdulai et al., 2023).

Enhancing digital infrastructure should be regarded not merely as a technology necessity but as a fundamental component of agricultural advancement.

6.4 Policy and Regulatory Hurdles

Governments around are contending with the regulation of modern agricultural technologies, striving to balance innovation, safety, and inclusivity. The regulatory landscape for agricultural automation is frequently vague, fragmented, or antiquated, resulting in uncertainty for investors, developers, and farmers (Bang & Han, 2025) .

In numerous places, the utilization of drones for agricultural applications is significantly constrained by airspace regulations, data privacy issues, or the absence of classification frameworks. Likewise, a definitive legal framework for responsibility concerning autonomous machinery malfunctions or data breaches is frequently absent(Tao et al., 2021).

In the absence of cohesive, progressive policies that promote innovation and safeguard stakeholders, adoption will continue to be sluggish and fragmented. Facilitating approvals, establishing certification criteria, and providing financial incentives are essential for fostering a conducive regulatory environment.

6.5 Ethical and Social Concerns (e.g., Job Displacement)

Automation engenders legitimate ethical and societal apprehensions, especially around job displacement. In numerous rural economies, agriculture sustains the livelihoods of a substantial segment of the population. Substituting manual labor with machinery may intensify unemployment, especially among landless laborers, seasonal workers, and women involved in informal agricultural activities (Wolfert et al., 2017). Concerns over data ownership and surveillance also exist. Who possesses the data gathered by automated systems? What is its application, and who benefits from it? These inquiries have ignited discussions around digital colonialism and agricultural autonomy. The societal acceptance of automation depends on inclusive development, transparent governance, and ethical protections. Technologies should seek to augment rather than replace human labor, boosting productivity while maintaining employment through task redefinition and skill enhancement(Martin, 2021).

The adoption of agricultural automation is impeded by a complex array of technological, economic, social, and policy-related difficulties. The transition to smart farming necessitates not only innovation but also systemic support, inclusive policy-making, and participatory implementation, given the challenges posed by the cost of advanced technology, the digital illiteracy of rural communities, and the lack of supportive infrastructure and regulations. (cheau et al., 2022).

Addressing these obstacles comprehensively is crucial to guarantee that agricultural automation transcends being a privilege of industrial farms and evolves into a transformational catalyst for all farmers, irrespective of location or economic standing.

7. FUTURE DIRECTIONS AND RECOMMENDATIONS

Agriculture, positioned at the intersection of innovation and necessity, requires a collaborative, inclusive, and sustainable strategy moving forward. The future of agricultural automation depends on advanced technologies that are also affordable, accessible, adaptable, and aligned with ecological and social concerns. The subsequent ideas delineate how stakeholders might together cultivate a resilient and equitable agricultural future.

7.1 Role of Education, Extension Services, and Public-Private Partnerships

To integrate automation into the mainstream, agricultural education and extension services must progress along with technology improvements. Universities, research organizations, and agricultural schools must incorporate courses on digital agriculture, IoT, AI, and robotics into their curricula to equip the forthcoming generation of agricultural professionals. Moreover, agricultural training must extend beyond fundamental mechanization to include practical sessions on drone operation, intelligent irrigation, sensor upkeep, and data analysis.

Extension services, especially in emerging nations, require rejuvenation. Field officers must be provided with digital tools and ongoing capacity-building to serve as facilitators of technology adoption. Mobile applications, content in regional languages, and demonstration farms can address the last-mile gap (Holm et al., 2024).

Public-private partnerships (PPPs) provide a scalable framework to integrate innovation with community engagement. Collaborations among government agencies, agri-tech firms, and NGOs can provide integrated services equipment leasing, agronomic consultancy, remote monitoring, and training specifically aimed at small and marginal farmers (Tagarakis et al., 2021) . Public-private partnerships are essential for infrastructure development, including the implementation of rural broadband and the coordination of logistics for automated machinery.

7.2 Affordable Tech Innovations for Smallholders

The forthcoming phase of agricultural automation must be economical, modular, and scalable. Contemporary trends predominantly emphasize premium, costly technologies designed for extensive agricultural operations, so largely marginalizing smallholders who represent over 80% of global farmers due to financial constraints and intricacy .

Innovations including affordable soil sensors, solar-powered irrigation systems, AI-driven mobile advisory applications, and tiny robotic weeders can facilitate the widespread adoption of automation. Shared service models, in which equipment such as autonomous seeders or drone sprayers is owned by cooperatives or agricultural service centers, can substantially alleviate the financial strain on individual farmers(Furendal & Jebari, 2023).

Local entrepreneurs and maker communities can be engaged to collaboratively develop frugal inventions suited to particular agricultural conditions. This necessitates the enhancement of rural innovation ecosystems via financial assistance, mentorship, and technical support. Utilizing open-source platforms and reverse innovation methodologies can facilitate the worldwide scaling of such solutions.

7.3 Integration with Sustainable Practices and Circular Agriculture

Automation must not function in isolation; it should be congruent with the ideas of sustainable agriculture and the circular economy. Smart technology can be utilized to mitigate the excessive application of fertilizers, pesticides, and water, thereby reducing environmental deterioration (Islam et al., 2024) . AI-driven pest detection systems can initiate targeted spraying, decreasing chemical usage by as much as 80%.

The integration of precision agriculture with organic methods, conservation tillage, and regenerative agriculture improves productivity and ecological equilibrium. Automated composting systems, real-time carbon footprint monitoring, and traceability instruments can facilitate closed-loop systems and climate-smart agriculture (Pal et al., 2025.) .

Circular agriculture, which prioritizes waste reutilization, nutrient cycling, and biodiversity, might additionally benefit from automation. Examples encompass robotic manure spreaders that optimally distribute nutrients and drone surveillance of cover crops for soil health evaluation.

7.4 Policies for Inclusive Digital Transformation

Governments are essential in creating a future where automation serves the interests of all. Inclusive digital transformation necessitates proactive policy formulation that fosters innovation while protecting farmer rights. Governments ought to develop national frameworks for digital agriculture policies, emphasizing digital literacy, rural infrastructure, open data standards, and cost-effective connection (Ferroni & Castle, 2011) .

Subsidies and tax incentives for sustainable automation technologies such as solar-powered pumps, intelligent irrigation systems, or bio-input applicators can promote adoption. It is essential that rules safeguard smallholders from digital marginalization and inhibit the monopolization of data or services by major agribusinesses(Mizik, 2023) . Data governance must guarantee that farmers maintain ownership and decision-making authority over their agricultural data. Protocols for ethical AI utilization, transparency in machine learning algorithms, and procedures for grievance redressal must be instituted(Kale et al., 2025) .

Interdisciplinary collaboration within ministries (agricultural, rural development, IT, education) and coordination with international organizations such as FAO, IFAD, and CGIAR is essential for standard harmonization and funding mobilization(Dara et al., 2022) . The future of agriculture relies not just on technology but also on the methods of its dissemination, governance, and sustainability. Automation provides robust solutions for labor shortages, climate issues, and productivity limitations, although it must be integrated within a framework of inclusion, affordability, and environmental awareness(Klerkx et al., 2019).

Education and training will enable farmers to adopt these tools. Frugal innovations will broaden their scope. Sustainability will guarantee enduring vitality. Enabling policies will integrate all these elements into a cohesive plan. The true problem resides not in constructing intelligent devices, but in developing an intelligent system that simultaneously benefits individuals, the environment, and production.

8. CONCLUSION

A new age has begun, one in which automation is no longer a luxury but rather a need. This new era was made possible by the convergence of technology breakthroughs and the issues that are faced in agriculture. It has been outlined in this chapter that the demands of climate change, decreased labor availability, and resource inefficiencies require a paradigm shift in the way that we cultivate food. Automation provides attractive solutions, such as precision farming and crop monitoring driven by artificial intelligence, as well as autonomous gear that is capable of doing jobs with an efficiency that is unparalleled.

Nevertheless, the successful deployment of these technologies is dependant upon a number of factors, including the cost of these technologies, the readiness of the infrastructure associated with them, and the digital literacy of farmers, particularly in developing nations. In order to ensure that small and marginal farmers are not left behind, it is necessary to formulate policies that provide support for automation solutions that are both scalable and inclusive.

REFERENCES

Abbes, A. B., Jarray, N., & Farah, I. R. (2024). Advances in remote sensing based soil moisture retrieval: Applications, techniques, scales and challenges for combining machine learning and physical models. *Artificial Intelligence Review, 57*(9), 224. DOI: 10.1007/s10462-024-10734-1

Abdulai, A. R., Tetteh Quarshie, P., Duncan, E., & Fraser, E. (2023). Is agricultural digitization a reality among smallholder farmers in Africa? Unpacking farmers' lived realities of engagement with digital tools and services in rural Northern Ghana. *Agriculture & Food Security, 12*(1), 11. DOI: 10.1186/s40066-023-00416-6

Alexander, P., Arneth, A., Henry, R., Maire, J., Rabin, S., & Rounsevell, M. D. (2023). High energy and fertilizer prices are more damaging than food export curtailment from Ukraine and Russia for food prices, health and the environment. *Nature Food, 4*(1), 84–95. DOI: 10.1038/s43016-022-00659-9 PMID: 37118577

Allen, R. G., Pereira, L. S., Raes, D., & Smith, M. (1998). Crop evapotranspiration-Guidelines for computing crop water requirements-FAO Irrigation and drainage paper 56. *Fao, Rome, 300*(9), D05109.

Ariesen-Verschuur, N., Verdouw, C., & Tekinerdogan, B. (2022). Digital Twins in greenhouse horticulture: A review. *Computers and Electronics in Agriculture, 199*, 107183. DOI: 10.1016/j.compag.2022.107183

Ayim, C., Kassahun, A., Tekinerdogan, B., & Addison, C. (2020). Adoption of ICT innovations in the agriculture sector in Africa: A Systematic Literature Review. *arXiv preprint* arXiv:2006.13831.

Bang, J., & Han, J. W. (2025). Factors Influencing Farmers' Motivation to Adopt Smart Farm Technology in South Korea. *arXiv preprint* arXiv:2504.01795.

Bazargani, K., & Deemyad, T. (2024). Automation's impact on agriculture: Opportunities, challenges, and economic effects. *Robotics (Basel, Switzerland), 13*(2), 33. DOI: 10.3390/robotics13020033

Benke, K., & Tomkins, B. (2017). Future food-production systems: vertical farming and controlled-environment agriculture. *Sustainability: Science. Practice and Policy, 13*(1), 13–26.

Binswanger, H. (1986). Agricultural mechanization: A comparative historical perspective. *The World Bank Research Observer, 1*(1), 27–56. DOI: 10.1093/wbro/1.1.27

Biswas, A., Sarkar, S., Das, S., Dutta, S., Choudhury, M. R., Giri, A., & Paul, D. (2025). Water scarcity: A global hindrance to sustainable development and agricultural production–A critical review of the impacts and adaptation strategies. *Cambridge Prisms: Water*, *3*, e4. DOI: 10.1017/wat.2024.16

Change, I. C. (2013). The physical science basis.

Cho, S. B., Soleh, H. M., Choi, J. W., Hwang, W. H., Lee, H., Cho, Y. S., Cho, B.-K., Kim, M. S., Baek, I., & Kim, G. (2024). Recent Methods for Evaluating Crop Water Stress Using AI Techniques: A Review. *Sensors (Basel)*, *24*(19), 6313. DOI: 10.3390/s24196313 PMID: 39409355

Dara, R., Hazrati Fard, S. M., & Kaur, J. (2022). Recommendations for ethical and responsible use of artificial intelligence in digital agriculture. *Frontiers in Artificial Intelligence*, *5*, 884192. DOI: 10.3389/frai.2022.884192 PMID: 35968036

Demestichas, K., Peppes, N., Alexakis, T., & Adamopoulou, E. (2020). Blockchain in agriculture traceability systems: A review. *Applied Sciences (Basel, Switzerland)*, *10*(12), 4113. DOI: 10.3390/app10124113

Duckett, T., Pearson, S., Blackmore, S., Grieve, B., Chen, W. H., Cielniak, G., . . . Yang, G. Z. (2018). Agricultural robotics: the future of robotic agriculture. *arXiv preprint* arXiv:1806.06762. DOI: 10.31256/WP2018.2

Etezadi, H., & Eshkabilov, S. (2024). A comprehensive overview of control algorithms, sensors, actuators, and communication tools of autonomous all-terrain vehicles in agriculture. *Agriculture*, *14*(2), 163. DOI: 10.3390/agriculture14020163

Evenson, R. E., & Gollin, D. (2003). Assessing the impact of the Green Revolution, 1960 to 2000. *Science*, *300*(5620), 758–762. DOI: 10.1126/science.1078710 PMID: 12730592

Fasciolo, B., Panza, L., & Lombardi, F. (2024). Exploring the Integration of Industry 4.0 Technologies in Agriculture: A Comprehensive Bibliometric Review. *Sustainability (Basel)*, *16*(20), 8948. DOI: 10.3390/su16208948

Ferroni, M., & Castle, P. (2011). Public-private partnerships and sustainable agricultural development. *Sustainability (Basel)*, *3*(7), 1064–1073. DOI: 10.3390/su3071064

Fountas, S., Carli, G., Sørensen, C. G., Tsiropoulos, Z., Cavalaris, C., Vatsanidou, A., Liakos, B., Canavari, M., Wiebensohn, J., & Tisserye, B. A. (2015). Farm management information systems: Current situation and future perspectives. *Computers and Electronics in Agriculture*, *115*, 40–50. DOI: 10.1016/j.compag.2015.05.011

Fuhrmann-Aoyagi, M. B., Miura, K., & Watanabe, K. (2024). Sustainability in Japan's Agriculture: An Analysis of Current Approaches. *Sustainability (Basel), 16*(2), 596. DOI: 10.3390/su16020596

Furendal, M., & Jebari, K. (2023). The future of work: Augmentation or stunting? *Philosophy & Technology, 36*(2), 36. DOI: 10.1007/s13347-023-00631-w

Garzón, J., Montes, L., Garzón, J., & Lampropoulos, G. (2023). Systematic review of technology in aeroponics: Introducing the technology adoption and integration in sustainable agriculture model. *Agronomy (Basel), 13*(10), 2517. DOI: 10.3390/agronomy13102517

Getahun, S., Kefale, H., & Gelaye, Y. (2024). Application of precision agriculture technologies for sustainable crop production and environmental sustainability: A systematic review. *TheScientificWorldJournal, 2024*(1), 2126734. DOI: 10.1155/2024/2126734 PMID: 39421732

Gupta, G., & Kumar Pal, S. (2025). Applications of AI in precision agriculture. *Discover Agriculture, 3*(1), 61. DOI: 10.1007/s44279-025-00220-9

Hasan, M. M., Islam, M. U., & Sadeq, M. J. (2022). Towards technological adaptation of advanced farming through AI, IoT, and Robotics: A Comprehensive overview. *arXiv preprint* arXiv:2202.10459.

Hebsale Mallappa, V. K., & Pathak, T. B. (2023). Climate smart agriculture technologies adoption among small-scale farmers: A case study from Gujarat, India. *Frontiers in Sustainable Food Systems, 7*, 1202485. DOI: 10.3389/fsufs.2023.1202485

Heikonen, S., Heino, M., Jalava, M., Siebert, S., Viviroli, D., & Kummu, M. (2025). Climate change threatens crop diversity at low latitudes. *. *Nature Food, 6*(4), 1–12. DOI: 10.1038/s43016-025-01135-w PMID: 40038529

Hemming, S., de Zwart, F., Elings, A., Righini, I., & Petropoulou, A. (2019). Remote control of greenhouse vegetable production with artificial intelligence—Greenhouse climate, irrigation, and crop production. *Sensors (Basel), 19*(8), 1807. DOI: 10.3390/s19081807 PMID: 31014024

Holm, S., Pedersen, S. M., & Tamirat, T. W. (2024). Robots in agriculture–A case-based discussion of ethical concerns on job loss, responsibility, and data control. *Smart Agricultural Technology, 9*, 100633. DOI: 10.1016/j.atech.2024.100633

Hoque, A., & Padhiary, M. (2024). Automation and AI in Precision Agriculture: Innovations for Enhanced Crop Management and Sustainability. *Asian J. Res. Comput. Sci, 17*(10), 95–109. DOI: 10.9734/ajrcos/2024/v17i10512

Islam, M. H., Anam, M. Z., Hoque, M. R., Nishat, M., & Bari, A. M. (2024). Agriculture 4.0 adoption challenges in the emerging economies: Implications for smart farming and sustainability. *Journal of Economy and Technology, 2*, 278–295. DOI: 10.1016/j.ject.2024.09.002

Kale, R. B., Dangi, P. A., Nikam, V., Khandagale, K., Gadge, S. S., & Mahajan, V. (2025). How are new collective models transforming support for smallholder farmers? Empirical evidence from thematic and sentiment content analysis. *Frontiers in Sustainable Food Systems, 8*, 1542676. DOI: 10.3389/fsufs.2024.1542676

Kaur, A., Bhatt, D. P., & Raja, L. (2024). Developing a hybrid irrigation system for smart agriculture using IoT sensors and machine learning in Sri Ganganagar, Rajasthan. *Journal of Sensors, 2024*(1), 6676907. DOI: 10.1155/2024/6676907

Kendall, H., Clark, B., Li, W., Jin, S., Jones, G. D., Chen, J., Taylor, J., Li, Z., & Frewer, L. J. (2022). Precision agriculture technology adoption: A qualitative study of small-scale commercial "family farms" located in the North China Plain. *. *Precision Agriculture, 23*(1), 1–33. DOI: 10.1007/s11119-021-09839-2

Khan, F. U., Nouman, M., Negrut, L., Abban, J., Cismas, L. M., & Siddiqi, M. F. (2024). Constraints to agricultural finance in underdeveloped and developing countries: A systematic literature review. *International Journal of Agricultural Sustainability, 22*(1), 2329388. DOI: 10.1080/14735903.2024.2329388

Klerkx, L., Jakku, E., & Labarthe, P. (2019). A review of social science on digital agriculture, smart farming and agriculture 4.0: New contributions and a future research agenda. *NJAS-Wageningen Journal of Life Sciences, 90*, 100315.

Lakhiar, I. A., Yan, H., Zhang, C., Wang, G., He, B., Hao, B., Han, Y., Wang, B., Bao, R., Syed, T. N., Chauhdary, J. N., & Rakibuzzaman, M. (2024). A review of precision irrigation water-saving technology under changing climate for enhancing water use efficiency, crop yield, and environmental footprints. *Agriculture, 14*(7), 1141. DOI: 10.3390/agriculture14071141

Lavelle, P., Decaëns, T., Aubert, M., Barot, S., Blouin, M., Bureau, F., Margerie, P., Mora, P., & Rossi, J. P. (2006). Soil invertebrates and ecosystem services. *European Journal of Soil Biology, 42*, S3–S15. DOI: 10.1016/j.ejsobi.2006.10.002

Lee, M. H., Yao, M. H., Kow, P. Y., Kuo, B. J., & Chang, F. J. (2024). An Artificial Intelligence-Powered Environmental Control System for Resilient and Efficient Greenhouse Farming. *Sustainability (Basel), 16*(24), 10958. DOI: 10.3390/su162410958

Li, D., Nanseki, T., Chomei, Y., & Kuang, J. (2023). A review of smart agriculture and production practices in Japanese large-scale rice farming. *Journal of the Science of Food and Agriculture, 103*(4), 1609–1620. DOI: 10.1002/jsfa.12204 PMID: 36069313

Liakos, K. G., Busato, P., Moshou, D., Pearson, S., & Bochtis, D. (2018). Machine learning in agriculture: A review. *Sensors (Basel), 18*(8), 2674. DOI: 10.3390/s18082674 PMID: 30110960

Martin, P. (2021). A future-focused view of the regulation of rural technology. *Agronomy (Basel), 11*(6), 1153. DOI: 10.3390/agronomy11061153

McLandsborough, A. (2023). Nothing Runs like a (Autonomous) Deere: How Government-Sponsored Reskilling Can Cushion the Blow of Technological Advancement for Kansan Farm Workers. *Kan. JL & Pub. Pol'y, 33*, 197.

Mehta, C. R., Chandel, N. S., & Dubey, K. (2023). Smart Agricultural Mechanization in India—Status and Way Forward. In *Smart Agriculture for Developing Nations: Status, Perspectives and Challenges* (pp. 1–14). Springer Nature Singapore. DOI: 10.1007/978-981-19-8738-0_1

Mizik, T. (2023). How can precision farming work on a small scale? A systematic literature review. *Precision Agriculture, 24*(1), 384–406. DOI: 10.1007/s11119-022-09934-y

Mohanty, S. P., Hughes, D. P., & Salathé, M. (2016). Using deep learning for image-based plant disease detection. *Frontiers in Plant Science, 7*, 215232. DOI: 10.3389/fpls.2016.01419 PMID: 27713752

Ngadi, N., Zaelany, A. A., Latifa, A., Harfina, D., Asiati, D., Setiawan, B., Ibnu, F., Triyono, T., & Rajagukguk, Z. (2023). Challenge of agriculture development in Indonesia: Rural youth mobility and aging workers in agriculture sector. *Sustainability (Basel), 15*(2), 922. DOI: 10.3390/su15020922

Pal, D., Mondal, R., Saha, S., & Chatterjee, S. (2025). INTEGRATION OF ADVANCED TECHNOLOGIES LIKE AI AND IOT INTO AGRICULTURAL EXTENSION.

Pingali, P. L. (2012). Green revolution: Impacts, limits, and the path ahead. *Proceedings of the National Academy of Sciences of the United States of America, 109*(31), 12302–12308. DOI: 10.1073/pnas.0912953109 PMID: 22826253

Qu, J., Zhang, Z., Qin, Z., Guo, K., & Li, D. (2024). Applications of autonomous navigation technologies for unmanned agricultural tractors: A review. *Machines (Basel), 12*(4), 218. DOI: 10.3390/machines12040218

Rai, N., Zhang, Y., Ram, B. G., Schumacher, L., Yellavajjala, R. K., Bajwa, S., & Sun, X. (2023). Applications of deep learning in precision weed management: A review. *Computers and Electronics in Agriculture*, *206*, 107698. DOI: 10.1016/j.compag.2023.107698

Rajwar, K., Pant, M., & Deep, K. (2024). Application of machine learning in agriculture: Recent trends and future research avenues. *arXiv preprint* arXiv:2405.17465.

cheau, M. C., Achim, M. V., Găbudeanu, L., Brici, I., & Vîlcea, A. L. (2022). Legal, economic and cyber security framework considerations for drone usage. *Applied Sciences (Basel, Switzerland)*, *12*(9), 4663. DOI: 10.3390/app12094663

Seto, K. C., Güneralp, B., & Hutyra, L. R. (2012). Global forecasts of urban expansion to 2030 and direct impacts on biodiversity and carbon pools. *Proceedings of the National Academy of Sciences of the United States of America*, *109*(40), 16083–16088. DOI: 10.1073/pnas.1211658109 PMID: 22988086

Sharma, H., Sidhu, H., & Bhowmik, A. (2025). Remote Sensing Using Unmanned Aerial Vehicles for Water Stress Detection: A Review Focusing on Specialty Crops. *Drones (Basel)*, *9*(4), 241. DOI: 10.3390/drones9040241

Skendžić, S., Zovko, M., Živković, I. P., Lešić, V., & Lemić, D. (2021). The impact of climate change on agricultural insect pests. *Insects*, *12*(5), 440. DOI: 10.3390/insects12050440 PMID: 34066138

Smith, P., Poch, R. M., Lobb, D. A., Bhattacharyya, R., Alloush, G., Eudoxie, G. D., & Hallett, P. (2024). Status of the World's Soils. *Annual Review of Environment and Resources*, •••, 49.

Soussi, A., Zero, E., Sacile, R., Trinchero, D., & Fossa, M. (2024). Smart sensors and smart data for precision agriculture: A review. *Sensors (Basel)*, *24*(8), 2647. DOI: 10.3390/s24082647 PMID: 38676264

Stanghellini, C. (2013, June). Horticultural production in greenhouses: efficient use of water. In *International Symposium on Growing Media and Soilless Cultivation 1034* (pp. 25-32).

Tagarakis, A. C., Dordas, C., Lampridi, M., Kateris, D., & Bochtis, D. (2021). A smart farming system for circular agriculture. *Engineering Proceedings*, *9*(1), 10.

Tao, W., Zhao, L., Wang, G., & Liang, R. (2021). Review of the internet of things communication technologies in smart agriculture and challenges. *Computers and Electronics in Agriculture*, *189*, 106352. DOI: 10.1016/j.compag.2021.106352

Tong, T., Ye, F., Zhang, Q., Liao, W., Ding, Y., Liu, Y., & Li, G. (2024). The impact of labor force aging on agricultural total factor productivity of farmers in China: Implications for food sustainability. *Frontiers in Sustainable Food Systems*, *8*, 1434604. DOI: 10.3389/fsufs.2024.1434604

Van Dijk, M., Morley, T., Rau, M. L., & Saghai, Y. (2021). A meta-analysis of projected global food demand and population at risk of hunger for the period 2010–2050. *Nature Food*, *2*(7), 494–501. DOI: 10.1038/s43016-021-00322-9 PMID: 37117684

Wolfert, S., Ge, L., Verdouw, C., & Bogaardt, M. J. (2017). Big data in smart farming–a review. *Agricultural Systems*, *153*, 69–80. DOI: 10.1016/j.agsy.2017.01.023

Yuan, X., Li, S., Chen, J., Yu, H., Yang, T., Wang, C., Huang, S., Chen, H., & Ao, X. (2024). Impacts of global climate change on agricultural production: A comprehensive review. *Agronomy (Basel)*, *14*(7), 1360. DOI: 10.3390/agronomy14071360

Zhang, X., Wang, C., & Li, Y. (2022). Practices, challenges, and future of digital transformation in smallholder agriculture: Insights from a literature review. *Agriculture*, *14*(12), 2193.

Zhang, Y. (2019). The role of precision agriculture. *Resource Magazine*, *26*(6), 9–9.

Zhao, C., Liu, B., Piao, S., Wang, X., Lobell, D. B., Huang, Y., Huang, M., Yao, Y., Bassu, S., Ciais, P., Durand, J.-L., Elliott, J., Ewert, F., Janssens, I. A., Li, T., Lin, E., Liu, Q., Martre, P., Müller, C., & Asseng, S. (2017). Temperature increase reduces global yields of major crops in four independent estimates. *Proceedings of the National Academy of Sciences of the United States of America*, *114*(35), 9326–9331. DOI: 10.1073/pnas.1701762114 PMID: 28811375

Chapter 6
From Data to Harvest:
How AI and IoT Are Shaping the Future of Farming

Lingala Thirupathi
https://orcid.org/0000-0003-0703-3860
Sreenidhi Institute of Science and Technology, India

Shyam Sunder Pabboju
https://orcid.org/0009-0002-0523-0537
Mahatma Gandhi Institute of Technology, India

Suman Dabbu
https://orcid.org/0000-0002-6749-2498
Osmania University, India

Deva Rajashekar
https://orcid.org/0000-0001-8871-0880
Anurag University, India

Manvitha Pola
https://orcid.org/0009-0008-4486-5862
Sreenidhi Institute of Science and Technology, India

Hari Chandra Prasad Cheerla
Saint Louis University, USA

ABSTRACT

The agriculture industry is at a crossroads, faced with rising global food demands, resource constraints, and the pressing effects of climate change. To move forward, farming must evolve, and that evolution is being powered by two transformative technologies: Artificial Intelligence (AI) and the Internet of Things (IoT). Together, they're creating what we now call Smart Farming. This chapter explores how AI and IoT are not only streamlining agricultural processes but are also helping reshape the very foundation of how farming operates. IoT brings the power of constant connectivity and real-time data collection through devices like soil sensors, drones, and weather stations. AI, on the other hand, processes that flood of data to uncover insights, predict outcomes, and drive automated decision making. We'll examine how these technologies are working together to enable precision agriculture, smart

DOI: 10.4018/979-8-3373-4207-8.ch006

irrigation, early pest detection, yield forecasting, and even supply chain optimization. The chapter will also address the challenges that come with digital transformation in agriculture.

1. INTRODUCTION

Agriculture, one of the oldest and most vital human industries, stands at a make-or-break threshold. With the world population approaching 10 billion in 2050, food demand is expected to increase by more than 60%. At the same time, available land for farming is dwindling, weather patterns are increasingly unpredictable, and conventional farming techniques are failing to meet the demand for efficiency and sustainability for what's to come. This requires an immediate shift not only in the way we farm but in the way we conceptualize farming. The advent of Smart Farming, driven by Artificial Intelligence (AI) and the Internet of Things (IoT), presents a viable route toward addressing these complex agricultural issues.

Smart Farming is not just a buzzword it is a paradigm shift that reimagines agriculture as a data-driven, networked, and smart process. With IoT technologies like soil moisture sensors, GPS-equipped tractors, weather stations, and aerial drones, farms can sense the environment in real-time and automate operations. These devices serve as the nervous system of contemporary agriculture, constantly taking vital data from the field. But information by itself is not intelligence and this is where AI comes in. AI technologies analyze this tremendous flow of information to identify patterns, make predictions, and inform decision-making with unprecedented accuracy.

The contribution of AI in agriculture goes well beyond automation. Machine learning is being employed to forecast crop disease before it infects, streamline irrigation schedules right down to the level of a single plant, and even suggest planting tactics based on climatic and soil conditions. With integration into IoT systems, AI makes possible closed-loop, autonomous learning systems that can dynamically react to changing circumstances tweaking fertilizer dosages, seeding with drones for precision pesticide application, or notifying farmers of pest infestations in real time. This convergence of sensing (IoT) and reasoning (AI) turns traditional reactive farming into proactive and predictive farming.

What makes this revolution stand out is not just its technological complexity but also its significant practical consequences. The marriage of AI and IoT is making possible Precision Agriculture an approach that uses real-time information and smart algorithms to guarantee each agricultural input, be it water, fertilizer, pesticide, or labour, is deployed in the most effective and efficient way. By reducing input waste and maximizing crop production, precision farming not only saves farmers operational expenses but also dramatically minimizes the ecological impact of farming.

In addition, this smart farming paradigm enables remote monitoring and control, enabling farmers to control large or scattered fields through mobile applications or cloud consoles. New capabilities like autonomous equipment, AI-driven crop health monitoring through drone photography, and blockchain-supported traceability systems are extending the limits of what contemporary agriculture can do. These technologies are not only increasing productivity but also promoting transparency, sustainability, and economic emancipation particularly for smallholder farmers in the developing world who now have access to affordable, scalable Agritech solutions.

This chapter explores how AI and IoT are transforming the face of agriculture. From field to cloud, from drones to data lakes, we will be looking at the architecture, applications, and synergistic working of these technologies. Along with success stories and live deployments, the chapter also explores adoption blockers infrastructure gaps, data privacy concerns, and cost factors. Finally, this chapter seeks to ensure that there is a complete perception of how AI and IoT not only modernize agriculture but reshape its future as well. The Figure 1 shows the Global population, food supply, and crop yield trends from 1960 to 2050, with data from FAOSTAT (2015).

Figure 1. Global population, food supply, and crop yield trends from 1960 to 2050, with data from FAOSTAT (2015).

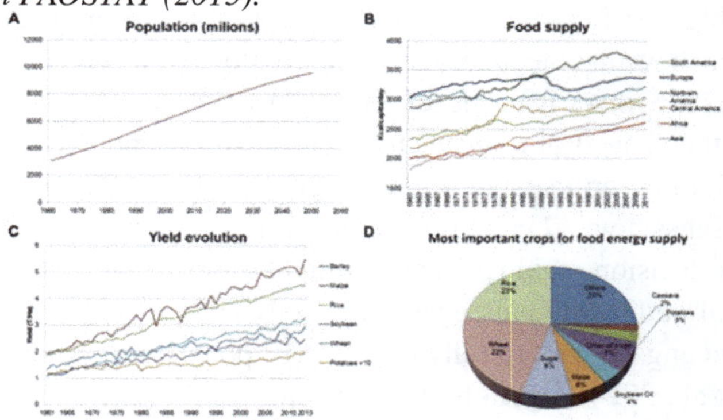

2. LITERATURE SURVEY

(Hussein et al., 2024) explore AI and IoT integration in agriculture, showing how technologies like drones, sensors, and big data enable precision farming and real-time decision-making. They discuss autonomous farming and energy-efficient IoT devices. The authors also examine the socioeconomic impact of smart agriculture. The study highlights the potential of these technologies in enhancing farming practices. They suggest that this transformation can lead to increased productivity and sustainability in agriculture.

(Sharma & Shivandu, 2024) discuss innovations such as AgroBots, remote sensing, and spectral imaging for crop monitoring. They highlight the role of AI in detecting fruits under varying conditions and its application in precision farming. The review includes case studies like PACMAN SCRI. The authors discuss future trends like the role of 5G in improving connectivity for farmers. They also address the need for more research in data integration and scalability.

(Issa et al., 2024) examine the role of AI and IoT in modern agriculture, with a focus on real-time monitoring and smart irrigation systems. They discuss the integration of machine learning, sensor networks, and drones in boosting agricultural productivity. The authors emphasize the use of SDSS for automated recommendations in developing regions. They highlight how these technologies can contribute to global food security. Their findings stress the importance of cloud-based farm monitoring solutions.

(Obaid et al., 2024) explore the use of AI, IoT, and blockchain in optimizing the agricultural supply chain. They discuss the role of RFID, sensor networks, and cloud computing in improving food production, safety, and distribution. The authors stress the importance of data security and transparency in farming practices. They point out how big data from sensors can aid in optimizing farming operations. The study also covers how these technologies support global food demand management.

(Aijaz et al., 2025) investigate the role of AI in overcoming agricultural challenges, particularly in the context of climate change and growing populations. They focus on AI's potential in crop management through automation and machine learning. The authors discuss how AI can provide real-time monitoring and predictive analytics for better decision-making. They highlight the barriers to AI adoption, such as technological and legal challenges. The study calls for comprehensive policy frameworks and engagement to fully unlock AI's potential in agriculture.

(Nautiyal et al., 2024) highlight the importance of AI and smart agriculture in addressing India's food security challenges. They propose using technologies like drones and precision farming to enhance agricultural productivity. The authors suggest that AI could significantly increase crop yields while reducing the environmental impact. They emphasize how these technologies can improve resource management and lower labour dependence. The paper argues that smart agriculture is crucial for global food security and sustainability. (Padmaja et al., 2021) Discussed the distributing of the message over the network using another frequency and timing technique to circumvent the jammers. Predicted (Gangula et al., 2021) ensemble machine learning based dengue disease with performance and accuracy elevation patterns.

(Qazi et al., 2022) discuss the role of IoT and AI in transforming traditional agriculture. They focus on how low-cost, low-power wireless sensors can optimize resource management, particularly water use. The authors also highlight the benefits of AI-driven predictive analytics for crop management and disease prevention. They address challenges in technology adoption, such as cost and connectivity issues. The paper explores future trends and technological advancements that could shape the future of agriculture.

(Bhangar & Shahriyar, 2024) explore AI and IoT in the cultivation of specific crops like onions and turmeric. They discuss how precision farming, automated irrigation, and disease detection can improve crop yield and quality. The authors highlight the importance of data from sensors for optimizing farming practices. They address barriers such as high initial costs, limited internet connectivity, and the need for farmer training. The study stresses the need for policy support and investment to overcome these challenges.

(Hussein et al., 2024) examine how AI and IoT technologies can address agricultural challenges like climate change and labor costs. The authors highlight applications such as crop yield forecasting, precision agriculture, and resource management. They discuss the role of AI in predicting diseases, automating operations, and enhancing decision-making. The study emphasizes the integration of AI and IoT with blockchain for better data security and supply chain transparency. These technologies are seen as key to making agriculture more sustainable and productive.

(Shaikh et al., 2022) explore the use of machine learning, robotics, and IoT in advancing precision agriculture. They discuss how AI and IoT technologies can optimize crop management, from monitoring to decision-making. The paper covers the use of drones and sensors in improving farming practices. The authors review the challenges these technologies face, particularly in rural settings. The study concludes with an outlook on emerging research directions in AI for agriculture.

(Thirupathi et al., 2023) predicted the disease in healthcare communities by machine learning and used IoT in the internet of lighting for smart cities. (Kumari &Nafchi, 2024) discuss the transformative role of AI, machine learning (ML), and deep learning (DL) in smart farming. The authors explore the integration of IoT and Variable Rate Technology (VRT) in modern agriculture. They highlight how advanced algorithms predict soil conditions and enhance crop yield forecasting. The paper emphasizes the use of image recognition for plant disease detection and water stress identification. Ultimately, the study points to AI's potential in optimizing farming efficiency and environmental sustainability.

(Holzinger et al., 2022) examine the role of human-cantered AI (HCAI) in the digital transformation of agricultural and forestry operations. The study emphasizes the integration of AI and ML with human expertise to enhance decision-making processes. The authors discuss the importance of explainability and robustness in AI models used in agriculture and forestry. They highlight the need for multimodal information fusion and the collaboration of AI systems with human knowledge. The paper identifies three key research areas for advancing HCAI: intelligent information fusion, robotics, and trusted decision support.

(Kumar et al., 2024) examine the role of technologies like AI, blockchain, IoT, remote sensing, and hydroponics in revolutionizing horticulture for smart farming. They discuss how these innovations optimize crop management, particularly in India, where land is limited. The paper highlights advancements in AI, robotics, and machine learning for monitoring plants, pests, and diseases. It also explores the challenges and future potential of these technologies to enhance food production. The review emphasizes the importance of digital tools for efficient farm management and sustainable practices.

(Bahari et al., 2024) explore key factors influencing IoT adoption in smart farming through a model based on the Technology-Organization-Environment (TOE) framework. Their study highlights the importance of governmental support, technological compatibility, and financial considerations in driving IoT integration. They also emphasize the role of the digital landscape in influencing smart farm adoption. Empirical validation using data from 179 managers provides practical insights for sustainable IoT adoption in agriculture.

(Shailaja, et al., 2023) explored the intrusion detection system for network applications in the real time. (Karunathilake et al., 2023) provide an overview of precision agriculture, focusing on how advanced technologies like IoT, big data, and machine learning can increase agricultural productivity while minimizing environmental impact. The paper discusses the latest innovations such as drones, sensors, and AI, highlighting their potential to improve crop yields and reduce waste. However, the authors also identify challenges in technology adoption, data management, and cost-effectiveness. They conclude by exploring the future prospects of smart farming in overcoming global agricultural challenges.

(Pratapagiri, et al., 2021) discuss the detection of plant leaf disease using convolutional neural networks. (Chataut et al., 2023) review the rapid growth of Internet of Things (IoT) technology, focusing on its applications in healthcare, agriculture, smart homes, smart cities, and Industry 4.0. The paper discusses how advancements in wireless networking and the increasing demand for automation are driving IoT proliferation. It highlights the unique challenges and opportunities in each sector while exploring the future potential of IoT. The authors emphasize the factors that have shaped IoT's current state and its ongoing advancements across industries.

(Padmaja et al., 2021) Distribute the Message over the Network Using another Frequency and Timing Technique to Circumvent the Jammers. (Fuentes et al., 2024) discuss the role of IoT, remote sensing, and AI in enhancing smart crop management. These technologies provide real-time data for better decision-making on irrigation, fertilization, and pest control. Drones and AI help improve monitoring and analysis, offering valuable insights for sustainable agriculture. The paper also highlights adoption challenges and the need for ongoing research and collaboration to advance these technologies globally.

(Mishra and Mishra, 2024) discuss the transformative impact of AI and ML on agriculture, addressing challenges like climate change, resource limitations, and population growth. These technologies enhance farming systems by improving productivity, precision agriculture, resource management, and pest control. AI and ML applications in crop and soil monitoring, agricultural robotics, and predictive analytics enable more efficient and sustainable farming practices. The integration of these technologies promises to revolutionize agriculture by optimizing resource use and fostering data-driven decisions.

(Balyan et al., 2024) discuss the role of technological innovations such as drones, smart sensors, robotics, and remote sensing in revolutionizing agriculture. These advancements support smart farming practices by enhancing crop modelling, weather forecasting, yield prediction, and disease detection, while also promoting sustainability and resilience. The integration of big data, improved connectivity, and computational power enables farmers to optimize productivity and minimize environmental impacts. However, challenges like high costs, data security, and digital literacy gaps remain, hindering widespread adoption.

(Wang et al., 2021) explore the concept of unmanned farms as a solution to the global shortage of agricultural labour due to population growth and decreased interest in farming. They highlight the integration of technologies like IoT, big data, AI, robotics, and 5G to create fully automated farms. These unmanned farms can perform agricultural tasks without human intervention, enabling more efficient and sustainable farming. The paper discusses the framework for unmanned farms, current technological applications, and challenges in implementing this mode of agricultural production. (Kumar et al., 2025) explores the transformative effects of smarter chatbots and text-based prediction, classification model of stress, anxiety, depression among the people.

3. IOT IN ACTION: CONNECTING THE FARM

The inclusion of the Internet of Things (IoT) in agriculture has been revolutionizing farming operations by enabling real-time monitoring, automation, and data-driven decision-making. With the rising global demand for food further intensified by the challenges of climate change and labour shortages IoT has emerged as a pivotal force in transforming how farms function. By enhancing efficiency, productivity, and sustainability, IoT is actively reshaping traditional agricultural methods. This section explores how IoT technologies are seamlessly connecting the farm ecosystem, the tangible benefits they offer, and the foundational role they play in driving the evolution toward Smart Farming.

3.1 The Basics of IoT in Agriculture

Essentially, IoT describes the collection of devices that are networked and can gather, exchange, and share information with minimal human effort. In farming, IoT devices go from the most basic of soil moisture gauges to high-tech autonomous tractors and drones. These devices gather information about the environment, machinery, crops, livestock, and so forth. This information is then routed to a centralized system for examination and action.

One example of a simple IoT application in agriculture is soil moisture sensors, which detect the quantity of water available in the ground. They feed real-time data, enabling farmers to observe their fields remotely and streamline irrigation activities. As the soil moisture levels fall below a predetermined level, automated systems can initiate irrigation so that crops will get adequate water, all automatically. Similarly, farm weather stations collect real-time information regarding temperature, humidity, and rain and enable farmers to expect weather activity and modify operations based on these forecasts.

3.2 The Role of IoT in Farm Automation

IoT has revolutionized farm automation by significantly reducing labour dependency and increasing operational efficiency. A prime example is self-driving tractors equipped with GPS, sensors, and AI, allowing them to perform tasks like ploughing, planting, and harvesting autonomously. These smart tractors use real-time data from IoT devices to adapt to soil conditions adjusting planting depth or speed based on terrain resulting in more precise and effective farming.

Drone technology is another major advancement, enabling high-resolution monitoring of crop health through aerial imaging. When integrated with IoT sensors in the soil and local weather stations, drones help detect pest infestations, diseases, or nutrient deficiencies early, allowing farmers to target interventions more accurately. Similarly, IoT-powered automated irrigation systems respond to real-time soil moisture and weather data, optimizing water usage and reducing unnecessary resource expenditure, ultimately enhancing sustainability and crop yields. The Figure 2 shows the role of IoT in precision farming.

Figure 2. Role of IoT in precision farming

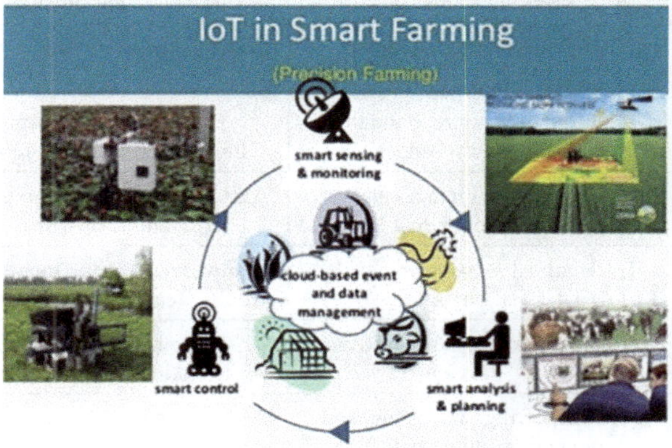

3.3 Enhancing Precision Agriculture with IoT

IoT is a cornerstone of Precision Agriculture, a modern approach that uses real-time data and analytics to optimize farming practices at a micro level. It allows farmers to tailor inputs like water, fertilizer, and pesticides based on the exact needs of specific plants or field zones. This precision results in improved crop yields, reduced input costs, and minimal environmental impact. For example, IoT-enabled smart irrigation systems adjust water supply based on soil moisture, crop type, and weather predictions, ensuring efficient water use and reducing waste especially critical in water-scarce regions.

The Variable Rate Technology (VRT), powered by IoT, fine-tunes planting, fertilizing, and pesticide application across different field sections. By leveraging data from soil sensors and local climate stations, VRT ensures that every part of the field gets exactly what it needs no more, no less. This targeted approach not only boosts productivity but also helps in conserving resources and minimizing environmental harm from overuse of chemicals. The Table 1 represents the IoT devices in smart farming.

Table 1. Common IoT devices in smart farming

IoT Device	Primary Function	Benefits in Farming
Soil Moisture Sensors	Measure soil water content	Enables precise irrigation; reduces water waste; improves crop health
Weather Stations	Monitor temperature, humidity, rainfall, wind speed, etc.	Helps with crop planning, pest/disease forecasting, and climate-resilient decisions
Drones	Capture aerial images and multispectral data	Aids in crop health assessment, plant counting, and early disease detection
Livestock Wearables	Track animal movement, health, and vitals	Improves herd health monitoring; detects illness early; enhances productivity
Smart Irrigation Systems	Automate watering based on sensor data	Optimizes water usage; ensures timely irrigation; supports sustainable farming
Remote-Controlled Tractors	Perform sowing, harvesting, and ploughing autonomously	Reduces manual labour; increases operational efficiency
RFID Tags & GPS Collars	Track livestock location and identity	Prevents theft/loss; helps with breeding and feeding management
Grain Bin Sensors	Monitor temperature and humidity in storage bins	Prevents spoilage; ensures post-harvest quality control
Crop Scouting Robots	Navigate fields to gather detailed crop data	Enhances decision-making through consistent field monitoring

3.4 Livestock Monitoring and Management

IoT's applications extend beyond crop cultivation into livestock farming, where it significantly enhances animal health and farm productivity. Smart wearables like sensor-enabled collars for cows and GPS tags for sheep allow real-time monitoring of animals' vital signs and movements. These devices collect data such as body temperature, heart rate, and activity levels, which are transmitted to cloud platforms or farmers' smartphones. Early detection of illness or stress through behavioural changes allows timely intervention, reducing veterinary costs and improving overall animal welfare.

Moreover, IoT helps optimize grazing patterns and land use by tracking the location and movement of livestock via GPS. This ensures animals graze in designated areas, preventing overgrazing and promoting pasture regeneration. By managing herd movement and monitoring environmental factors, farmers can implement efficient rotational grazing systems, leading to healthier livestock and more sustainable land management practices.

3.5 Supply Chain Optimization and Traceability

Outside the boundaries of the farm, IoT is transforming the agricultural supply chain by enhancing traceability, quality control, and operational efficiency. With the use of RFID tags and smart containers, produce can be tracked in real-time from the farm to retail shelves. IoT sensors monitor key parameters like temperature and humidity during transportation. Any deviation from optimal conditions triggers alerts to farmers or distributors, allowing immediate action such as adjusting cooling systems or re-routing the shipment. This minimizes spoilage, ensures food safety, and maintains high product quality.

Moreover, integrating IoT with blockchain technology offers full traceability across the supply chain. Consumers can scan QR codes on packaging to access detailed information about the product's origin, farming practices, and sustainability measures. This level of transparency builds trust between producers and consumers and supports eco-conscious choices. As consumer demand for responsible sourcing increases, such technologies play a crucial role in ensuring accountability and promoting sustainable food systems.

4. AI'S EXPANDING ROLE IN AGRI-DECISION MAKING

The adoption of Artificial Intelligence (AI) in farming represents a paradigm shift in the way farms are run and make vital decisions. With the increasing amounts of data from IoT sensors, drones, satellite images, and equipment, AI is the intelligence that takes all this data, learns from it, and converts it into actionable information. This ability allows farmers to shift from legacy reactive decision-making into predictive and prescriptive decision-making. AI technologies like machine learning, deep learning, and computer vision are central to this revolution, enabling the farming industry with more precision, efficiency, and sustainability.

4.1 Machine Learning for Agricultural Intelligence

Machine learning models are especially effective in managing the varied and intricate data sets created in agriculture. Machine learning models can learn from past and real-time data to enable predictive analytics for important decisions like planting, irrigation, fertilization, or harvesting. For instance, supervised learning models may be trained using past crop yield data, soil types, and climate variables to predict future yields or pinpoint lagging regions of a farm. Unsupervised learning methods, such as clustering, may be employed to classify similar crop health zones for enhanced input management.

In plant disease and pest management, ML models scan plant images or environmental data to predict early warning signs of infestations or infections. These can signal farmers early enough for them to apply intervention specifically, reducing crop damage and lowering the use of pesticides.

4.2 Deep Learning and Computer Vision in Precision Agriculture

Deep learning, a machine learning technique motivated by the human brain's neural networks, excels at working with image and video data. Convolutional Neural Networks (CNNs) are widely employed in agriculture to classify crops, detect weeds, count fruits, and identify diseases from leaf images. Deep learning models, when combined with drone or satellite images, can chart vast agricultural lands, detecting changes in vegetation health (through NDVI and other indices) and assist in prioritizing areas for attention.

Computer vision technology is also helpful for automation in farming. For example, AI harvesters employ computer vision systems to recognize and harvest ripe fruits without any human intervention. Automated weeders also distinguish between weeds and crops, minimizing the use of herbicides and manual weeding. The table 2 shows the comparison of AI Techniques and Their Agricultural Applications.

Table 2. Comparison of AI techniques and their agricultural applications

AI Technique	Key Applications in Agriculture	Advantages
Machine Learning (ML)	- Crop yield prediction - Soil health analysis - Pest and disease forecasting	- Learns from structured data - Good for trend and pattern recognition
Deep Learning (DL)	- Disease detection from leaf images - Weed identification - Fruit counting	- Excels with large image datasets - High accuracy in visual classification
Computer Vision (CV)	- Real-time plant monitoring - Automated harvesting - Growth stage analysis	- Enables visual automation - Reduces labour costs - Real-time insights

4.3 AI in Livestock Management

AI is not just for crops; its application in livestock management is also revolutionary. Through video monitoring and sensor-attached animals, AI software can track health metrics like feeding habits, activity levels, and body temperature. Anomaly detection systems notify farmers of any departures from baseline patterns, which may signal disease or stress. This allows for early diagnosis and lower veterinary bills while enhancing general animal well-being. In dairy farming, AI technologies process milking machine and wearable sensor data to monitor milk production, cow health, and breeding cycles. These details inform optimal feed formulations, allow for early mastitis detection, and simplify breeding programs.

4.4 Predictive Analytics and Decision Support Systems

One of the most beneficial contributions of AI to agriculture is its predictive analytics capability to predict future conditions. By combining real-time environmental observations with past trends, AI can predict weather patterns, foresee pest infestations, or project yield potential with uncanny accuracy. Such abilities enable farmers to plan ahead and make fact-based decisions that minimize risk and maximize profitability. AI is also the focus of decision support systems (DSS), which provide farmers with dashboards and notifications that translate complicated information into actionable, timely advice. They can suggest the optimal type of crop to sow based on soil information and weather forecasts or suggest ideal times to sow and harvest. In accessible interfaces, DSS applications make sophisticated analysis available to plain farmers, putting smart technologies in their hands.

5. MAKING IT WORK: HOW AI AND IOT INTERACT IN SMART FARMS

In the fast-changing world of agriculture, the convergence of AI and the IoT is building a robust digital environment that facilitates precision, efficiency, and automation throughout the entire cycle of farming. While IoT is tasked with gathering real-time information through a range of connected sensors and devices, AI takes this constant flow of data and turns it into valuable insights and action. Collectively, they are a synergistic system wherein data-driven decision-making occurs in a matter of seconds, building really smart farms.

5.1 The Data Journey: From the Field to the Cloud

At the centre of this integration is the information flow among devices, platforms, and intelligence layers. The process normally starts on the field with IoT devices like soil moisture sensors, temperature and humidity sensors, weather stations, GPS-tractors, and drones with cameras. These devices are deployed strategically over land or fixed to livestock to take data about the environment and livestock in real-time. This sensor information is then sent to edge computing nodes or centralized cloud platforms, based on network capabilities. In remote locations, where connectivity may be a problem, edge computing enables preliminary processing at the source through local microcontrollers or gateways. Otherwise, cloud systems gather, organize, and store large amounts of sensor information in structured formats, awaiting analysis.

5.2 AI as the Decision-Making Core

Once the data arrives at the processing platform, AI takes over. Various layers of AI machine learning models, deep neural networks, and pattern recognition algorithms process the incoming data for trends, anomalies, and predictions. For example, temperature and humidity values from IoT sensors might be processed by an AI model to predict the likelihood of fungal disease outbreaks. Also, real-time soil information can assist in deciding the optimum irrigational timing and quantity, tailored to every parcel of land. AI models also correlate different types of data such as satellite imagery with in-field sensor data to make more accurate predictions and localized insights. This capability to correlate different datasets improves decision-making precision, minimizing wastage of resources and maximizing productivity.

5.3 Real-Time Feedback Loop and Automation

As perhaps the most influential part of the AI-IoT loop is real-time feedback, once AI has received the sensor data and made a decision, it sends the resulting insight back to the user interface or farm equipment for execution. In most smart farms, this will be automated. For instance, if the AI senses that a specific patch of the field is too dry compared to what is ideal, it can trigger an automated watering system instantly to provide only as much water as needed to the area. Drones could be automatically sent out to spray pesticide or fertilizer where crop stress has been identified using AI-powered image analysis. Such a closed-loop system guarantees prompt action with the least amount of human intervention.

AI systems usually display their suggestions or choices on user dashboards or mobile applications in a ready-to-understand format so that farmers can manually override, approve, or schedule activities as and when required. This combination of automation along with human intervention is essential while adapting to changing farm conditions and establishing farmer trust in technology.

5.4 Interoperability and Data Integration

Smart farming is dependent on the easy interoperability of varied devices, platforms, and AI tools. New-generation IoT agricultural ecosystems are created with interoperability in their design, whereby information from different sources soil sensors, weather stations, drone imagery, and livestock wearable's is channelled into one data platform. AI modules process this converged dataset to build farm intelligence as a whole.

For instance, while one set of sensors monitors plant health, another tracks irrigation usage, and a third follows livestock movement patterns. AI algorithms can link this information to uncover complex relationships: reduced plant growth may be correlated with poor water supply or pest activity, which in turn could be traced back to nearby livestock movements or weather changes. These multi-dimensional insights are only possible through the combined power of IoT's sensing and AI's reasoning.

5.5 Examples of Synergistic Applications

Across the agricultural value chain, AI and IoT are being used together in several impactful ways:

- Precision irrigation: Soil moisture sensors (IoT) send real-time data to AI models that calculate optimal irrigation schedules, conserving water while maximizing yield.
- Pest and disease detection: Cameras and environment sensors detect early signs (IoT); AI models identify patterns and trigger warnings or treatment actions.
- Livestock monitoring: Wearable IoT devices track movement and health metrics; AI identifies anomalies and alerts for sickness.
- Autonomous machinery: GPS and computer vision sensors (IoT) guide tractors and harvesters, while AI interprets terrain data for optimal paths and planting depth.
- Greenhouse management: IoT sensors track temperature, humidity, and light; AI adjusts climate control systems in real-time for optimal crop growth.

This harmonious interaction between AI and IoT not only optimizes inputs like water, fertilizers, and energy but also reduces labour dependency and enhances sustainability through data-driven automation.

6. KEY APPLICATIONS AND USE CASES

The fusion of AI and IoT in smart farming has enabled a variety of high-impact applications that go beyond traditional agricultural practices. These use cases exemplify how technology is reshaping agriculture from soil to supermarket.

6.1 Precision Agriculture

Precision agriculture refers to the site-specific management of crops and resources using real-time data collected by IoT sensors and analyzed through AI models. Instead of uniform application of inputs across entire fields, farmers now apply water, fertilizers, and pesticides only where and when needed, maximizing efficiency while minimizing waste.

IoT-enabled soil sensors continuously monitor moisture, nutrient levels, and pH, while drones and satellite imagery offer spatial data on crop health. These datasets feed into AI systems that determine zones of stress, pest vulnerability, or nutrient deficiencies. AI models provide farmers with prescriptive recommendations, enabling targeted interventions that reduce chemical usage and environmental impact. This data-driven approach improves crop quality, yield, and profitability, while promoting sustainability by conserving vital resources.

6.2 Smart Irrigation

Water scarcity is a growing challenge in agriculture, making smart irrigation systems critical. IoT devices such as soil moisture probes, flow meters, and weather sensors provide continuous feedback on soil and atmospheric conditions. AI models analyze this data in real-time to predict optimal watering schedules, considering factors like evapotranspiration rates, upcoming rainfall, crop stage, and soil composition. The system automatically activates irrigation valves or sprinklers only when necessary, ensuring precise water distribution. Farmers can monitor and control irrigation remotely via mobile apps or dashboards, receiving alerts if anomalies such as leaks or excessive water use are detected.

This integration of AI and IoT in irrigation reduces water waste by up to 50%, lowers costs, and improves plant health by preventing overwatering or drought stress.

6.3 Pest and Disease Detection

One of the most promising use cases of AI in agriculture is early pest and disease detection. Traditionally, pest outbreaks were identified only after visible damage occurred. With AI and IoT, detection happens in real-time and at earlier stages, enabling timely responses. IoT-enabled devices like camera-equipped drones or in-field imaging sensors capture high-resolution images of leaves and plants. These images are analyzed by AI-based computer vision models trained to recognize symptoms such as discoloration, wilting, spots, or fungal patterns. Other sensors may track temperature, humidity, and CO_2 levels that foster disease development. Once an issue is identified, farmers are alerted instantly, often with automated recommendations for pesticide or fungicide application. This targeted approach minimizes chemical usage and avoids large-scale crop loss, supporting sustainable and cost-effective pest management.

6.4 Crop Yield Prediction

Accurate crop yield forecasting is essential for agricultural planning, logistics, pricing, and policy-making. AI algorithms can predict future yields based on a range of input data including historical yield records, soil quality, weather patterns, irrigation schedules, and satellite imagery. IoT systems feed real-time data into these AI models, which are continuously refined with machine learning techniques. These models identify patterns and trends that indicate expected output and potential risks, such as heat stress or nutrient deficiency, well before harvest time. Farmers use these insights to make informed decisions about resource allocation, harvesting times, and market engagement. On a larger scale, governments and cooperatives can plan food distribution, storage, and exports more efficiently.

6.5 Livestock Health Monitoring

IoT and AI are also revolutionizing animal husbandry, particularly in large-scale livestock farms. Wearable IoT devices such as smart collars, ear tags, or ingestible biosensors collect continuous data on body temperature, heart rate, activity levels, rumination patterns, and GPS location. AI algorithms analyze this data to detect early signs of illness, reproductive cycles, or behavioural anomalies. For example, a drop in activity or food intake may indicate fever or lameness, triggering alerts to veterinarians or farm managers.

This proactive health monitoring reduces mortality, prevents disease outbreaks, and improves animal welfare and productivity. AI systems also support feeding optimization, breeding cycles, and milk yield prediction, enhancing operational efficiency across livestock farms.

6.6 Supply Chain Management

Beyond the field, AI and IoT play a critical role in agricultural supply chains. IoT sensors attached to packaging or transport vehicles track location, temperature, humidity, and spoilage risk of perishable produce during transit. This ensures traceability and compliance with food safety regulations. Simultaneously, AI models analyze market trends, weather forecasts, storage levels, and consumer behaviour to forecast demand and optimize logistics. This prevents overproduction, reduces post-harvest loss, and ensures timely delivery to markets. Farmers and agribusinesses benefit from real-time visibility across the supply chain, allowing better inventory management, reduced waste, and enhanced profit margins.

7. CASE STUDIES FROM THE FIELD

As AI and IoT technologies continue to mature, numerous real-world implementations offer compelling evidence of their transformative potential in agriculture. This section explores three diverse case studies from large-scale farms to tech startups that demonstrate how these innovations are reshaping traditional practices, enhancing efficiency, and driving sustainability.

Case Study 1: John Deere – AI and IoT in Precision Farming

Location: United States
Technology Used: Computer Vision, Machine Learning, IoT-enabled equipment

John Deere, one of the world's most well-known agricultural machinery manufacturers, has pioneered smart farming tools through its integration of AI and IoT in machinery. One standout innovation is the See & Spray™ technology, which uses computer vision and machine learning to detect weeds in real time and precisely spray herbicides only where needed.

IoT-enabled tractors collect GPS data, crop health information, and soil condition metrics using various sensors embedded in the equipment. This data is transmitted to a centralized platform where AI models analyze trends, monitor anomalies, and optimize decisions like planting density and irrigation schedules. Figure 3 shows the John Deere's See & Spray™ cuts herbicide use by up to 90% with AI precision.

Outcomes:

- Up to 90% reduction in herbicide usage
- Increased crop yields through targeted resource use
- Decreased operational costs and environmental impact

Lessons Learned:

- Integrating AI into legacy farm equipment can deliver immediate ROI.
- Real-time data feedback loops are key to maximizing AI efficiency.

Figure 3. John Deere's See & Spray™ cuts herbicide use by up to 90% with AI precision.

Case Study 2: Fasal – Smart Farming for Smallholder Farmers in India

Location: India

Technology Used: IoT sensors, Predictive AI models, Mobile-based dashboards

Fasal, an Indian agritech startup, provides affordable IoT and AI solutions tailored for small and medium-sized farms. It deploys climate-resilient sensors that monitor microclimatic conditions like humidity, temperature, rainfall, and soil moisture. This data is fed into AI algorithms that generate actionable insights including personalized irrigation schedules, disease alerts, and harvest time predictions.

Farmers receive real-time alerts and recommendations through a mobile app in vernacular languages, increasing adoption and usability in rural areas. Fasal's AI models are trained on region-specific data, improving their relevance to local farming conditions. Figure 4 represents the Fasal boosts yield and saves water with AI-powered farming.

Outcomes:

- Water usage reduced by up to 40% through smart irrigation
- Yield increase of 15-20% observed in crops like grapes and chillies
- Early disease warnings reduced crop losses by over 20%

Lessons Learned:

- Localized, language-friendly AI interfaces enhance farmer trust and engagement
- Affordable IoT deployment can scale smart farming to low-income regions

Figure 4. Fasal boosts yield and saves water with AI-powered farming.

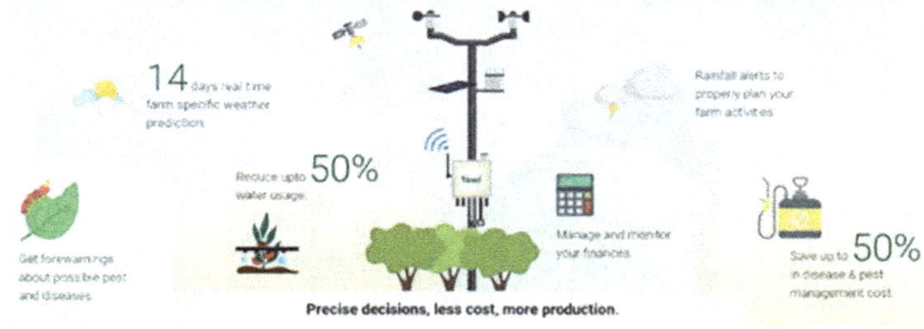

Case Study 3: SmartAgriHubs – EU-Wide Collaboration on Digital Agriculture

Location: European Union

Technology Used: Distributed IoT networks, AI platforms, Digital Innovation Hubs (DIHs)

Smart Agri Hubs is an ambitious European Commission-funded initiative connecting over 2000 agricultural stakeholders including start-ups, research institutions, and farms to accelerate digital innovation in agriculture. Through its network of Digital Innovation Hubs, the project facilitates the deployment of AI and IoT technologies across multiple farming use cases like precision livestock monitoring, soil nutrient analysis, and smart greenhouse control.

One sub-project in the Netherlands deployed IoT sensors in dairy farms to monitor cow movement, temperature, and feed intake. AI models then analyzed this data to detect early signs of lameness or health issues, allowing for timely intervention and reducing vet costs. Figure 5 represents the Livestock management: traditional methods vs. AI and sensor-powered monitoring.

Outcomes:

- Improved animal welfare and milk quality
- Reduced greenhouse gas emissions through optimized feeding
- Empowered SMEs to test and scale AI innovations

Lessons Learned:

- Public-private collaborations accelerate the adoption of AI/IoT
- Shared infrastructure (like DIHs) reduces barriers for small tech providers

Figure 5. Livestock management: Traditional methods vs. AI and sensor-powered monitoring.

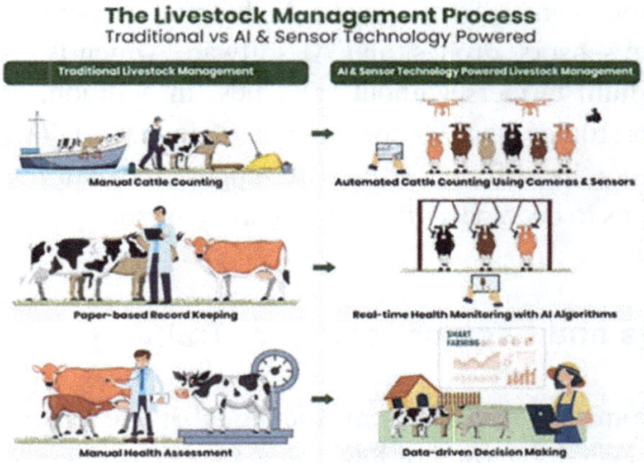

8. CURRENT BARRIERS AND LIMITATIONS

Despite the transformative potential of AI and the IoT in agriculture, several key challenges and limitations continue to hinder widespread adoption. While smart farming technologies promise higher efficiency, greater yields, and environmental sustainability, their real-world implementation is far from seamless. The barriers span across infrastructure, economics, digital literacy, data handling, and social acceptance.

8.1. Technological Challenges

Smart farming faces key challenges, primarily due to poor rural connectivity, which hampers real-time data transfer and cloud-based AI processing. Without stable internet or cellular networks, technologies like automated irrigation and predictive analytics lose effectiveness. Additionally, interoperability issues arise when devices from different manufacturers use incompatible platforms and data formats, limiting seamless integration. Unreliable electricity further complicates operations, as IoT devices need consistent power, and while solar options help, dependence on stable energy remains a barrier.

8.2. Economic Barriers

A key limitation of smart farming is the high initial investment required for IoT infrastructure like sensors, drones, and AI software, which is often unaffordable for small or medium farmers without subsidies. In addition, ongoing costs for maintenance, data storage, and AI model updates, along with the need for technical expertise, make adoption difficult. Even with support, the return on investment may take several seasons to become evident, depending on crop type and environmental factors.

8.3. Skill Gaps and Lack of Technical Training

Despite the promise of smart farming, the digital divide remains a major barrier. Many farmers, especially in rural or low-literacy areas, struggle to use complex AI and IoT tools due to a lack of training and user-friendly interfaces. The shortage of Agri tech experts and support staff further limits the adoption of these technologies. Language barriers, non-intuitive dashboards, and inaccessible designs make it harder for farmers to benefit from data-driven insights.

8.4. Data Privacy and Ownership Concerns

The rise of IoT and cloud platforms in agriculture has made data privacy and ownership a key concern. While devices gather detailed farm data, farmers often lack clarity on who owns this data and how it's being used or shared. Without transparent policies, vendors may claim data rights or use it commercially, leading to mistrust. Data leaks could expose sensitive information, impacting market access or pricing. Strong data governance and ethical practices are essential to safeguard farmer interests.

8.5. Social and Cultural Resistance

Traditional mindsets and resistance to change remain key barriers in smart farming adoption. Many older farmers are sceptical of digital tools, fearing job loss and distrusting new methods over traditional practices. Wider adoption is also hindered by social inequality and unclear government policies. Without collective participation, robust regulations, and incentives, the full potential of AI and IoT in agriculture remains difficult to achieve. Table 3 shows the Challenges, causes, and solutions for overcoming barriers in smart farming.

Table 3. Challenges, causes, and solutions for overcoming barriers in smart farming.

Challenge	Cause	Suggested Mitigation
Poor Connectivity	Lack of high-speed internet in rural areas	Implement satellite-based internet, improve rural broadband infrastructure
High Initial Cost	High costs of smart farming devices and infrastructure	Subsidies, government grants, or low-interest loans for farmers
Skill Gaps and Lack of Training	Lack of digital literacy and technical training in rural areas	Develop localized training programs, mobile apps with simple interfaces
Data Privacy Concerns	Unclear ownership and usage policies for farm data	Establish transparent data ownership policies, secure encryption for data storage
Cultural Resistance to Change	Deep-rooted traditional farming practices	Promote success stories, organize community-based workshops and demonstrations
Fragmented Technologies	Incompatibility between different IoT devices and platforms	Standardize IoT devices, encourage open-source solutions, and industry collaboration
High Maintenance and Operational Costs	Regular maintenance of IoT devices, data storage costs	Develop low-cost or modular IoT solutions with user-friendly interfaces
Lack of Skilled Workforce	Shortage of Agritech experts and advisors	Increase investment in agritech education, university programs, and internships

9. WHAT'S NEXT: EMERGING TRENDS AND RESEARCH AVENUES?

The future of agriculture is being shaped by a combination of emerging technologies that promise to revolutionize farming practices. Key trends like Generative AI, collaborative robots (cobots), blockchain with AI/IoT, and climate-smart agriculture are paving the way for a more efficient, sustainable, and productive agricultural sector. These technologies are set to address the challenges of rising food demands, climate change, and labour shortages, providing smarter solutions for farmers worldwide.

9.1. Generative AI in Simulating Agricultural Scenarios

Generative AI is a powerful tool that can simulate agricultural environments and predict outcomes of various farming practices. These simulations allow farmers to test different strategies without the risk of trial and error in real-world conditions. For instance, AI can generate models of crop growth based on soil quality, weather patterns, and irrigation methods, helping farmers optimize their practices and improve crop yields. This ability to predict outcomes under varying conditions can assist in making informed decisions for resource management, pest control, and water conservation.

9.2. Collaborative Robots (Cobots) in Field Tasks

Collaborative robots, or cobots, are designed to work alongside human workers, enhancing productivity while reducing the physical strain of labour-intensive agricultural tasks. Cobots can be used for planting, harvesting, sorting, and packaging, and they work autonomously or in tandem with farmers. These robots are equipped with advanced sensors and AI, allowing them to perform tasks like identifying ripe crops or monitoring livestock health. As cobots become more autonomous and efficient, they will help address labor shortages and improve the precision of agricultural practices.

9.3. Blockchain + AI/IoT for Secure Food Traceability

Blockchain combined with AI and IoT is emerging as a game-changer in food supply chain management. Blockchain provides a transparent, secure, and tamper-proof record of each step in the food production process, from farm to table. IoT devices, such as temperature and humidity sensors, collect real-time data, while AI analyzes this information to ensure that food remains safe and fresh throughout the supply chain. This integration will enhance food safety, reduce waste, and increase consumer confidence by offering verifiable proof of the origins and quality of food products.

9.4. Climate-Smart Agriculture and Autonomous Systems

As the impacts of climate change continue to affect agriculture, climate-smart agriculture is becoming increasingly important. This approach involves using technology to adapt to changing weather patterns, optimize resource use, and reduce environmental impact. Autonomous systems such as drones, robots, and AI-powered sensors are already being deployed to monitor soil conditions, weather, and crop health in real-time. These systems can make adjustments automatically, such as optimizing irrigation or applying fertilizers, helping farmers reduce their environmental footprint and cope with extreme weather events like droughts or floods.

10. CONCLUSIONS

The convergence of AI and the IoT has emerged as a powerful force transforming the landscape of modern agriculture. By enabling real-time data collection, precise decision-making, and autonomous systems, these technologies are not only optimizing farm operations but also laying the foundation for Smart Farming. As global food demand continues to rise and environmental challenges like climate change intensify, the role of AI and IoT in shaping the future of farming becomes increasingly critical. AI's ability to process and analyze vast amounts of agricultural data is enhancing decision-making in areas such as crop management, livestock monitoring, pest control, and yield prediction. IoT devices, on the other hand, provide the connectivity and data collection necessary for continuous monitoring and real-time feedback, driving more efficient and sustainable practices. Together, AI and IoT form an integrated ecosystem that can adapt to evolving challenges, optimize resource usage, and improve overall productivity.

The synergy between AI and IoT holds transformative potential not only for increasing agricultural efficiency but also for addressing critical global issues such as food security, resource conservation, and climate resilience. Through precision agriculture, smart irrigation systems, and supply chain optimization, these technologies enable farmers to achieve higher yields with fewer resources, making farming more sustainable and environmentally friendly. The journey towards widespread adoption of AI and IoT in agriculture is not without its challenges. Connectivity issues in rural areas, high initial costs, and the need for skilled labour are some of the barriers that need to be addressed for successful implementation. Yet, as technological advancements continue and the agricultural industry adapts, these challenges can be mitigated, paving the way for more inclusive and accessible smart farming solutions.

In conclusion, the integration of AI and IoT offers unparalleled opportunities for revolutionizing agriculture, enhancing food security, and fostering sustainability. By embracing these innovations, the agricultural sector can meet the growing demands of a global population while ensuring that future generations have access to nutritious food produced in an environmentally responsible manner. The future of farming is undoubtedly smart, and its transformative potential will continue to unfold as AI and IoT continue to evolve.

REFERENCES

Aijaz, N., Lan, H., Raza, T., Yaqub, M., Iqbal, R., & Pathan, M. S. (2025). Artificial intelligence in agriculture: Advancing crop productivity and sustainability. *Journal of Agriculture and Food Research, 10*, 101762. DOI: 10.1016/j.jafr.2025.101762

Bahari, M., Arpaci, I., Der, O., Akkoyun, F., & Ercetin, A. (2024). Driving Agricultural Transformation: Unraveling Key Factors Shaping IoT Adoption in Smart Farming with Empirical Insights. *Sustainability (Basel), 16*(5), 2129. DOI: 10.3390/su16052129

Balyan, S., Jangir, H., Tripathi, S. N., Tripathi, A., Jhang, T., & Pandey, P. (2024). Seeding a Sustainable Future: Navigating the Digital Horizon of Smart Agriculture. *Sustainability (Basel), 16*(2), 475. DOI: 10.3390/su16020475

Bhangar, N. A., & Shahriyar, A. K. (2024). *IoT and AI for Next-Generation Farming: Opportunities, Challenges, and Outlook*. Creative Commons International License.

Chataut, R., Phoummalayvane, A., & Akl, R. (2023). Unleashing the Power of IoT: A Comprehensive Review of IoT Applications and Future Prospects in Healthcare, Agriculture, Smart Homes, Smart Cities, and Industry 4.0. *Sensors (Basel), 23*(16), 7194. DOI: 10.3390/s23167194 PMID: 37631731

Fuentes, F., Gutter, K., Vega, R., & Carrasco Silva, G. (2024). Transformative Technologies in Digital Agriculture: Leveraging Internet of Things, Remote Sensing, and Artificial Intelligence for Smart Crop Management. *J. Sens. Actuator Netw., 13*(4), 39. DOI: 10.3390/jsan13040039

Holzinger, A., Saranti, A., Angerschmid, A., Retzlaff, C. O., Gronauer, A., Pejakovic, V., Medel-Jimenez, F., Krexner, T., Gollob, C., & Stampfer, K. (2022). Digital transformation in smart farm and forest operations needs human-centered AI: Challenges and future directions. *Sensors (Basel), 22*(8), 3043. DOI: 10.3390/s22083043 PMID: 35459028

Hussein, A., Jabbar, K., Mohammed, A., & Alzubaidi, L. (2024). Harvesting the Future: AI and IoT in Agriculture. E3S Web of Conferences, 477. DOI: 10.1051/e3sconf/202447700090

Hussein, A. H. A., Jabbar, K. A., Mohammed, A., & Al-Jawahry, H. M. (2024). AI and IoT in Farming: A Sustainable Approach. E3S Web of Conferences, 491, 01020. DOI: 10.1051/e3sconf/202449101020

Issa, A., Majed, S., Ameer, S., & Al-Jawahry, H. (2024). Farming in the Digital Age: Smart Agriculture with AI and IoT. E3S Web of Conferences, 477, 00081. DOI: 10.1051/e3sconf/202447700081

Karunathilake, E. M. B. M., Le, A. T., Heo, S., Chung, Y. S., & Mansoor, S. (2023). The Path to Smart Farming: Innovations and Opportunities in Precision Agriculture. *Agriculture, 13*(8), 1593. DOI: 10.3390/agriculture13081593

Kumar, M., Kumar, A., Sajjan Kumar, S., Patle, G. T., Kumar, S., & Tanwar, A. (2024). Role of cutting-edge technology in horticulture crops for shaping the future smart farming ecosystem: A comprehensive review. *Environment and Ecology, 42*(4B), 1933–1942. DOI: 10.60151/envec/KSOK1056

Kumari, K., & Nafchi, A. M. (2024). Sustainable agriculture with AI, machine learning, deep learning, and IoT for future farming. 2024 ASABE Annual International Meeting, 2400906. DOI: 10.13031/aim.202400906

Mishra, H., & Mishra, D. (2024). *Artificial Intelligence and Machine Learning in Agriculture: Transforming Farming Systems* (Vol. I). Research Trends in Agriculture Science.

Nautiyal, C., Nautiyal, P., Papnai, G., Mittal, H., Agrawal, K., Shivani, , Vishesh, , & Nandini, R. (2024). Importance of Smart Agriculture and Use of Artificial Intelligence in Shaping the Future of Agriculture. *Journal of Scientific Research and Reports, 30*(3), 129–138. DOI: 10.9734/jsrr/2024/v30i31864

Obaid, M. K., Abood, B. S. Z., Alazzai, W. K., & Jasim, L. (2024). From Field to Fork: The Role of AI and IoT in Agriculture. E3S Web of Conferences, 491, 02006. DOI: 10.1051/e3sconf/202449102006

Padmaja, P., Sophia, I. J., Hari, H. S., Kumar, S. S., & Somu, K.. (2021). Distribute the Message over the Network Using another Frequency and Timing Technique to Circumvent the Jammers. *J Nucl Ene Sci Power Generat Techno, 10*, 9.

Padmaja, P., Sophia, I. J., Hari, H. S., Kumar, S. S., & Somu, K.. (2021). Distribute the Message over the Network Using another Frequency and Timing Technique to Circumvent the Jammers. *J Nucl Ene Sci Power Generat Techno, 10*, 9.

Pratapagiri, S., & Gangula, R. R. G., Srinivasulu, B., Sowjanya, B., & Thirupathi, L. (2021). “Early Detection of Plant Leaf Disease Using Convolutional Neural Networks,” 2021 3rd Interna-tional Conference on Electronics Representation and Algorithm (ICERA), , pp. 77-82, doi: DOI: 9.1109/ICERA53111.2021.9538659

Qazi, S., Khawaja, B. A., & Farooq, Q. U. (2022). IoT-Equipped and AI-Enabled Next Generation Smart Agriculture: A Critical Review, Current Challenges and Future Trends. *IEEE Access : Practical Innovations, Open Solutions, 10*, 21219–21235. DOI: 10.1109/ACCESS.2022.3152544

Gangula, R., Thirupathi, L., Parupati, R., Sreeveda, K., & Gattoju, S. (2021). Ensemble machine learning based prediction of dengue disease with performance and accuracy elevation patterns, Materials Today: Proceedings, , ISSN 2214-7853, DOI: 10.1016/j.matpr.2021.07.270

Shaikh, T. A., Rasool, T., & Lone, F. R. (2022). Towards leveraging the role of machine learning and artificial intelligence in precision agriculture and smart farming. *Computers and Electronics in Agriculture*, *199*, 107119. DOI: 10.1016/j.compag.2022.107119

Shailaja, K., Srinivasulu, B., Thirupathi, L., Gangula, R., Boya, T. R., & Polem, V. (2023). An Intelligent Deep Feature Based Intrusion Detection System for Network Applications. *Wireless Personal Communications*, *129*(1), 345–370. DOI: 10.1007/s11277-022-10100-w

Sharma, K., & Shivandu, S. (2024). Integrating artificial intelligence and Internet of Things (IoT) for enhanced crop monitoring and management in precision agriculture. Scientific Intelligence, Advance online publication. DOI: 10.1016/j.sintl.2024.100292

Thirupathi, L., Rajesh, A., & Sandeep, R. (2023). Internet of Lighting for Smart Cities. In Tiwari, M., Ismail, Y., Verma, K., & Garg, A. K. (Eds.), *Optical and Wireless Technologies. OWT 2021. Lecture Notes in Electrical Engineering* (Vol. 892). Springer., DOI: 10.1007/978-981-19-1645-8_41

Thirupathi, L., Srinivasulu, B., Khanapurkar, U., Rambabu, D., & Preeti, C. M. (2023). Heart Disease Prediction in Healthcare Communities by Machine Learning Over Big Data. In Bhateja, V., Sunitha, K. V. N., Chen, Y. W., & Zhang, Y. D. (Eds.), *Intelligent System Design. Lecture Notes in Networks and Systems* (Vol. 494). Springer., DOI: 10.1007/978-981-19-4863-3_38

Wang, T., Xu, X., Wang, C., Li, Z., & Li, D. (2021). From Smart Farming towards Unmanned Farms: A New Mode of Agricultural Production. *Agriculture*, *11*(2), 145. DOI: 10.3390/agriculture11020145

Chapter 7
Role of AI and IoT in Smart Farming

G. Muthupandi
https://orcid.org/0009-0002-8822-0157
Chennai Institute of Technology, India

N. Rathinavel Subramanian
https://orcid.org/0009-0006-2064-8957
Chennai Institute of Technology, India

S. Jawagar
https://orcid.org/0009-0006-9899-8658
Chennai Institute of Technology, India

Angamuthu A.
http://orcid.org/0009-0001-9877-1545
Chennai Institute of Technology, India

S. Pavithra
Chennai Institute of Technology, India

ABSTRACT

This chapter explains how Artificial Intelligence (AI) and the Internet of Things (IoT) can change present-day farming practices through smart farming. Dealing with problems like climate change, shortage of resources and labor issues, using AI and IoT could help the agriculture sector provide secure, sustainable and efficient food supplies. This chapter describes how smart farming has advanced, focusing on how IoT devices provide real-time data and AI helps to make future decisions easier. In this chapter, examples and case studies are used to explain how artificial intelligence is changing the methods of farming, covering irrigation, monitoring

DOI: 10.4018/979-8-3373-4207-8.ch007

crops, managing pests and boosting the supply chain. It also covers the difficulties preventing many from using it such as infrastructure, budget and data protection and suggests ways to make it available to more people. In essence, the chapter describes how the future of farming is being affected by the combination of AI and IoT.

1. INTRODUCTION

Humans have depended on agriculture to support their survival from the beginning of civilization. All the way from ancient settlements to today's industrial world, the role of farming in economic, cultural and everyday growth has been very important. Despite being an ancient practice, agriculture today is dealing with problems so big that it needs to undergo a fundamental change (Thilakarathne, et,al (2025)). Because of rapid population change, climate shifts, soil damage, lack of water, dropping biodiversity, outbreaks of pests and not enough workers, agriculture worldwide is under a lot of stress. In order to feed over 10 billion people in 2050, we will need both more food and environmentally friendly ways of farming.

Traditional farming which depends on human effort, experience and timing linked to seasons, is having a hard time responding to current food system needs. Traditional farming may have strong roots in older knowledge and tested skills, but it usually is not very accurate, efficient or large-scale. For this reason, new ideas are required that make farming more effective, reduce harm to nature and help farms stay strong during changing climates and economies.

Using AI and IoT, automation, intelligence and connectivity are being brought to every step of farming. Soil sensors, weather stations, drones and automated machinery in the Internet of Things collect information about temperature, humidity, soil moisture, pH and whether crops are healthy. They let you see the details of soil conditions that were not continuously possible to check earlier. At the same time, machine learning, deep learning and data analytics are used by Artificial Intelligence to analyze the data and make useful recommendations (Shen, Z. (2025)). AI tools are able to look for patterns, make predictions and suggest the best actions for planting, watering, fertilizing, dealing with pests and harvesting crops.

AI and IoT work together for more than merely tracking and handling tasks automatically. It causes farming to be managed in a different way. For example, using AI, image recognition systems can detect diseases in plants using images captured by drones or phones, so the problem can be solved fast and crop losses minimize. Mirikon's smart irrigation system only supplies water when needed by checking the soil's condition which reduces water and energy consumption (Devarajan, Y. 2025). Using GPS and AI algorithms, autonomous farming machines guide themselves through fields with high accuracy, making it easier to reduce how much manual work is involved.

Smart Farming is responsible for strengthening sustainability, increasing productivity, sharing data and making better decisions. Cloud-based platforms and mobile applications now allow farmers and particularly smallholders in developing countries, to get instant information and notices (Sagar, S (2025)). Now, farmers have access to information and resources that used to be owned by the biggest farming businesses. Table 1 illustrates the stark contrasts between traditional and smart farming methods in terms of decision-making, input usage, and sustainability.

Table 1. Comparative overview of traditional vs. smart farming

Criteria	Traditional Farming	Smart Farming
Decision Basis	Experience-based	Data-driven and automated
Input Usage	Uniform across fields	Precise and site-specific
Technology Dependency	Minimal or none	High – sensors, AI, cloud, automation
Environmental Impact	Often high due to overuse	Reduced due to optimized input application
Scalability	Limited	Highly scalable with remote monitoring

Using AI and IoT in farming does come with its own set of issues. Limits to infrastructure, high initial expenses, not having enough specialists, data security threats and laws are the biggest issues that must be overcome. At the same time, the positives are much greater than the negatives. With proper backing, teamwork between public and private groups and upgrading skills, these challenges can be met and farmers supported for better results.

It's seeks to describe how agricultural AI and IoT play an important role in changing the process from traditional farming to Smart Farming. First, it looks at how agriculture has developed and how digital tools have appeared. Later sections cover IoT and AI use cases in agriculture, the results from using both technologies, case studies from real life, the advantages and potential problems (Assimakopoulos, F. (2025)). This chapter also points out upcoming changes, for example, using robotics, blockchain and climate-friendly methods, giving a clear picture of how farming will change as the world goes digital.

2. SMART FARMING: CONCEPT AND EVOLUTION

With the development of people's economy, agriculture was changing to match it, with such activities as simple farming and mechanized methods, agriculture was developing. The modes of working the farm that the indigenous people passed down to each new generation were the most important element when it comes to providing enough food. Urban growth and increase in the number of people, as well as climate change, soil depletion, and depletion of water, are the problems that farming now has to address if it is to remain efficient and productive (Long, S. P. (2025)). Still, the methods used from by-gone generations, while rich in environmental awareness and protection, are inadequate for the present-day challenges. They will have to undergo a massive transformation focusing on innovation and productivity to still continue to be the engine that drives the balance between profitability and sustainability of the future society.

Agriculture that requires much hand labor, experience from older generations and simple tools, has long been the main support of food supplies for the world. Nevertheless, these methods, though belonging to a historic tradition, are not keeping up with the requirements of today's farming (Dinis, I (2025). During this century, the world is dealing with climate change and not enough resources, while more and more people are settling in cities and the demand for food rises. These factors require us to change our approach to farming which is motivating a focus on innovative practices.

The name Precision Agriculture is another way to express the idea of Smart Farming which represents a major change needed for modern farming. It consists of using innovative technologies—for example, the IoT, AI, robotics, drones, remote sensors and cloud services—in farming to improve how decisions are made and how work is carried out. The main purpose is to collect data, examine it and use it to take smart actions about crop management, irrigation, fertilization and pest control (Ajith, S (2025)). Using data to guide farming maximizes the use of water, fertilizers and energy which helps save money, lessens the harm to nature and improves both the quantity and quality of crops. In place of using the same blanket strategy, Smart Farming responds to the different conditions in each part of a farm.

It's use of sensors, automation, data analytics, robotics and cloud computing helps manage and optimize both what goes into and what comes out of agricultural systems. The main goal of Smart Farming is to provide farmers with real-time and past data for decision-making so they use resources better, reduce losses and improve how much they produce (Awais, M (2025)). Instead of the same one-size-fits-all treatment for the entire field, Smart Farming helps with targeted measures based on requirement. It provides a bigger harvest and also decreases the negative effects of agriculture on the environment.

Besides being technological, Smart Farming also involves a change in how farming is viewed and done. Today's Digital Agriculture was made possible thanks to changes like the mechanization era which introduced machines and the Green Revolution which focused on chemical fertilizers and high-producing plant varieties.

Data is now seen as the essential information needed for precise actions in agricultural fields (Gracia-Rodriguez, C (2025)). Machines, maps and monitoring tools that use GPS, GIS, satellite data and AI have given farmers the ability to measure crop health, see weather predictions and decide what to do in advance. Keeping an eye on entire fields is possible for drones armed with multispectral imaging and the irrigation systems themselves get information from sensors to better save water. At first, farming was mechanized with tractors and combines and the following Green Revolution introduced new seeds that generated high yields and chemicals. Today, data is the main element that drives farming more than anything else. Using Global Positioning Systems (GPS), Geographic Information Systems (GIS), remote sensing and now advanced analytics from AI and IoT devices, Precision Agriculture has grown (Rana, R (2025).). Farming now includes using drones for pictures from the sky, soil sensors to track nutrients and satellite data to predict the weather.

Smart Farming can connect tech-savvy farms with resource-poor smallholder farmers. With the help of modern mobile apps, cloud services and easy-to-use sensor kits, farmers of all social backgrounds can obtain useful information about farming. For farmers, mobile notifications can give advice on the right time to plant, any pest-related problems or weather that might be important. There are efforts by governments, research centers and agri-tech companies to help these technologies be more usable, easy to scale and open to all (Batan, A. (2025)). Slowly, building the skills of farmers, teaching them new ideas and grouping them into cooperatives through technology sharing are reaching rural areas, particularly in developing countries. Even those farmers with little money in developing countries can take advantage of real-time information using mobile apps and new sensors. Different groups are launching platforms to help with this, so that technology is open to all and no one is left out of digital agriculture.

Smart Farming is not just about enhancing crop yields but it is rather the concept of trying to develop food production that is resilient, adaptive, and can be ecological stabilized. Real-time monitoring and predictive analytics make it easier to not only determine but also to prevent the impacts of weather changes like global warming that lead to disasters such as droughts or floods. Environment-friendly methods such as precision irrigation, integrated pest management, and targeted nutrient application are the sources of health in the ecosystem.

3. INTERNET OF THINGS (IOT) IN AGRICULTURE

Smart Technologies or IoT (Internet of Things) has been the reason of the great shift in the industry different from its usage in agriculture. Described as a "Connected Agriculture" IoT in farming is the practice of using a network of sensors, and systems that can transfer, process, and analyze data instantaneously. These technologies give farmers the power to do things, like watch the environmental conditions, automate several processes, and arrange farming operations with unmatched accuracy, through deep analysis and insightful data. In its essence, IoT in agriculture is a system that consists of interconnected physical objects that are embedded with units of electronics, software, and connectivity that operate in a controlled network. These things help in the capture of data for further transformation through the internet or a local network so they are processed in a computer or other machines, making agriculture decisions more accurate through data analytics or pre-defined rules.

IoT allows farmers to replace decisions based on instinct with those based on collected data. In traditional farming, knowledge and what farmers could see guided the choices about watering, fertilizing, pest control and when to harvest. But thanks to IoT, real-time information on soil, crops, weather and equipment can make farmers' decisions more accurate. With sensors and automated tools, farmers now can precisely time when to use water and fertilizers which helps them save water and increase what they produce from the same amount of land.

Sensor devices, connectivity to the Internet, data gathering platforms and user interfaces are the main elements of an agricultural IoT system. IoT in agriculture depends heavily on sensors. Such systems consist of soil moisture and thermometers, sensors for weather monitoring (humidity, rainfall, wind speed), devices for analyzing soil nutrients (pH and minerals), sensors mounted on animals (livestock GPS collars and RFID tags) for observing their health and cameras (optical, multispectral) from drones used to watch over crops. Smart Farming function As shown in Table 2, smart farming technologies serve distinct roles from real-time sensing to predictive analytics.

Table 2. Key technologies used in smart farming and their functions

Technology	Function in Agriculture	Example Use Case
IoT Sensors	Real-time monitoring of soil, moisture, temperature	Soil moisture sensors in drip irrigation
Drones	Aerial imaging, crop health diagnostics	Pest infestation detection from above
GPS and GIS	Location mapping and spatial analysis	Precision planting and spraying
AI-Based Analytics	Predictive modeling and decision-making	Yield prediction and disease forecasting
Cloud Platforms	Data storage and remote access	Centralized farm management systems

For data to transfer smoothly from devices to central systems, connectivity and communication protocols are very important. Depending on where they are and the technologies they can access, farmers may link their devices using LoRaWAN, Zigbee, NB-IoT, cellular networks (3G/4G/5G) and Wi-Fi. When a rural area lacks strong internet connections, low-power wide-area networks (LPWAN) are often used for communication over big distances with low energy use (Ganaou, N. I (2025)). For agricultural devices that rely on batteries, remote access is very important. Once sensors gather the data and send it over networks, the information is collected, managed and processed on the cloud or edges of the network. Storing, analyzing and presenting the data collected from sensors is made possible using IBM Watson, Microsoft Azure FarmBeats and AWS IoT Core which are all cloud services. They use machine learning and artificial intelligence to identify useful insights, show where unusual activity happens and suggest actions. As an example, a system may see that one part of the field is not getting enough water and will turn on an irrigation valve to correct it automatically.

Mobile apps, web dashboards or software systems used for farm management make up the user interface of the IoT system. The methods show data in an easy-to-understand way, providing data on soil moisture, the weather, possible pests and details on livestock. In addition, platforms often provide automated updates and advice, so farmers can respond quickly to the data provided.

A typical IoT system for agriculture consists of the four layers - the perception layer, the network layer, the processing layer and the application layer. In the perception level, the sensor and the actuator will be responsible for detecting the emergencies and gathering information and there are also other field hardware such as drones that play a great role in the collect of information process. The next piece is the network layer which connects the devices and the processing units through wired and wireless links. The processing layer, on the other hand, is generally equipped with a server, database and the necessary analytics tools to turn the raw data into meaningful insights. The highest layer, the application layer, offers the possibility of human-system interaction through various access points, like software applications, dashboards and application programming interfaces. IoT technology is widely employed in agriculture. Precision irrigation is one of the most striking examples. When farmers use soil moisture sensors, weather forecasts and crop evapotranspiration models, they decide the correct time for the watering (Ali, A., Tajamul, H., & Azlan, Z. (2025)). In this way, they obtain the maximum use of water while minimizing the water wastage and at the same time, guaranteeing high-quality crop growth and abundant harvest. Companies like Netafim and CropX have each come up with different crop/crop types and regions of applications where they increased yield by better water use.

Collecting localized information at field weather stations makes it much easier to monitor current climate and make future predictions. While other climate services give regional averages, IoT sensors give specific, up-to-date data on the farm's own particular weather conditions. Thus, farmers have better ideas about when to sow, apply chemicals and harvest their crops, so they are less likely to lose crops because of sudden weather changes (Arularasan, A. N.(2025)). Livestock are also cared for and watched using IoT technology. Smart collars, sensors eaten by animals and RFID tags check on pets' movements, what they eat, their body heat and a range of other physical signs.

Figure 1. Overview of IoT devices used in smart agriculture

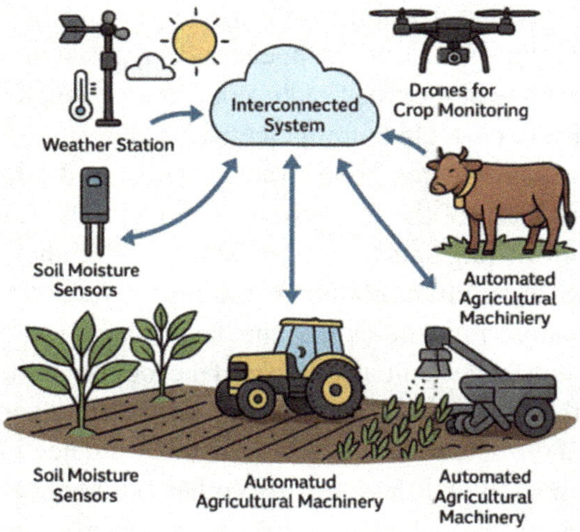

This figure illustrates various IoT devices such as soil moisture sensors, weather stations, drones, and livestock trackers, forming the interconnected system in modern agriculture.

Greenhouse automation has been greatly changed by the introduction of IoT. Using sensors for CO_2, humidity, light and temperature, greenhouses have the ability to function on their own with current data. Optimal growing conditions are achieved by actuators that handle fans, windows, irrigation pumps and the shade for the crops. Because of this, workers are needed less and crops will still be of good quality regardless of changing weather conditions.

Another very critical need is to monitor pests and diseases. etc Traps integrated with sensors and cameras are placed in the fields to watch insects and quickly diagnose the plants. The drones equipped with imaging sensors have the capacity to cover the huge area of fields with high resolution in order to detect the crops under stress, the occurrence of pests or the spots where nutrients are less. (Agnor et al. (2019)) showed that the deployment of technology for pest surveillance benefits the environment by lowering the cost of pesticides and reducing losses from pests. It is essential to conduct regular soil and crop observation to monitor the changes in nutrient status, the amount of water that the crop can hold and the health of the plants. The soil moisture can be measured with the help of IoT sensors which also enables us to quantify the degree of soil compaction and the volume of soil organic matter. Through these data, farmers can decide on the appropriate fertilizer, the proper crop plants, and the right time to carry out the fieldwork. Consequently, soil productivity will be enhanced, less input is needed, and the possibility of achieving sustainable yields will be higher.

4. ARTIFICIAL INTELLIGENCE IN AGRICULTURE

Artificial intelligence (AI) has been rapidly changing the agricultural landscape by converting older methods into modern, efficient, and environmentally friendly methodologies. The AI solutions that are in use now mostly depend on machine learning, deep learning, computer vision, and predictive analytics to facilitate better decision making and to replace the different steps of the process with automated operations. The utilization of devices by farmers helps them optimize the various processes of breeding, growing crops, and the animals and get more as well as good quality harvest, while at the same time minimizing the use of resources and causing the least possible environmental impact.

AI in agriculture is the use of machine learning and other technologies to gather, organize, and interpret data from the past that can be very helpful in automating processes of different kinds and the prediction of yield and it is the most important work of AI in agriculture. Machine learning does this job quite well by taking information from the past, present and assists the algorithms in recognizing patterns and predicting precise outcomes. The disease management systems in the cases of plant diseases and pests are definitely dependent on the skills and expertise of supervisors and a good example that comes to mind is the one where the model is trained with many pictures of different plants (labeled set) where there are healthy plants and the plants that have diseases such as powdery mildew, blight or leaf rust so that the disease symptoms can be identified. The observance of small changes in the leaf structure is a major improvement that will allow for an early diagnosis, which can start the treatment of affected plants before they grow. With the help of AI, smart solutions are helping agriculture in the early prevention of the catastrophe and the use of less pesticide, which leads to sustainable farming practices.

Advancements in agriculture oplanions oplanments using AI have come from deep learning and neural networks. Drone or satellite images taken in aerial surveys can be investigated by Convolutional Neural Networks (CNNs), helping to assess the condition of crops on a large scale. The software can tell the difference between crops, look at their growth stages, find nutrient shortages and notice pests (Kamilaris., A 2018). Unlike going manually to observe crops, using AI for aerial imaging quickly and accurately replicates farm monitoring, so farmers can respond quickly. When used together, multispectral imagery and CNNs can find chlorophyll variation and water stress that humans cannot see directly.

AI allows farmers to use predictions to manage choices for their crops, dealing with pests and managing when to harvest. Leveraging a wide spectrum of data — including previous crop yields, nature of the soil, weather history and occurrence of pests — allows machine learning models to predict things like the crop's expected yield which pests may come and the optimal window to harvest. Such predictions help protect against misuse of resources and allow better use of existing resources. Should the model project frequent pest infestation along with current environmental data, farmers would be able to take preventive actions for pests instead of only using a lot of pesticides after an infestation.

AI has introduced robots and self-governing systems into farming today. More robotic solutions such as drones, harvesters and self-driven tractors are adopted for sowing, spraying, weeding and harvesting. They check the land by polling regular data from the sensors and GPS and later, they handle their work according to the farmer's directions regarding crops and soil. Agricultural drones fitted with AI perform precise spraying only where pests or weeds are found, so less chemical is applied and it's safer for the environment. Using precision agriculture, farmers can be more productive, work with less effort and protect the environment.

AI shows in smart irrigation systems how it can save resources. They regularly check soil moisture with sensors, check weather forecasts and calculate how much water the crops should get using AI methods. Depending on what's happening in the field, sensors guide the system to give just the right amount of water to plants. In many water-short places, smart irrigation can help save water and still provide good crop yields. Just as AI is used in food processing such technology is also used to examine soil for nutrients and advise on the right amounts of fertilizer to apply. More precise fertilization prevents fields from receiving too much fertilizer, helps stop nutrients from going into rivers and reduces expenses.

AI has an influence on livestock farming by keeping animals healthier and more productive. Data on animals' motion, habit of eating, temperature and squeaking is collected throughout the day with wearable sensors and cameras. These inputs are examined by the models which use them to spot the first signs of health problems, stress or reproduction. Should computer vision display unusual gait in a pet, this might show signs of lameness, encouraging owners to see their vet quickly. AI can predict which animals will be healthy or productive which allows scientists to choose better breeds and quicken breeding. Because of these advances, farmers can handle their livestock better and in a more ethical way.

Figure 2. Communication architecture for IoT in agriculture

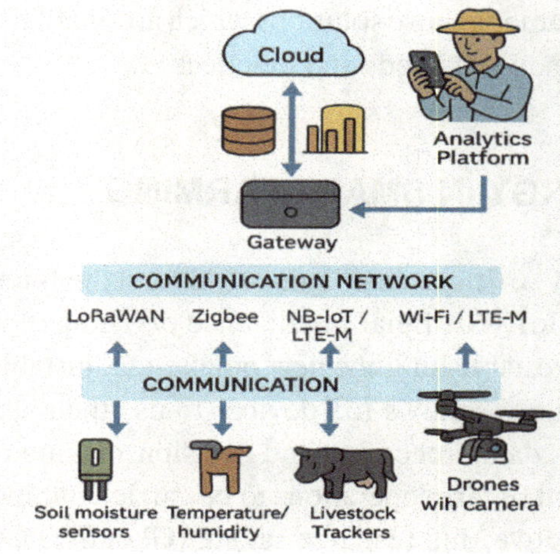

This figure shows how different IoT devices communicate through various network layers to cloud or edge platforms for real-time processing and analytics.

Advances in artificial intelligence could be very useful for agricultural purposes, but some problems still need to be managed. Because of a lack of technology and network, getting access to information is still problematic in poor regions. It is important to use data that is real, correct and from numerous sources to avoid problems, as this allows for good AI results; but if data is limited, unreal or flawed, results will not be as expected. Due to the many issues affecting agro-ecosystems, widely useful models are hard to put together. The AI system stays dependable in various situations by continuously updating the data used and regularly updating the model.

Ethical and socio-economic matters are also requiring deep involvement. The accessibility of AI technologies for many small farmers is limited by such problems as high expenses, lack of digital skills, or fear of losing work. There is a fear that data privacy and ownership will not be well taken care of, since the information is valuable to agriculture and at the same time sensitive. A complete and equitable embracing of citizens is only possible when the AI solutions are also designed in a way that they are cheap, user-friendly, and in accordance with the different cultures (Nkwo, M (2025)). It is essential besides to provide appropriate training to farmers and support services in the process of adopting AI tools to effectively utilize them in their operations. Governments, research institutions, and the private sector are cooperating to solve these issues. Open data initiatives as well as cloud computing platforms promote data sharing and reduce computation costs. Through the availability of affordable sensors and mobile connectivity, AI can be accessed by even the rural areas. The greater the transparency, reliability, and explainability AI has the more farmers will build confidence using it. Pilot projects and farmer-centric design approaches are among those solutions which are enabling the personalization of AI applications to the local needs and practices

5. AI + IOT SYNERGY IN SMART FARMING

The combination of Artificial Intelligence (AI) and the Internet of Things (IoT), better known as the AIoT (Artificial Intelligence of Things), is reshaping the face of modern agriculture establishing the new quickest, adjustable, and precise farming systems. This synthesis allows IoT devices to use their sensing capabilities in combination with AI's data processing and decision-making capabilities and thus getting the management of farm operations to be carried out in such a way that they are sustainable, productive, and resource saving (Gkoulis, D. (2025)). A structure provided by IoT devices, such as sensors, motors, and connected machinery, those which are processing the data through the clouds and those which are also able to take decisions on their own, should be the most appropriate for data collection from different farmlands on a continuous basis. These devices mainly utilize the concepts of IoT such as the collection of environmental data, processing of such data, and the establishment of general information surfaces. Specifically, these gadgets are used for measuring such things as soil moisture, temperature, humidity, pH levels, nutrient content, pest activity, and animal health. It is these devices that mostly own the power to ensure that every aspect of the farming that is at length, the control of the moisture, the protection of the products, and finally the monitoring of the animal kingdom. Through the use of a sensor network distributed in the close vicinity of the fields or in the same spot with greenhouses, farmers are able to get a 3D picture of the status of the crop and/or the animal at that particular time. This high-resolution data source provides decision-makers the opportunity to have a live and updated complete print of the crop field in all its physiological and visual aspects at the exact moment.

However, just the mere collecting of data is not the key to agricultural problems resolution. By utilizing powerful machine-learning and deep-learning approaches, AI handles the large amount of collected data, which makes it possible to find patterns, make forecasts as well as offer actionable advice (Fasal. 2023). For instance, AI can estimate the appearance of pests by assessing the climate data together with historical circumstances or it can figure out the perfect time for the irrigation by relating soil-water-content to weather forecasts and crop water needs. Such intelligent data interpretation changes the raw information from the sensors into the current insights which help implement automatic or semi-automatic farm management decisions.

One of the most far-reaching uses of AIoT that come to our minds is in precision irrigation. The conventional methods of irrigation usually result in over-watering or under-watering which in turn leads to the loss of water as well as the health of crops. With the use of Io T soil moisture sensors that are interconnected with a computer AI control system, the sensors are able to provide data to the AI that then determines the quantity and time of water application in the fields. This method is eco-friendly and costs only a fraction of the traditional method in the long run. The AI enabled system also updates a website where the information is easy to access. The system, also, sends status updates, email alerts, notices through the app, and on phones and computers. It uses less water and they are sustained and even improved.

Another field of precision agriculture that is changed by AIoT fusion is the control of pests and diseases. The sensors can detect the presence of certain environmental factors that make it suitable for pests to grow and for the disease to spread; in addition, the AI model can also be engaged in envisioning risks of an attack through the use of data that it has gathered so far and is being collected real-time. Furthermore, unmanned aerial vehicles and ground robots equipped with imaging sensors and treatment delivery devices are used to target the specific areas that are under pest attacks and their location as well as apply the correct pesticide or biocontrol agent. This strategy curtails the occurrence of a pest attack, decreases the cost of production, and preserves the current natural ecosystem.

Figure 3. AI+IoT synergy in smart farming

The above Figure is showing the concept of AIoT synergy on smart farming. AIoT is also greatly exploited in livestock-breeding. Through these multimedia IoT devices, worn on animals, among others, AI recognizes and records the life indicators, movement, food preferences, and position of the animals. Thus learned, the AI skill is health monitoring, timely diagnosis of fever or anxiety without waiting for obvious symptoms to develop first. Feeding and climate systems run by AI regulate themselves in real time, according to the environmental conditions detected by the sensors, thus switching to the best status immediately. For example, smart barns which use AI methods and are also furnished with the temperature, humidity, and ventilation sensors ensure the AI control of comfort and reduction of the incidence of disease is further to which the weight or milk performance gets improved.

Usually, AIoT systems in agriculture come with several layers. The perception layer is the source of sensors and devices that are responsible for collecting environmental and biological data. The network layer transfers data securely and at the same time saves energy via wireless technologies such as Low Power Wide Area Networks (LPWAN), 5G, or Wi-Fi. The data processing layer, which is mainly cloud-based, makes use of edge computing, i.e. it can be done on the terminal (e.g., IoT edge device) to process data, and also easy to use AI algorithms for analysis. The application layer is the part that comes into direct interaction with the farmers, wherein mobile apps or dashboards give actionable insights and automated control over devices such as irrigation valves or drones.

AIoT makes it possible for processes to continuously be controlled in real time. Unlike with regular monitoring and changes by staff, AIoT systems keep an eye on everything and fine-tune crop management automatically. If the sensors notice the soil is becoming drier, the irrigation system will start immediately. If pests are found by sensors, self-driving vehicles can start spraying right away. In this way, less waste occurs, crops are better protected and labor needs fall. AIoT also allows for full decision support systems that use past information, climate predictions, market trends and understanding of crops. They can help choose when to plant, the crop rotation, when to harvest and which supply chain method to use, supporting better field control and the farm business as a whole.

6. REAL-WORLD CASE STUDIES OF AIOT IN AGRICULTURE

Using Artificial Intelligence (AI) and the Internet of Things (IoT) in agriculture is really changing the way farmers operate worldwide. Because of the merging of AI and IoT, also called AIoT, new innovations can be seen today helping farmers increase yields, protect the environment and include everyone in the industry. AIoT is being used in both small or large agricultural settings such as in developing and developed nations, to manage issues related to climate, slimming resources and highly volatile markets. This part looks at real-life projects in agriculture that show how AIoT is used and how it affects farming, discussing key innovations, successes and obstacles.

AIoT-powered Fasal,et.al shows how advanced technology in farming can be given to resource-poor farmers. A lot of the world's farmers, relying on smallholder agriculture, do not have access to quick and correct information needed for optimizing what they grow. As an example, Fasal uses affordable IoT sensor kits to watch over soil moisture, temperature, humidity and nutrient levels in the fields of small farmers. Real-time data from the sensors goes to AI platforms in the cloud which review it and offer detailed advice for growing crops. Farmers can use the mobile applications to get helpful suggestions on when to irrigate, put out fertilizer, fight pests and diseases and plan the right time to harvest. With the help of Fasal, farmers can respond to the needs of their land and experience which lowers expenses, increases their yields and lessens the threat of weather or pests. Similarly, one of the key features of Fasal's model is how inexpensive IoT technologies and AI analytics give access to tools and techniques that most farmers were unable to use before. Because the startup supports local languages and uses mobile access, remote rural areas can enjoy the benefits of AIoT as well.

John Deere is recognized as a pioneering company on bigger farms primarily because of its AIoT-enabled technology that brings better accuracy and efficiency to farm equipment. John Deere stands out as one of the superior agricultural machinery manufacturers with its tractors, combines and sprayers having a lot of IoT sensors and up-to-date communication technology. By using AI, the data collected by these machines from the soil, crop and equipment is sent to a farm management software that uses AI. AI after processing the data helps significantly in the usage of the equipment for different purposes, e.g., such as control of how deep to plant, the number of seeds and pesticide application, which are done throughout the year. In addition to that, Deere has a major commitment to the development of robotic farming machines in its portfolio enabling machines to perform tasks regularly hence reducing human labor and making farming more efficient. Monsanto's acquisition of Blue River Technology, makers of intelligent sprayers which use AI and vision to find out the plants that are in need of care and therefore, be able to reduce chemical use and consequently the impact of pollution on nature is also a notable one. By means of 'Artificial Intelligence of Things' technologies, Deere proves the effects of automation, data analysis, and connectivity on productivity, sustainability and resource utilization on farms. Table 3 summarizes successful implementations of smart agriculture technologies across different geographic and socioeconomic contexts.

Table 3. Real-World case studies in AI and IoT-Based agriculture

Case Study	Technology Used	Country/ Region	Outcome Achieved
Fasal	AI + IoT sensors + weather API	India	40% water savings, 20% increase in yield
John Deere	Autonomous tractors + GPS + data cloud	United States	Precision harvesting, reduced fuel consumption
Microsoft AI4Earth	Machine learning + satellite data	Global	Pest forecasting, improved crop planning
e-Choupal	Digital kiosks + weather + market info	Rural India	Enhanced farmer access to market prices & info

Through its AI for Earth project, Microsoft highlights the ways AIoT is helping towards better farming and more conservation for the environment worldwide. AI for Earth gives grants and collaborates with efforts that apply AI and Internet of Things to address issues in agriculture, including increasing results, adequate water use, pest detection and resistance to climate changes. The main highlight in this program is FarmBeats which was designed with the help of Microsoft Research (Ghetiya, R.(2025)). By using simple IoT sensors, drones and data from satellites, FarmBeats gathers a huge amount of data from farming activities. AI algorithms study the information to produce models that can forecast about crop development, irrigation, pests and weather. A main feature of FarmBeats is that it strives to be accessible and affordable to farmers in areas where both fast internet and advanced farming technologies are rare. Local processing of data by FarmBeats with edge computing helps cut the need for uninterrupted internet access which in turn brings down the cost of running the service. It makes it possible for both local and overseas farmers to use methods of precision agriculture that were unreleased to them before. Because of FarmBeats, AIoT is being used in different socio-economic sectors to make sure technology for agriculture is available for everyone around the world.

Tamil Nadu Agricultural University (TNAU), in India, is an eminent case of how government and academic institutions employ AIoT for their agricultural sectors. IoT sensor networks are established across the research farms of TNAU to continuously monitor a number of variables such as soil moisture, temperature, plant indicators, and meteorological data. This data is then used in AI models created by TNAU scientists which regulate the irrigation schedules, the amount of fertilizers, and the pesticide management protocols in order to produce more crops on less land and at the same time save resources. TNAU also utilizes mobile advisory services that distribute AI-generated knowledge to Tamil Nadu farmers through SMS and local language smartphone apps (Ghetiya, R. (2025)). These programs have resulted in measurable gains, such as precise irrigation, which led to significant water savings, increased yields, and reduced pesticide use. The attempts of TNAU have shown the critical role played by the public sector to carry out research, technology development, and farmer education that makes AIoT solutions accessible. Through the combination of scientific and extension services, TNAU is building a more resistant and sustainable agricultural sector in the area.

Furthermore, AIoT is being used in agriculture in many other emerging ways. Bayer's Climate FieldView platform gathers information from different machines, sensors and weather monitoring stations, then AI technology is used to personalize both planting and pest control methods for various zones within a field. In Indonesia, the smart machine on eFishery which uses IoT and AI, makes sure aquaculture fish are fed at the right times, lessening chances of wasting feed and enhancing the fish's overall health. Autonomous robots from the Small Robot Company in the UK contain AI and sensors from the IoT, monitoring crops and soil all the time and thus enable easy and careful farming that limits the company's impact on the environment. Various cases like these show how AIoT has proven useful in farming crops, operating in aquaculture environments and is suitable for different locations and scales.

7. REAL-WORLD CASE STUDIES OF AIOT IN AGRICULTURE:

Artificial Intelligence (AI) and the Internet of Things (IoT) being used in agriculture is not far off anymore and is already starting to reshape global food production. DA in agriculture is transforming many parts of the world, since it makes farming more accurate, efficient and eco-friendly. This part provides examples of how AIoT is helping startups, big corporations, government organizations and research institutes address the issues in agriculture and support farmers.

Fasal, an Indian start-up in the agri-tech sector, has introduced helpful AIoT solutions at lower costs for small farmer groups which are the key defenders of agriculture. Fasal uses IoT sensors which are easy to install, to monitor on a daily basis the moisture, temperature, humidity and nutrient levels in the soil. They send information about crops and the environment to AI models in the cloud which then use the data to offer practical advice. On their mobile app, farmers can get notices and advices that suit their farmland and at the right time for crop irrigation, handling pests and applying fertilizer. By localizing precision agriculture solutions and adding predictions about weather and pests, Fasal allows less wealthy farmers to decide on the best actions, use less and produce more. How AIoT is being made accessible to all, brings inequalities in digital access to the spotlight and increases chances for inclusive success in farming.

John Deere, a leading company in industrial business, has used AIoT technology to upgrade its agricultural equipment, now making them smarter and easier to connect. The equipment from John Deere today uses many Internet of Things (IoT) sensors and GPS devices to monitor and record real-world and environmental data. This information is entered into farm management software which helps AI correctly determine the rates for seeds, depth of soil to plant and pesticide usage. By using AI, autonomous tractors and robotic harvesters take care of repetitive, labor-intensive jobs with high accuracy and need very little help from people, so their use leads to better results and lesser labor costs. As seen in Table 4, AIoT contributes positively to both economic productivity and ecological sustainability.

Table 4. Potential benefits of AIoT in smart farming

Benefit Area	Description	Technology Enabler
Resource Efficiency	Optimal use of water, fertilizer, and energy	IoT Sensors, AI analytics
Increased Yield	Better decision-making and monitoring	ML models, drones
Environmental Gains	Reduced chemical runoff, improved soil conservation	Precision agriculture platforms

continued on following page

Table 4. Continued

Benefit Area	Description	Technology Enabler
Labor Reduction	Automation of repetitive and dangerous tasks	Robotics, autonomous machinery
Market Access	Price forecasting, e-commerce channels	Cloud-based mobile apps

Besides private businesses, large technology companies and research teams are promoting AIoT applications in farming with initiatives on a worldwide scale. An example of this is Microsoft's AI for Earth which is a main program using AI and IoT to solve environmental and agricultural problems. Among all the projects in this program, FarmBeats stands out by bringing together low-cost Internet of Things (IoT) sensors, drones and satellite photos to collect precise farming information. Using the data, AI models provide ideas about crop development, the need for water and how to deal with pests. FarmBeats puts a lot of attention on being affordable and accessible, so it can help large farms as well as smallholder farmers in less-developed countries. Instead of using the internet for data analysis, the platform works by doing analysis at the edge. This lowers the need for high-speed internet which many rural communities cannot get. FarmBeats does this by allowing low-cost, scalable automation in farming and showing how AIoT is useful in different places to make smart farming more popular worldwide.

Development and applications of AIoT in a region are greatly supported by government and academic institutions. In Tamil Nadu Agricultural University (TNAU), IoT sensor networks have been set up on experimental farms to monitor the condition of the soil, climate and the growth of crops. Collected data is input into AI models made by university researchers to perfect schedules for watering, fertilizing and controlling pests that are right for the local crops and weather. Farmers receive AI advisories sent by SMS and smartphone apps which helps TNAU ensure that more people can understand and make use of the information. Experiments by TNAU have brought about improvements in the use of water, increase in crop yields and lower use of pesticides. Here, it is evident that work done by research institutions goes a long way in getting new scientific discoveries to benefit farmers and promote eco-friendly farming in the region.

AIoT applications used in agriculture around the globe show the range of ways this technology may be employed. Using the help of AI-generated analytics, the Climate FieldView platform supported by Bayer relies on data from farming machines and local weather stations to assist with precision farming choices such as varying the rate of planting and using crop protection. The platform makes sure that the AI's analytics are accurate. When there is variation in the field, farmers will quickly realize it affects the standard rate. They can then apply their method accurately in the field which raises their yield and minimizes waste. An example is eFishery from Indonesia that has used AI to figure out how to feed fish without harming the environment. An AI system which is set up within online networks, takes care of the feeders by setting optimized feeding schedules to reduce the amount of food that is wasted. There is a simple example of how AI is used in IoT. Also, the company has come up with a way using robots that have Internet of Things sensors and artificial intelligence to meet these needs. They are built to watch both crops and soil during the growing season and carry out remedial local actions that almost do not contain chemicals and help improve the quality of the soil. Thus, these achievements emphasize that AIoT serves many agricultural areas, including aquaculture, demonstrating its important effect across the industry.

8. BENEFITS OF AIOT IN FARMING

The conjoining of AI and IoT, referred to collectively as AIoT, in agriculture has led to a new wave of transformation. AIoT not only equips the agriculture system with automation and data collection but also affects the way the system is designed, operated, and optimized. With the aid of AIoT the problems of today's agriculture can be solved to a great extent, thus making it economic, green, and humane. This chapter focuses particularly on these aspects of AIoT implementations and technology supporting modern farming in efficiency, sustainability, and adaptability.

One of the most immediate and quantifiable advantages of AIoT in agriculture is the significant arising of the plant's yield and quality. AIoT systems collect detailed data from many different sources like soil sensors, drones, weather stations, and satellite imaging that are available - building the crop health and field conditions picture precisely (v Pintus, M., Colucci, F., & Maggio, F. (2025)). These data are then analyzed by AI algorithms streaming--in real-time--to create a chain that can be used to produce very accurate decision-making on various agronomic operations such as irrigation, fertilization, seeding, and pest control. The said accuracy of the implementation by the AI eliminates any doubts which in turn makes it possible for crops to get the only necessary inputs at the most appropriate time, thus allowing better growing conditions and losses reduction due to mistreatment. As an example, real-time AIoT-based irrigation that adapts with soil moisture data can easily avoid both water stress and waterlogging and at the same time lead to healthier plants and more yield. In particular, the use of predictive analytics enriches the capability of agriculturalists to foresee occurrences of crop diseases or pest crises with great accuracy even before they really break out and start causing damage, facilitating the prompt response.

AIoT, apart from boosting productivity, favours the reduction of the expenses connected with the usage of resources. Conventional farming used to depend on standard practices, like spreading chemicals equally throughout a big field or irrigating at a regular pace which did not take into account the localised situation of the crop or soil. Such an inefficient technology not only leads to the waste of inputs but also increases operational costs and causes harm to the environment. AIoT is suitable for precision agriculture, a system that utilizes resources selectively in specific areas and specific times. For instance, smart irrigation systems save water consumption by 30–50% by determining the amount of water the soil needs at that moment. Moreover, the variable application of fertilizers and pesticides also results in the reduction of chemical usage and the drop of input costs. AIoT-driven automation of activities like weeding, sowing, or harvesting not only decreases the reliance on manpower but also can provide particular unrivalled benefits to the places which encounter an immense shortage of agricultural workers. In simple terms, all the abovementioned efficiencies result in the increment of the relative profitability and sustainability of the farm economy.

AIoT-enabled cultivation practices are the main cause behind the immense development of environmental sustainability. The exploitation of natural resources for agriculture is the biggest contributor to global warming, deforestation, and the depletion of biodiversity. AIoT technology is, nonetheless, in a position to effect a change towards more environmentally friendly methods. A typical AI example is one that can reduce the soil nitrous oxide emission by limiting the dosage of the fertilizer used by the farmer that is a major contributor to the pollution of water. In a similar way, AI-powered systems of pest detection along with targeted pesticide application are technology examples that will ensure no virulent pesticides are misplaced and through that spread into untargeted ecosystems. Drones and robots that are AI-driven are the ones that can be controlled so that they do not pass through the fields that are most vulnerable to soil compaction, thus maintaining soil structure and decreasing erosion. Besides that, AIoT has a great potential to assist in the efficient utilization of such valuable natural resources as water and energy. The smart irrigation systems powered by renewable energy and the solar sensors are the technologies which prove to be the alternatives to the traditional, inefficient, energy-intensive practice of irrigation. These improvements have together initiated a better interpersonal relationship between agriculture and the environment and at the same time set the foundations of a sustainable environment. Table 5 outlines some of the key barriers hindering widespread adoption of AIoT in agriculture.

Table 5. Common challenges faced in smart farming adoption

Challenge	Description	Affected Stakeholders
High Initial Cost	Devices, installation, and maintenance are expensive	Smallholder farmers
Data Privacy Concerns	Risk of misuse or unauthorized access to farm data	All farmers and governments
Technical Skill Gap	Lack of digital literacy in rural populations	Rural communities
Infrastructure Barriers	Inadequate connectivity, electricity, or logistics	Developing nations
Interoperability Issues	Incompatible devices or software standards	Tech providers, farmers

One more way in which AIoT technology in farming brings a lot of benefits is that it enables intelligent decision making and risk management. Agriculture has always been considered a risky business due to a number of factors – from the changeable weather to variable market prices. The adoption of AIoT is bringing the much-needed relief as far as this uncertainty is concerned, and doing it services that go far beyond the existing state. In the process, it not only alleviates their current predicament but also plays a role in the coordination of the various steps to improve their performance. A great example of this is where a farmer is faced with weather forecasts so that he can determine the best time to sow his crops and to check the field conditions. In the same way, the farmers can be aware of the seasons of high prices and the places in which they can market their products effectively through AI-based crop price forecasting. Predicting post-harvest losses also make good use of risk management tools in the agriculture sector, as they can know about the spoilage and decide on the proper temperature and humidity settings for storage facilities. These risk management tools then assist the farmers to make better decisions sooner and not when the issue has already happened, in the long run, reducing the total losses and safeguarding food production. Such tools that give warning and support to the farmer through the problem-solving process.

AIoT systems are indeed amazing in a supply chain and not only do they help in better the ability to trace and transparency the process, this new technology is beneficial to everyone starting from the farmer to the consumer and the regulators. By taking information at each step, the process becomes transparent and thus a digital audit trail is created that not only serves the needs of the counsellor but also safeguards the whole production. For example, if there's an issue of contamination, the stakeholders will know the cause very quickly and will be able to solve the problem. With this level of transparency, consumers can become more cautious and pick organic foods when shopping. Furthermore, AIoT systems can be teamed up with blockchain technologies in order to give an unchangeable record of all agricultural transactions, thus lowering the opportunity of fraud and providing a fair trading environment, particularly in commodity markets.

Figure 4. Benefits of AIot in smart farming

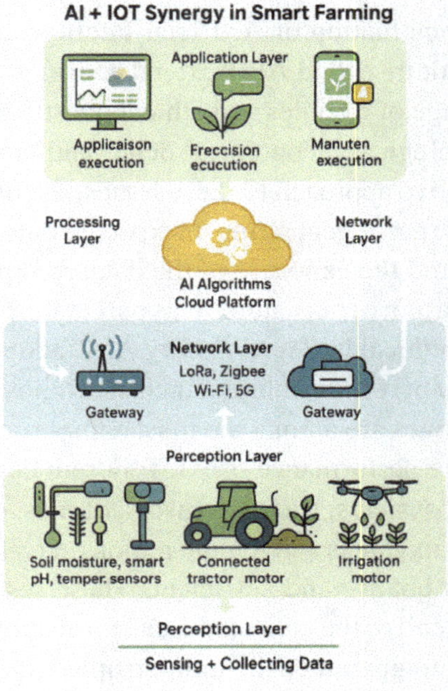

Figure 4 has cleanly explained the benefits of the AIoT in smart farming. AIoT offers a slew of possibilities that could not only help people with disabilities but also contribute to global agriculture. Great part of the farming population in developing countries is made of smallholder farmers, who usually have no access to adequate information, financial services, or modern machinery. If the marginalized of the marginalized can get the benefit from AIoT applications that run on mobile networks and are very cheap, this will give them the possibility to get real-time weather forecasts, pest alerts or the current market price. A lot of AIoT tools are developed with poor connectivity in mind and thus they rely on edge computing to be operational without an internet connection (Alakkari, K. (2025)). Such tools are programmed to wait for the moment when internet accessibility is on the flip side to upload data. Moreover, governmental and non-governmental organizations (NGOs) have committed themselves to the very same thing as they are purposely favoring these AIoT solutions by giving out mostly free/discounted sensor kits, digital literacy training sessions, and open-source platforms. AIoT shortens the digital divide and hence makes the whole agricultural society more equal.

9. CHALLENGES AND ETHICAL ISSUES:

Not only does the combination of Artificial Intelligence (AI) and the Internet of Things (IoT) in agriculture called AIoT extend enormous capabilities, but it also puts the sector into a range of troubles and ethical questions. These uncomfortable issues reflect the technological, economic, social, and environmental aspects. In the absence of a preemptive approach to these issues, the utilization of AIoT in the agriculture industry is at risk of generating unexpected outcomes such as the aggravation of digital divide and the exposure of the dataset to potential danger and the destruction of the equilibrium of out nature. This part of the paper critically discusses the main challenges and ethical issues caused by AIoT adoption in farming and also underscores the necessity of responsible and inclusive innovation and policy setting.

The infrastructural issues are among the most complex challenges that the AIoT technology pulls off. The performance of AIoT system depends on power and internet stability of the ecosystems, such as sensors, drones, edge devices, and cloud platforms. However, in rural areas and remote regions such resources are not always available or not of high quality and sufficient. These resource constraints make farmers' work more difficult or the collected data less accurate because of the nature of AI technology failing to guarantee the uninterrupted data collection. Thus, they are at risk of receiving not fully, or even worse, incorrect, AI-prompted suggestions for better farming practices. Additionally, the connection of different devices and platforms usually needs common standards of interfaces and staff professionalism, most of which are not available in small and medium-sized farms. No user-friendly interfaces, especially for those who are not technology-savvy, make the use of the technology inconvenient and thus leads to low rates of its implementation.

Another important obstacle to the implementation of AIoT solutions is their cost. The substantial investment needed for the acquisition of such devices as sensors, drones, machine learning software, and connected infrastructure can be an insurmountable obstacle for small farmer families or low-income agricultural communities. Even when the economic benefits in the long run are beyond any reasonable doubt, the necessary capital for the realization of the AIoT project can be very high without the help of third-party financial institutions. Besides, the extra costs are likely to be a burden on farmers who are also dealing with flimsy income sources owing to weather, pests, or market changes. The cost of maintenance, software upgrades, cloud storage subscriptions, and the new equipment needed for the replacement of the worn technology become the farmer's inherent expenses. The gap that exists between the adoption of technologies in the farms and those lacking financial resources can increase the agricultural disparities already existing.

Data privacy and security are basic ethical issues in the context of AIoT-powered agriculture. Through the AIoT systemic approach, the matter of data from and to the smart environment becomes important as these data is collected, transferred, and saved—earth content, and crop health from the field to the geographical position cannot be missed. The large volume of data collected can be of concern if it is not kept secure or shared without the right kind of consent, as it may fall victim to unauthorized use by third parties, including agribusiness companies, insurance agencies, and possibly political entities. When the data are out of farmers' control this can obviously lead to issues of data ownership, informed consent, as well as sensitive information being misused. The other way security might be violated is through the introduction of cyber threats that can cause such big and harmful damages as the shutdown, theft, or distortion of the most important sub-areas of the economy. In addition to this, the more prone to cyber-attack, ransomware, and hidden surveillance your agriculture is, the more it's digital.

One more drawback that is strongly under discussion is the potentially caused displacement of agricultural labor. AIoT technologies with their autonomous machinery, robotic harvesters, and automated spraying systems are ways of making work more efficient and at the same time, kinds of work that hardly need human resources. On the positive side, it can be the solution for labor scarcity problems, but on the other hand, poverty may soon face the workers without the daily wages they depend on. In countries where agriculture is a major part of the population employment and at the same time, it is no longer viable; machines are required to carry out the entire process expeditiously, but human workers meanwhile have lost their jobs and have to find other sources of income that are still hardly available, which in the medium term such a situation can provoke tensions and even violent acts in these societies. Ensuring that technology advances' leading to the sacrifice of people's well-being can be considered the most central AIoT transition ethics responsibility.

AI algorithms and bias in AI systems raise other questions about fairness. The functioning of AI programs basically depends on the data they have been trained with. In cases where the training datasets are incomplete, outdated, or biased toward a particular or a few geographies, or crop types, the suggestions of the AI model may turn out to be faulty or useless in other environments. For instance, with the weather and soil data of North America as training data, a prediction model may not be effective in sub-Saharan Africa or Southeast Asia. This may result in acts that endanger crops or the waste of resources. What is more, there is a danger of AI systems favoring large-scale industrial farming methods in comparison to traditional or indigenous practices, which in turn may result in the neglect of cultural and global agricultural homogenization enhancement.

10. FUTURE TRENDS IN AIOT-BASED SMART FARMING:

Given that global food demand is ever-increasing, and the issue of sustainability is now more of an emergency, it is evident that Artificial Intelligence and the Internet of Things- termed AIoT - are going to meet one another with more frequency, resulting in the future of farming getting more and more dependent on them. In other words, smart farming will be the driving force for a sustainable agriculture sector, thanks to the combination of technological innovations, changing consumer behavior, different policies, the issue of climate change, etc. Throughout the next ten years, farmers will experience a swift change in the way the food chain starts from the use of self-driving vehicles and secure digital ledgers in the supply chains to managing the information sent by drones and extremely local data. An era is coming that will be dominated by the effective strategies employed to grow, manage as well as distribute food. In this chapter, we look into the not so distant future, and consider what are likely to be the leading trends that will define AIoT in the agriculture sector and consequently transform the whole industry.

In the modern era, significant changes have been seen with the revolution in the world of farming machines and robots. The use of AI-based navigation systems, vision systems, and edge computing in the process has made it possible to have self-operating tractors, the robotic harvesters, and even the ganzers unmanned air vehicles (like drones) along with the planting, weeding, spraying, and harvesting of highly precise quality. This kind of technology not only results in the disadvantage of using manual labor but also shows that the machines can run error-free and be in constant working condition, thereby improving and increasing productivity on the whole. Both startups and leading industry players are joining the movement and investing in the production of farming technologies that are fully autonomous with just minimal human input(Mowat, K. M. (2025)). For example, swarm robotics is a new technology that is aimed at performing multiple tasks that can be coordinated to cover large agriculturally occupied areas In large agricultural fields, where there is a need to perform complex agricultural tasks simultaneously, the technology of coordinated operation among many robots, i.e., the smart robots working as team, could be the best choice. A group of robots with wireless communication and coordination capabilities, and designed to work together, could be deployed, covering large areas, to deliver the required level of service on time.

10.1 Evaluation Points

- Security & Privacy: Insert the discussion on data security, encryption and privacy issues in IoT-enabled agriculture to overcome the risk of data breaches and ownership challenges.

- Policy & Government Support: Add the role of the governmental policies, subsidies, and public private partnerships in promoting the use of IoT, particularly in the developing world.
- ROI & Cost-Benefit Analysis: Add the economic analysis of IoT systems to outline the Return on Investment to both the small and large-scale farmers.
- Smallholder Challenges: Remove the hindrances of cost, poor digital literacy, and lack of infrastructure among the small or marginal farmers.
- Emerging Trends – Blockchain: Describe how blockchain enhances traceability, food safety and transparency of the supply chains in smart agriculture.
- Emerging Trends – Edge Computing: Mention the real time, on device data processing capability to bypass connectivity challenges in the rural or remote agricultural areas.
- Agriculture 6.0: Which is the next level of hyper-connected, AI-powered, and sustainable agriculture, mention the future vision of agriculture, going beyond precision farming.
- AI + IoT Combination: Demonstrate the predictive analytics and automated decision making where AI and IoT are combined in crop surveillance, animal control and prediction.
- Excessive Focus on Devices: There are too many IoT tools stated; therefore, it will be more appropriate to make it a table or figure to make it less wordy and obvious and eliminate repetitiveness.
- Interoperability Discussion: Discuss the issues of having separate devices and/or systems of various vendors to be used in one single platform, and how difficult it is to integrate them into a singular platform.
- Ethics & Data Ownership: Debate the ethical aspect surrounding the case of who owns the farm data, how the farm data is used, go as far as speaking with the technology providers.
- Future Proof: Expand on the topic of the environmental impact of agriculture and effective usage of the resources and how the IoT fits into the sustainable farming procedures.
- Redundant Descriptions: It is unnecessary to describe the technical descriptions here again- a briefer and more altered version, or a presentation of comparative visuals where possible would suffice.
- Scarcity of Regional Examples: They can use many more cases, of various type as related to various counties, various crops, various organizations, etc. This may help in making the work globally relevant and applicable.
- Lacks Analysis: The argument/conflict can be the controlling theme of the piece and hence the writer ought to have added more criticism on the numerous challenges, trade-offs and system constraints under various use-case situations.

- Suggestions on Improvement: It would be good to add some references to scholar articles, utilize some diagrams, insert some transition between various sections, and, finally, have a clear introduction that would appear as the roadmap guiding the reader through the text.

One of the major trends in the agricultural business at the moment is the merging of blockchain technology into the supply chain. Blockchain provides a secure, transparent, and unmodifiable ledger system that is particularly useful when tracing the origin of food products from farm to table. With the help of Internet of Things (IoT) sensors placed in the supply chain nodes like storage facilities and transportation units, the data about the temperature, humidity, and handling conditions are tracked and written onto a blockchain network immediately. This allows for the trust among stakeholders to be built up and the responsibility to be put in place, certifies that the resulting foodstuff complies with safety standards and permits the rapid removal of the contaminated product. Through the use of blockchain technology, small farmers can be guaranteed fair prices, as there will be no middlemen that take the lion's share, and they also will have access to direct market-specific information, as well as transactions.

The future of AIoT in agriculture is going to be marked with extremely customized, data-enabled decision making. Due to the progress of machine learning algorithms and edge AI, farmers will get very local, plant-specific information, which is based on the current condition of the environment. AI models will use past climate data, satellite pictures, the information about nutrients in the soil, and the extent of pest activity to give precise instructions, which can be followed in every piece of land. Decision support systems will be transformed into the clever consultants who are constantly learning and reforming according to the conditions in the field, hence allowing small farms to be as precise and strategic as the large farms.

The 5G network rollout is another powerful enabler that will speed up the AIoT uptake in the countryside. 5G's very low latency and high data rate capacity are perfect for real-time tasks such as drone surveillance, remote-controlled irrigation, and AI-guided machinery. Due to faster data transfer, farmers will be able to keep an eye and control their activities even when they are not at the farm, with mobile dashboards and cloud-integrated apps (Thampi, S. M., Thomas, T (2025)). In addition, scratch communication between machines will facilitate distributed, decentralized smart farming ecosystems where edge devices carry out their operations with minimum interventions from centralized processing hubs.

11. CONCLUSION

The utilization of Artificial Intelligence (AI) and the Internet of Things (IoT) in agriculture - which is also called AIoT - is a milestone in a new era of the history of the whole planet's farming practices. To mention just a few, the world is currently dealing with a number of challenges such as climate change, population growth, resource depletion, and food insecurity. Thus, traditional farming methods have become insufficient. Hence, the AIoT-based smart farming is a way of using data and technology that characterizes the new tools of the agri-food sector that aim at efficiency, sustainability, and resilience. Taking advantage of the intelligent automation, the real-time data analytics, and the interconnected sensing systems, smart farming not only makes the productivity higher but also it leads to fewer environmental pollution, it uses the resources in a more efficient way, and it creates climate resilience.

The progression of Smart Farming starting from the mechanization and the Green Revolution up to the present time of digital agriculture, where precision and flexibility are the most important factors. We have discussed the involvement of IoT in the human role of environmental sensing and automation. We also dwelt upon the role of AI in providing insights, prediction, and autonomous decision-making. We further investigated the cooperation of AI and IoT that facilitates the development of intelligent agricultural systems which are capable of handling everything from irrigation and fertilization to pest control and crop forecasting. These technologies have thus far been proven and realized in different real-world use cases, including large agribusinesses and smallholder farms—showing the scalability, the affordability, and the ability to democratize the agricultural innovation of these technos.

However, the combination of AIoT with agriculture is not without problems. Problems such as very high expenses for implementation, digital illiteracy in the countryside, data security issues, and the digital gap, among others, have to be solved by means of inclusive policy frameworks, farmer training programs, and public-private partnerships. The ethical questions raised in relation to the AI decisions, the ownership of data, and the transparency of the algorithms should be dealt with in a very cautious way so that the transformation in the field remains fair and just.

REFERENCES

Ajith, S., Vijayakumar, S., & Elakkiya, N. (2025). Yield prediction, pest and disease diagnosis, soil fertility mapping, precision irrigation scheduling, and food quality assessment using machine learning and deep learning algorithms. *Discover Food*, *5*(1), 1–23. DOI: 10.1007/s44187-025-00338-1

Alakkari, K., & Ali, B. (2025). Artificial Intelligence of Things: A Review. *Babylonian Journal of Internet of Things*, *2025*, 113–120. DOI: 10.58496/BJIoT/2025/006

Ali, A., Tajamul, H., & Azlan, Z. (2025). Smart Irrigation Technologies and Prospects for Enhancing Water Use Efficiency for Sustainable Agriculture. *AgriEngineering*, *7*(4), 106. DOI: 10.3390/agriengineering7040106

Arularasan, A. N., Kalaiyarasi, D., Ashokkumar, P., Suresh, K., Rajasekar, M., & Prasad, A. R. (2025). Utilization of Advanced IoT and High-Performance Computing for Enhanced Agricultural Systems. In *Integrating Artificial Intelligence Into the Energy Sector* (pp. 467–492). IGI Global Scientific Publishing. DOI: 10.4018/979-8-3693-7112-1.ch022

Assimakopoulos, F., Vassilakis, C., Margaris, D., Kotis, K., & Spiliotopoulos, D. (2025). AI and related technologies in the fields of smart agriculture: A review. *Information (Basel)*, *16*(2), 100. DOI: 10.3390/info16020100

Awais, M., Wang, X., Hussain, S., Aziz, F., & Mahmood, M. Q. (2025). Advancing precision agriculture through digital twins and smart farming technologies: A review. *AgriEngineering*, *7*(5), 137. DOI: 10.3390/agriengineering7050137

Batan, A. (2025). Assessing the Role of Public-Private Partnerships in Advancing Sustainable Development Goals Through Agri-Tech and Renewable Energy Solutions. *Transactions on Social Innovation, Digital Inclusion, and Ethical AI*, *15*(2), 1–11.

Devarajan, Y. (2025). Investigation of Emerging Technologies in Agriculture: An In-depth Look at Smart Farming, Nano-agriculture, AI, and Big Data. *Journal of Biosystems Engineering*, •••, 1–23.

Dinis, I., Santos, D., & Mendes-Moreira, P. (2025). Revitalising Traditional Cereals in Portugal: Challenges, Opportunities, and Strategies for Value Chain Development. Sustainability (2071-1050), 17(6).

Fasal. (2023). Smart agriculture platform: Revolutionizing farming with AI and IoT. https://www.fasal.co

Ganaou, N. I., & Salaou, A. I. (2025). Communication Technologies and Protocols in IoT Systems: Principles, Standards, and Security. In Enhancing Data-Driven Electronics Through IoT (pp. 323-390). IGI Global Scientific Publishing.

Ghetiya, R., Kumari, P., Vekariya, P., & Kumar, M. S. (2025). Artificial Intelligence in Achieving Sustainable Development Goals: Models and Approaches. In AI in Agriculture for Sustainable and Economic Management (pp. 12-24). CRC Press.

Gkoulis, D. (2025). Creating interpretable synthetic time series for enhancing the design and implementation of Internet of Things (IoT) solutions. *Internet of Things : Engineering Cyber Physical Human Systems*, *30*, 101500. DOI: 10.1016/j.iot.2025.101500

Gracia-Rodriguez, C., Martínez-Medina, A. E., Torres-Cosio, L., Lopez-Ortiz, C., Nimmakayala, P., Luévanos-Escareño, M. P., Hernández-Almanza, A. Y., Castro-Alonso, M. J., Sosa-Martínez, J. D., Reddy, U. K., & Balagurusamy, N. (2025). Can the molecular and transgenic breeding of crops be an alternative and sustainable technology to meet food demand? *Functional & Integrative Genomics*, *25*(1), 1–32. DOI: 10.1007/s10142-025-01594-1 PMID: 40205022

Kamilaris, A., & Prenafeta-Boldú, F. X. (2018). Deep learning in agriculture: A survey. *Computers and Electronics in Agriculture*, *147*, 70–90. DOI: 10.1016/j.compag.2018.02.016

Long, S. P. (2025). Needs and opportunities to future-proof crops and the use of crop systems to mitigate atmospheric change. Philosophical Transactions B, 380(1927), 20240229. DOI: 10.1098/rstb.2024.0229

Mowat, K. M. (2025). A Case Study on Readiness of Engineers in Agricultural Technology Start-Ups (Doctoral dissertation, The University of Nebraska-Lincoln).

Nkwo, M., Ikwunne, T., Adejoro, C., & Anuyah, O. (2025). Exploring Usability of AI Systems in the Global South—Toward Responsible Human-Centered AI for Sustainable Cities and Communities. In *Usability for the World: Building Better Cities and Communities* (pp. 125–165). Springer. DOI: 10.1007/978-3-031-88326-2_5

Pintus, M., Colucci, F., & Maggio, F. (2025). Emerging Developments in Real-Time Edge AIoT for Agricultural Image Classification. *IoT*, *6*(1), 13. DOI: 10.3390/iot6010013

Rana, R., & Bhambri, P. (2025). The Role of IoT in Shaping the Future of Geospatial AI. In *Recent Trends in Geospatial AI* (pp. 177–216). IGI Global Scientific Publishing.

Sagar, S., & N Birje, M. 2025. Precision Agriculture Using Internet of Things and Cloud Computing:Review, A. (•••).. . *SN Computer Science*, *6*(4), 1–18.

Shen, Z., Wang, Z., Chew, J., Hu, K., & Wang, Y. (2025). Artificial intelligence empowering robo-advisors: A data-driven wealth management model analysis. *International Journal of Management Science Research, 8*(3), 1–12. DOI: 10.53469/ijomsr.2025.08(03).01

Thampi, S. M., Thomas, T., & Mukherjee, P. (Eds.). (2025). *Securing the Connected World: Exploring Emerging Threats and Innovative Solutions*. Springer Nature. DOI: 10.1007/978-3-031-82826-3

Thilakarathne, N. N., Bakar, M. S. A., Abas, P. E., & Yassin, H. (2025). Internet of things enabled smart agriculture: Current status, latest advancements, challenges and countermeasures. *Heliyon, 11*(3), e42136. DOI: 10.1016/j.heliyon.2025.e42136 PMID: 39959477

ADDITIONAL READING

FAO. (2020). *The State of Food and Agriculture 2020: Overcoming water challenges in agriculture*. Food and Agriculture Organization of the United Nations. https://www.fao.org/3/cb1447en/cb1447en.pdf

Fasal. (2023). *Smart agriculture platform: Revolutionizing farming with AI and IoT*. https://www.fasal.co

Kamble, S. S., Gunasekaran, A., & Sharma, R. (2019). Analysis of the driving and dependence power of barriers to adopt Industry 4.0 in Indian manufacturing industry. *Computers in Industry, 101*, 212–237. DOI: 10.1016/j.compind.2018.11.002

Kamilaris, A., & Prenafeta-Boldú, F. X. (2018). Deep learning in agriculture: A survey. *Computers and Electronics in Agriculture, 147*, 70–90. DOI: 10.1016/j.compag.2018.02.016

Liakos, K. G., Busato, P., Moshou, D., Pearson, S., & Bochtis, D. (2018). Machine learning in agriculture: A review. *Sensors (Basel), 18*(8), 2674. DOI: 10.3390/s18082674 PMID: 30110960

Qin, J., Liu, Y., & Grosvenor, R. (2016). Digital twin and big data towards smart manufacturing and industry 4.0: 360 degree comparison. *IEEE Access : Practical Innovations, Open Solutions, 6*, 3585–3593. DOI: 10.1109/ACCESS.2018.2793264

Wolfert, S., Ge, L., Verdouw, C., & Bogaardt, M.-J. (2017). Big data in smart farming – A review. *Agricultural Systems, 153*, 69–80. DOI: 10.1016/j.agsy.2017.01.023

KEY TERMS AND DEFINITIONS

Artificial Intelligence (AI): Refers to the capability of a machine to imitate human cognitive functions such as understanding language, learning, reasoning, and problem-solving. Intellectual faculties emerge in a computer.

Autonomous Farming Equipment: Tractors, drones, robots, and other machinery that are either fully autonomous or semi-autonomous make use of AI and IoT for going through activities such as planting, harvesting, and monitoring.

Big Data Analytics: Massive sets of organized and unorganized data are being analyzed to reveal hidden patterns or relationships, and they can provide useful signals to decision-making engagements.

Digital Twin: Such a computerized model of an actual entity or procedure gives one the opportunity to carry out the operation virtually to get the actual effects without any mistake and to optimize performance.

Internet of Things (IoT): A system of interconnected devices with sensors and softwre, thus they are able to collect and exchange data through the internet for continuous monitoring and control.

Machine Learning (ML): AI tech that includes ML techniques and statistical models that allow software to improve its performance on certain tasks from experience and without explicit programming.

Precision Irrigation: Agrotechnology utilizes monitoring devices and measuring the results of the data to apply water in the most appropriate way to the crops thus lowering water wastage and at the same time improving the health of the crops.

Remote Sensing: The process of gathering data about an object or a place without any physical contact, usually by using sensors installed on satellites or drones. This data can be used to detect changes in the crops, soil, or environment.

Smart Farming (Precision Agriculture): Implementing digitaltechnology in agriculture an IoT sensors, AI and data analytics to improve production of crops, usage of resources and environmental sustainability.

Chapter 8
Generative AI in Practice:
From Core Principles to Real-World Use

Basudha Dewan
https://orcid.org/0000-0001-6513-4374
Manipal University Jaipur, India

Sabyasachi Bhattacharyya
Barak Valley Engineering College, Residential Girls' Polytechnic, Golaghat, India

Ishu Sharma
Chandigarh Group of Colleges Jhanjeri, Mohali, India

Ehsan Sheybani
University of South Florida, Tampa, USA

ABSTRACT

Generative AI is a subset of machine learning focused on generating new, coherent data samples that mimic a given dataset. This chapter introduces core principles and theoretical foundations of generative modeling, emphasizing probabilistic approaches and models such as GANs, VAEs, and transformers like GPT. It explores design strategies, training techniques, and trade-offs involved in quality and efficiency. Practical applications are discussed across domains including natural language generation, synthetic media, drug discovery, medical imaging, education, and content recommendation. Real-world case studies support the theoretical insights. The chapter also addresses ethical considerations like deepfakes, data privacy, and algorithmic fairness, promoting responsible AI development. Emerging trends such

DOI: 10.4018/979-8-3373-4207-8.ch008

as diffusion models and multimodal generation are also covered, offering a concise yet comprehensive guide for researchers and practitioners in academic, industrial, and cross-disciplinary settings.

1. INTRODUCTION

Generative Artificial Intelligence (Generative AI) refers to a class of machine learning techniques designed to create new data samples that resemble a given dataset. Unlike traditional AI systems that are designed to recognize patterns or make predictions based on historical data, generative AI learns the underlying probability distribution of the input data and uses it to generate new, coherent, and meaningful content. These outputs can be texts, images, audio, video, 3D models, etc. (Rishi Bommasani et al., 2021; Goodfellow, 2016). Generative AI is an umbrella term for a variety of models and algorithms, namely, Generative Adversarial Networks (GANs), Variational Autoencoders (VAEs), and Transformer-based architectures like the Generative Pre-trained Transformer (GPT) (Ian Goodfellow et al. 2014, Kingma et al. 2014, Vaswani et al., 2017). Such models have been adopted by a number of different areas including natural language processing, computer vision, drug discovery, education, which has from the ground level transformed the way in which synthetic content is created and utilized (Ramesh et al., 2021).

In order to understand the contribution of the generative models, it is helpful to compare them with discriminative models (Ng et al., 2001). Discriminative models focus on modeling the boundary between different classes of data (M Christopher Bishop et al., 2006). They are trained to answer questions like: "Given this data point, to which class does it belong?" Examples include logistic regression, support vector machines (SVMs), and most classification-focused deep learning networks.

Generative models, on the other hand, attempt to model the joint probability distribution of inputs and outputs (i.e., P(x,y)) or just the data itself (i.e., P(x)) (Murphy, 2012; Tipping, 2003). Their goal is to generate entirely new instances that could have plausibly come from the training data. While discriminative models are optimal for classification and regression tasks, generative models excel in content creation, data augmentation, and simulation-based learning.

Generative AI has rapidly become a cornerstone of modern artificial intelligence, revolutionizing how machines interact with data, create content, and support human creativity. Its importance lies in its ability to not only understand patterns in data but also to generate entirely new and meaningful information that mirrors the characteristics of the original dataset. This capability has made it instrumental in various high-impact domains. People in the creative fields use generative models to make images and videos that look real, original music, and write text that can't be told apart from something made by hand. Generative AI is used in drug discovery by helping to simulate how molecular structures behave and investigate their qualities, speeding up the process and making it more efficient (Karras et al. 2019; Jumper et al. 2021). Also, in education and personalized settings, generative AI helps create different types of content, tutoring interactions, and simulations suited to the needs of each student. Generative AI is essential in business, helping to build better recommendation systems, power customer chatbots, and supply synthetic data for training AI in areas with few examples. Apart from these uses, generative AI is making a huge impact on basic AI research by improving unsupervised learning, self-supervised learning, and multimodal understanding (Floridi & Cowls, 2019). Now that the technology is improving, its potential to transform industries, automate creative work, and boost innovation makes generative AI important to the current AI space, not only the future.

The main purpose of this chapter is to give the readers a solid yet informative knowledge of what generative artificial intelligence is based on and what areas of application this concept has. It starts by debunking the theoretical underpinnings that sustain generative modeling – probabilistic reasoning and learning of data distribution which are needed to know how the generative models operate. The chapter next examines the star architectures – Generative Adversarial Networks (GANs), Variational Autoencoders (VAEs), and Transformer-based models like GPT – exploring their philosophies behind their design, their training pipelines, the strengths and weaknesses, and their practical deployment. One of the main aims is to close the gap between theory and practice and introduce real-world applications from different spheres, for example, natural language processing, the synthetic media creation, drugs discovery etc., and education technology. In the process, the chapter also touches on fundamental ethical issues with generative AI, such as risks of misinformation, deep fakes, violations with data privacy, and algorithmic bias. The expressing of these concerns through the chapter promotes responsible innovation and critical thinking regarding AI systems' development and implementation. In addition, the chapter presents readers with state-of-the-art developments like diffusion models and multimodal generation practices that shed a light on the further direction of generative AI. In the end, the chapter is aimed at providing the conceptual arsenal, practical knowledge, and moral awareness for students, researchers, and professionals that would enable them to effectively interact with generative AI in an academic environment, in industry, or even across disciplines.

2. THEORETICAL FOUNDATIONS OF GENERATIVE MODELING

Probabilistic modeling is the concept that is at the base of generative AI, which consists in learning the probability distribution of a set of data. Rather, than just learning the limits between various classes (as in discriminative models), the generative models try to understand how the data was produced, which enables them to produce new samples that are statistically consistent with the original dataset (Bacciu et al., 2015; Zhang et al., 2022). In a formal sense, for a dataset X= ={x1,x2,...,xn}, a generative model would attempt to estimate the probability distribution P(x), which presents the likelihood of obtaining a given piece of data point xxx. After this distribution is acquired, the model can take random samples x'∼P(x).

2.1 Types of Probabilistic Models in Generative AI

1. Explicit Density Models

These models try to approximate P(x) directly and explicitly, often trying to maximise the likelihood of the data observed.

Examples:

- Autoregressive Models (e.g., PixelRNN, GPT):

Model the joint distribution as a product of conditional distributions:

$$P(x) = \prod_{i=1}^{n} P\left(x_i \middle| x_1, \ldots.., x_{i-1}\right) \tag{1}$$

- Normalizing Flows:

Transformation of a simple known probability density function (e.g., Gaussian) to data probability density function using an invertible function.

2. Implicit Density Models

These models do not give an explicit definition of P(x). Rather, they become good in creating samples that cannot be distinguished from real data.

Examples:

- **Generative Adversarial Networks (GANs):** Use a discriminator to distinguish real from fake samples while the generator improves its ability to fool the discriminator.
- **Diffusion Models**: Gradually transform noise into a data sample via a learned denoising process.

Generative AI relies on the modeling of the data distribution as a fundamental capability, since it facilitates many sophisticated behaviors quite beyond a simple classification or prediction. One of the most convincing arguments for mastering the distribution P(x) is the capability to generate new high-quality synthetic data, which shares a lot of similarities with the training dataset. The ability is especially helpful when there is a limited number of data, or when the sensitivity is high, such as in medical imaging or auto vehicle simulations. Additionally, probabilistic models have uncertainty estimations established in them and therefore provide certainty of the outputs they generate. This functionality increases not only the interpretability but also trustworthiness of AI systems that makes them more appropriate within the decision-critical context such as finance, healthcare. Also, a well-modeled distribution helps in detecting an anomaly. the samples can be determined as these that are considered unusual or potentially faulty for the individual cases as the part of cybersecurity, fraud detection and system monitoring, which belong to a low-probability region under P(x). Finally, probabilistic generative models allow Bayesian inference, which allows integrating prior information and transformed data improve the predictions even with noise or uncertainty. Such capability of doing reasoning in uncertainty is a strong point of scientific research, robotics as well as other disciplines relying on flexible and intelligent systems. On the whole, proper representation of the data distribution not only makes the generative AI capable of generating realistic outputs but also provides tools for making sense of, explaining, and acting within the world in a complex way.

Figure 1. Simulated imaging of training generative model matching to true data distribution (Ian Goodfellow et al. 2014).

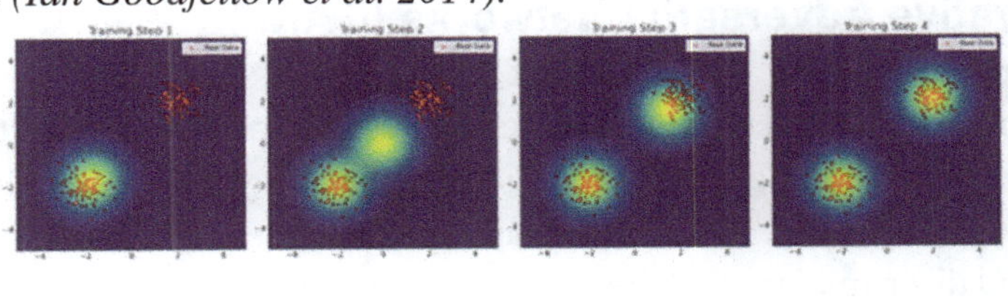

As it can be seen from Figure 1, a generative model revises its comprehension of data distribution in an iterative manner through training. The figure shows how the distribution of the probability density estimated by the model comes to resemble the actual distribution of data as time progresses, showing the convergence behavior that enables generative learning to be effective.

In generative modeling, likelihood estimation refers to the process of assigning probabilities to data points under a model's learned distribution. This is fundamental to many generative models, as it enables both the generation of new data and the evaluation of how well the model captures the underlying structure of the training data. At its core, likelihood estimation involves selecting model parameters that maximize the likelihood function—the probability of observing the given data under the current model (Blei et al., 2017). The most common approach is Maximum Likelihood Estimation (MLE), where the objective is to find parameters θ that maximize the likelihood P(x;θ) across all data samples. In practice, it is often more convenient to work with the log-likelihood, which transforms the product of probabilities into a sum and improves numerical stability. The other approach is variational inference that is applied in the models such as Variational Autoencoders (VAEs), in which the exact likelihood is intractable. Instead, a lower bound – Evidence Lower Bound (ELBO) is maximized as a lower bound measure to the true log-likelihood (Nowozin et al., 2016; Papamakarios et al., 2019). For other models such as Generative Adversarial Networks (GANs) that do not give a defined explicit density function; this traditional likelihood estimation is not possible; instead, these models rely on the use of adversarial loss functions to indirectly align the generated distribution with actual distribution of the data. Moreover, the score-based diffusion models have brought new paradigms of likelihood estimation by modeling the gradient (or score) of the data distribution (Song et al., 2019). In summary, likelihood estimation is a vital part of assessing and training generative models, and plays a direct role in their ability to generate realistic high-fidelity samples, and generalize to data which they have never seen before.

3. CORE ARCHITECTURES OF GENERATIVE AI

3.1 Generative Adversarial Networks (GANs)

This generative architecture in the generative AI community has the form of Generative Adversarial Networks (GANs) introduced by Ian Goodfellow in 2014 (Goodfellow et al., 2014). A GAN is made up of two neural networks including a generator and discriminator that are trained concurrently by an adversarial learning process. The generator generates synthetic data which is consistent with the real data distribution while the discriminator checks if a data is real or fake. Through time, the generator becomes better at deceiving the discriminator, while the discriminator becomes better at identifying actual data and fake. This min-max game drives both the networks toward the iterative improvement in their performance (Salimans et al., 2016). It is however not easy to train GANs. One of the most popular is mode collapse, the case when the number of the output is very limited, and the generator is unable to represent all the varieties of the data distribution. Other problems include instability of training and the sensitivity of a hyperparameter. The training dynamics have been proposed to be improved by researchers with such approaches as feature matching, Wasserstein loss, and spectral normalization. In spite of all these problems, GANs have been used for various applications such as image synthesis, super-resolution, style transfer, data augmentation among others (Miyato et al., 2018; Radford et al., 2016). Nevertheless, they suffer from the need for huge datasets, hardness to control generated outputs, and their non-interpretability. Such restrictions reflect the necessity of a proper design and tuning in the real-world GANs deployments.

Figure 2. The architecture of generative adversarial networks (GANs), where generator and discriminator interact with each other with supporting elements.

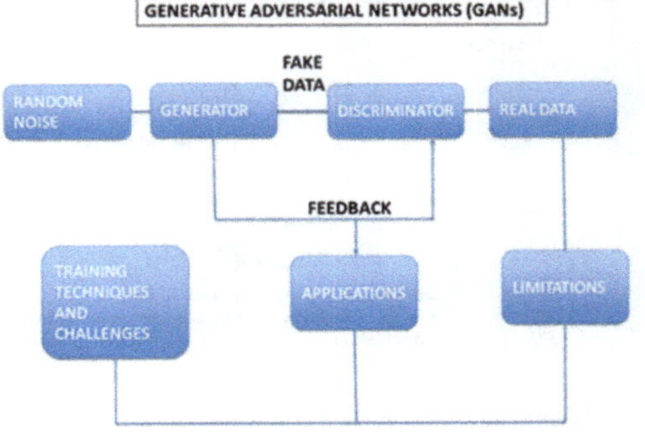

3.2 Variational Autoencoders (VAEs)

Variational Autoencoders (VAEs) form a kind of generative models which are based on the concepts of both deep learning and probabilistic inference in learning complex data distributions. The architecture of VAE is of the form of an encoder-decoder where the encoder reduces the input data to a latent space by learning a probability distribution (usually Gaussian) while the decoder retrieves the data from a set of samples from the latent space (Doersch 2016; Zhao et al., 2017). Unlike conventional autoencoders that train the encoding phase to produce fixed latent vectors, VAEs approximate the posterior distribution of latent variables using the variational inference. This entails optimization of the Evidence Lower Bound

(ELBO), which depends on two important aspects. the reconstruction loss function, which determines how well the decoder can reconstruct the input and Kullback-Leibler (KL) divergence that regularises the latent distribution to be similar to a standard normal distribution. This equilibrium makes VAEs capable of producing diverse yet significant samples from the latent space. VAEs are themselves strong in image generation, anomaly detection and representation learning because of having smooth and interpretable latent spaces (Kingma et al., 2014; Gregor et al., 2015). As compared to GANs, VAEs are, in general, more stable to train and deliver more stable outputs, but, owing to the use of probabilistic reconstruction, they output images containing more blur than the GAN outputs. Conversely, GANs produce sharper images while suffering more from instability in the training process and the lower structured latent spaces.

Figure 3. Architecture of variational autoencoders (VAEs), in which the encoder-decoder structure is illustrated

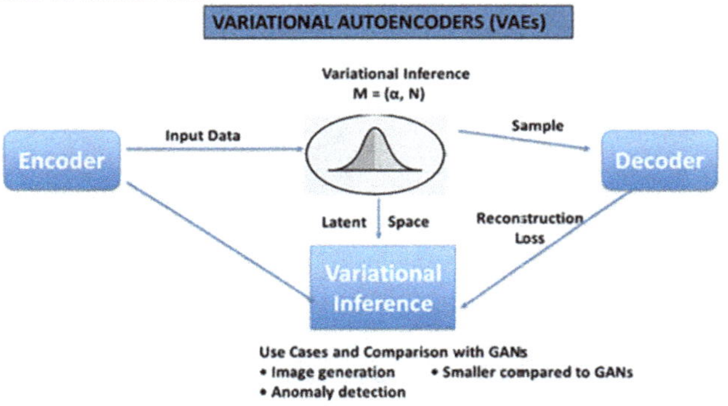

3.3 Transformer-Based Models (e.g., GPT)

Transformer-based models have emerged as the dominating architecture in modern generative AI propelled by the scalability, the flexibility, and superiority of performance. The Generative Pre-trained Transformer series (GPT) is one such example. In the centre of these models, there is self-attention mechanism that enables each token in input sequence to weigh and attend to all other tokens dynamically; thus, modeling long-range dependencies at the exponential cost without sequential restrictions of previous models such as RNNs, or LSTMs. Such an architecture allows for a high level of parallelism while training and also supports superior scalability for large datasets and large models. Transformer based models are effective for text generation where it predicts the next token in a sequence from previous context

(Radford et al., 2018; Brown et al., 2020). First, these models are trained on large corpora using unsupervised learning and then fine-tuned by such methods as supervised learning on task-specific datasets or reinforcement learning from human feedback (RLHF). In comparison with the previous generative models such as GANs and VAEs, Transformers provide a set of benefits such as broader applicability to various domains (e. g. language, code, protein structures), better coherence and fluency in generated content, and ability to perform multiple downstream applications with merely a few modifications (Long Ouyang et al. 2022). Not only did it change natural language processing, but it has also encouraged innovation in multimodal and cross-disciplinary AI research.

4. APPLICATIONS OF GENERATIVE AI

Generative AI has been on a rapid expansion on a variety of applications in various industries changing the way in which content is created, optimized, and personalized. In the imaginative field, it enables apps of making realistic images, videos, music, and even entire paintings, changing the face of digital design and amusement. Generative AI models are applied to text generation, machine translation, summarization and conversational agents like chatbot and virtual assistant in natural language processing. It has also significant contribution in healthcare including drug discovery, medical imaging synthesis and personalised treatment recommendations. In engineering and production, the generative design algorithms suggest innovative designs that provide maximum performance with the minimum cost. Additionally, in the areas of cybersecurity, finance, and education, the generative AI is utilized for data simulation, fraud detection, as well as adaptive learning systems. These apps not only improve productivity and creativity but also provide new opportunities for automation and solving different issues in any field.

4.1 Natural Language Generation

Natural Language Generation (NLG) is one of the major areas of applying generative AI related to automatic generation of human-like text from data, which is either structured or unstructured. It empowers machines with the ability to transform information into sensible, appropriate to the context narratives, summaries, explanations, or responses. Underlying core of modern NLG systems are the transformer-based architectures such as GPT that rely upon huge datasets and self-attention mechanisms in order to comprehend and produce fluent language. These systems are trained on large corpora and fine-tuned for specific tasks such as report generation, conversational dialogue, content creation, and personalized communication. NLG is widely used in applications like automated news writing, customer service chatbots, real-time translation, and assistive technologies for accessibility (Maestre et al., 2024). A major strength of NLG lies in its ability to produce high-quality, context-aware output that closely mimics human expression. However, challenges persist in ensuring factual accuracy, managing biases in training data, and controlling tone or style across diverse domains (Wani Bisen et al. 2022). As models and training methods continue to evolve, NLG is expected to play an increasingly central role in human-AI interaction and automated content generation.

4.2 Synthetic Image and Video Generation

Synthetic Image and Video Generation refers to the process of creating artificial visual content using computer algorithms, particularly through advanced techniques such as deep learning, generative adversarial networks (GANs), and diffusion models. Such technology allows producing extremely realistic pictures and videos which can show non-existing units, things, or even people which are not real. The synthetic assets can be applied in a wide variety of industries such as those related to entertainment, VR, gaming, advertisement, fashion, and even in the scientific simulations. For example, filmmakers apply synthetic video generation, to develop visual effects as well as realistic characters without depending on physical set or actors completely. In the field of AI studies, models, such as StyleGAN and DALL·E can create credible portraits or fantastical scenarios from text commands, while models for video synthesis can create fluid motion sequences based on static images or command text. Although synthetic media has the potential to be creative and commercial, ethical issues associated with it, include misinformation, deep fakes and altering of public opinion. As such, in spite of the power of synthetic image and video generation for innovation purposes, there is a need for regulating its terms and appropriate usage to strike a balance with creativity, trust and transparency.

4.3 Drug Discovery and Molecular Design

Drug Discovery and Molecular Design is an integrated field that relates to identifying new therapeutic compounds for use in the treatment of diseases. It integrates principles from chemistry, biology, pharmacology, and computer science to study diseases mechanisms at molecular level and construct molecules that can interact with particular biological targets. The process usually starts from target identification: the search for a biomolecule (usually a protein or an enzyme) that is critical to some disease. Researchers proceed to utilize such techniques as high-throughput screening and structure-based drug design, as well as molecular docking to find or design small molecules that can bind to this target and control its function. Advances in computational tools, such as artificial intelligence and machine learning, have significantly accelerated drug discovery by enabling the rapid analysis of vast chemical spaces, prediction of molecular properties, and optimization of drug candidates (Oluwafemi Shittu Bakare et al. 2024). Molecular design also includes considerations of a compound's pharmacokinetics, toxicity, and bioavailability. This approach not only shortens the development cycle but also increases the precision and effectiveness of new drugs, making it a cornerstone of modern pharmaceutical innovation.

4.4 Medical Imaging

Medical Imaging as an Application of Generative AI involves the use of advanced generative models—such as Generative Adversarial Networks (GANs), Variational Autoencoders (VAEs), and diffusion models—to enhance, synthesize, or interpret medical images like X-rays, CT scans, MRIs, and ultrasound images. Generative AI can augment medical imaging workflows by improving image quality, generating high-resolution images from low-resolution inputs (super-resolution), reducing noise, and even synthesizing missing or corrupted parts of scans. In scenarios where obtaining large datasets is challenging, synthetic images generated by AI can be used to augment training data, helping to develop more robust diagnostic models (Musalamadugu et al., 2023; Chataut et al., 2024). Moreover, generative AI enables the simulation of rare conditions, aiding in the training of medical professionals and machine learning systems. It also supports precision medicine by enabling patient-specific modeling, such as predicting disease progression through simulated future scans. Notably, it is possible to use generative models for segmentation of anatomical structures and detection of anomalies with a high accuracy. Although, these technologies provide infinite promise for enhancing diagnostic precision and efficiency, these also pose problems related to validation, data bias, and regulatory approvals, which require strict clinical testing and ethical guidelines before their implementation.

4.5 Personalized AI in Education

Using Personalized AI in Education as an Application of Generative AI involves the application of generative models to develop personalized learning experiences appropriate for an individual student's needs, preferences, and learning style. An analysis of a learner's performance, behavioral trend and progress over a period can help generative AI generate dynamic educational content such as personalized quizzes, explanations, summaries or even interactive simulations/storytelling exercises. Such AI-based tools evolve in real time, providing targeted feedback, changing the level of difficulty, and filling the gaps in knowledge in a more effective manner than was previously provided by the one-size-fits-all teaching methods (Mittal et al., 2024; Maity et al., 2024). For instance, it can reword complicated ideas to a struggling student or provide enrichment materials for a competent learner. Simulation of tutors or tutoring partners for language learning is also possible with language models, helping learn how to be fluent by generating dialogue. Also, generative AI allows to create the multilingual educational content, which increases access to quality education internationally. The outcome is a livelier, a more inclusive, and thus more effective educational environment. However, for equity and accountability, it is essential to address issues such as concern for data privacy, content bias, and tendency in over-reliance in pedagogy on automated systems.

4.6 Content Recommendation Systems

Generative AI Application- Content Recommendation Systems apply Generative AI models to improve how these digital platforms raise relevant and personalized content to its users in the domains of entertainment, e-commerce, education, and social media. Previously based on collaborative filtering and statistical approaches, the contemporary recommendation systems incorporate generative AI (e.g. language models and deep neural networks) to approach the identification of user preferences and the creation of content recommendations more dynamically and subtly. Generative AI is able to simulate user patterns, interests, and signals (such as time of the day or current trends) of sorts to create personal recommendations that make use of more than "like other users, you may like this" reasoning (Aggarwal, 2025). For example, on streaming platforms, it can provide movie or show suggestions based on a user's history of watching by creating narratives or summaries and then matching with intent of the user. In e-learning, it is capable of suggesting such resources as tutorials, articles, or exercises that suit the current progress of the student. In news apps, generative models can even generate custom headline or summary to meet reader preference. These systems learn and adapt all the times, providing ever more relevant and varied suggestions. However, ethical considerations such as filter bubbles, content diversity, and algorithmic bias must be managed carefully to ensure a balanced and fair user experience.

5. ETHICAL CONSIDERATIONS IN GENERATIVE AI

Ethical Considerations in Generative AI are crucial to ensuring the responsible and fair use of these technologies, as they can have profound impacts on society, privacy, and trust (Lasker, 2024; Nag, 2024). Below are some key ethical concerns:

- **Deepfakes and misinformation**

Generative AI models, particularly GANs (Generative Adversarial Networks), are capable of creating highly realistic fake images, videos, and audio clips—referred to as deepfakes. While these can have creative applications in entertainment and education, they pose a significant risk in the spread of misinformation and manipulation. Deepfakes can be used to create convincing but entirely fabricated content, potentially leading to harm in political, social, or personal contexts. The ease of generating such content has made it more difficult to distinguish between real and fake media, leading to concerns about public trust, privacy violations, and the undermining of democratic processes. All the negative effects need to be compensated for using the technological solutions (e.g. deepfake-detection algorithms) and legislative efforts.

- **Data privacy and model transparency**

While generative AI models need a lot of data for training and providing accurate outputs. This creates great concerns in terms of data privacy and unfair usage of personal information. Specifically, models that can be applied to the creation of synthetic media or prediction of people's behavior can accidentally disclose sensitive information. The transparency of such models is another problem – many AI models are being used as "black boxes", so that the people that are using them cannot understand how the decisions are being made. This lack of transparency may lessen the trust in AI systems, especially, in the areas where it will predominantly be used, i.e., in the sphere of healthcare, finance, or legal decisions. There has to be clear guidelines on collection of data, consenting of the users, and explainability of the model to uphold the rights of the users in ethical AI development.

- **Algorithmic fairness and responsible use**

Fairness and safety of the algorithms is one of the critical ethical concerns when it comes to the matter of generative AI. AI systems, such as generative models, are usually trained with huge datasets that can have an inherent bias. These biases can also be carried on and even intensified in the output of the model. For instance, in the case where a generative model is trained using biased data (for example gender, racial or socioeconomic bias), the model may generate outputs that support stereotypes or class-based discrimination. The responsible use of generative AI will require to put checks regarding fairness during the training of the models, inclusion of varied sets of data and continuous monitoring to ensure that the models do not spread destructive biases or inequalities.

- **Societal implications and regulatory frameworks**

With the generative AI technology continuously advancing, it is important to think about its larger consequences for society. For example, there are fears of job displacement in sectors that may potentially see creative work to be replaced by generative models, such as content creation, design, or even journalism. Moreover, democratization of AI tools implies that the AI tools with the positive and with the negative outcomes are available for a large number of people. Therefore, there is need to develop regulatory frameworks to regulate responsible development and deployment of generative AI. These frameworks should address issues like accountability, transparency, the prevention of misuse (e.g., in the case of deepfakes), and the protection of public interest. Laws and regulations must evolve to ensure that generative AI technologies are used ethically, safely, and in ways that benefit society as a whole, without compromising privacy or fairness.

6. EMERGING TRENDS AND ADVANCED TECHNIQUES

Emerging Trends and Advanced Techniques in Generative AI are shaping the future of artificial intelligence by introducing more powerful, flexible, and context-aware models capable of generating high-quality and multimodal outputs (Yadav et al., 2024; Batchu et al., 2024). Below is a detailed exploration of the most impactful innovations:

- **Diffusion models and their advantages**

Diffusion models represent a significant leap forward in generative modeling. These models work by learning to reverse a gradual process of noise addition to data, effectively learning to denoise data back into a coherent image, audio, or text. Compared to traditional GANs (Generative Adversarial Networks), diffusion models offer greater training stability, higher output fidelity, and better diversity in generated samples. Prominent models like DALL·E 3, Stable Diffusion, and Imagen have demonstrated the ability to generate incredibly detailed and realistic images from text prompts. Their structured and iterative generation process also makes them more interpretable and tunable, allowing for finer control over the final output. These advantages are positioning diffusion models as the new state-of-the-art in many generative tasks, especially in visual and creative content creation.

- **Multimodal generative systems (text-image, text-audio, etc.)**

Another transformative trend is the development of multimodal generative AI, which integrates multiple types of data—such as text, images, audio, and video—into unified models. These systems are capable of generating one modality from another, such as:

Text-to-image (e.g., DALL·E, Midjourney),
Text-to-audio (e.g., AudioLM, MusicLM),
Image-to-text (e.g., captioning tools),
Text-to-video (e.g., Sora by OpenAI).

By combining different types of sensory inputs and outputs, multimodal models allow for richer, more intuitive AI systems. For example, a model can generate a short educational video with narration and animations based on a simple text description. These systems are opening up new possibilities in education, accessibility (e.g., generating descriptions for visually impaired users), gaming, creative arts, and human-computer interaction.

- **Recent research directions and future scope**

Recent Research Directions and Future Scope of Generative AI reflect a dynamic and rapidly evolving landscape, where the focus is shifting from generating high-quality content to building more intelligent, controllable, and ethically aligned systems. One of the most significant trends is the development of foundation models—large, pre-trained models that can be fine-tuned for a variety of generative tasks across modalities such as text, image, audio, and video. These models, like GPT-4, DALL·E 3, and Gemini, represent a move toward general-purpose AI, capable of understanding and generating content in complex, real-world scenarios. Alongside this, researchers are exploring multimodal architectures that can simultaneously process and generate different types of data, enabling applications like AI assistants that can see, speak, and understand context across multiple channels.

Another primary line of research is fine-grained control of generation, such that users or programmers can guide the result of generative models with prompts, constraints, or reinforcement learning. This is important in real world of professional applications such as those in drug design, architecture and education where precision and domain relevance are important as creativity. Further still, real-time and interactive creation is emerging as more concentrated area, making it possible for AI systems to provide answers to user input immediately, meaning that these are used in live storytelling, virtual reality, gaming, and shared designing scenarios. As with any systems, as they develop in complexity and outreach, so there is growing tendency to the energy-efficient and lightweight models. Study is being applied in enhancing the efficiency of models in better training algorithms, pruning techniques and use of edge computing in a way that generative AI can be utilized at scale without prohibitive costs or environmental impact. In addition, neuro-symbolic AI – which involves the combination of neural networks with logic-based reasoning – is picking up momentum to add explainability and trustworthiness for generative outputs where it matters most, such as healthcare, finances, and legal services. Diving into the future, the future horizon of generative AI is wide. It consists of development of digital humans and AI companions, hyper-personalized content, AI-advanced scientific discovery and next-generation education platforms. However, all this is hinged on simultaneous development in ethical AI research, such as frameworks for bias mitigation, factual consistency, protection of intellectual property, and regulatory compliance. While generative AI will only continue in its role of obscuring the lines between human and machine creativity, its responsible maturation will be what makes it a true thing in society.

7. HUMAN-AI CO-CREATION AND CREATIVITY

Human-AI co-creation means a creative partnership between human and artificial intelligence that generates new things. Unlike automation which often results in replacing human labor, co creation is about augmenting human capability. The focus is on usage of AI in interactive, iterative and dynamic workflows using AI as a creative partner or a collaborator, not as the final decision maker.

In this AI relationship to humans, we often bring context, emotion, intent and values and the AI provides speed, memory, pattern recognition and generation of variations. It is this complementary balance that permits both scalable and expressive co creative processes. Within co creation, there are several interaction paradigms. Generative prompting involves a human input (text prompt, sketch, style, emotion) as the input and options returned by the AI. A loop where the human and AI pass back end forth in attempting to modify, select or refine each other's content. Human sets the boundary or set of goals and AI goes out looking to find solution in that space. Generative AI is natively woven into design tools, writing assistants, game dev engines and music composition platforms and these frameworks already exist.

Visual arts has been one of the earliest and most visibily transformed fields of Artificial Intelligence assisted creativity. The complex, generative models StyleGAN, DALL·E, Midjourney and Stable Diffusion have helped artists, illustrators and designers explore a new aesthetic, create concept art or mimic the style of past masters. As a concept artist working either in game or film, time is at a premium. Instead of spending the hours necessary to create dozens of environments by hand, an artist can simply prompt an AI with "cyberpunk city at dusk with flying cars" and the AI will provide some visual interpretation. These results can be refined further or used to serve as starting points for traditional work, facilitating faster prototyping and more rich exploration by the artist.

Not least of all, artists view AI not as replacement, but as a new brush, as a new kind of medium, as charcoal or Photoshop. The human determines your theme, emotion and message. Or structural composition, texture ideas or possible lighting sourced through AI, that can be curated, layered and edited. This is a co creative process of hybrid workflows which generate parts of the image with artificial intelligence, parts hand drawn and parts digitally painted. Therefore, AI is a creative amplifier that empowers non artists to create and professional artists to go further.

One of the most mature domain of generative AI is text generation. The advent of GPT 4, Claude, Gemini and other models of such great skill now allow us write essays, dialogue, poetry and entire novels. However, it is they really shine when they're used for collaborative writing. ReWrite stories in other tones, give suggestions for character traits or for plot ideas. For instance, you might give a writer a prompt such as "describe a battle scene in the style of Homer," in which the words they produce will inspire their own poetry that they will decide is either worth adaptation or not. It provides us with an ever available co writer to get us out of the sludge of creative inertia.

8. CONCLUSION

Generative artificial intelligent is a vast field that has a probabilistic foundation and observes different real-world applications and ethical fronts. This chapter has, therefore, presented a holistic delve into this field. By differentiating generative models from discriminative counterparts, the discourse clarified how generative procedures model underlying distribution of data to generate novel and high fidelity content. The key architecture such as Generative Adversarial Networks (GANs), Variational Autoencoders (VAEs), and Transformer-based models like GPT were examined as regards its design philosophy, strength and weakness respectively. The idea of generative AI was given practicality of representation across such fields as natural language processing, synthetic media creation, drug discovery, medical imaging, and personalized education. These use cases highlight the interdisciplinary utility of the generative AI that has the ability to not only automate content but also improve decision-making, creativity, and accessibility. Trends that are emerging – such as diffusion models, multimodal systems, or foundation models – are pointing towards a future where generative AI is becoming more generalizable, controllable and interactive. At the same time, the spread of these technologies calls for the ethical considerations of deepfakes, privacy, bias, and algorithmic transparency. In summary, the advancement in generative AI denotes technology revolution and societal inflection point. As the capabilities evolve, so should the responsible innovation – the commitment to making the generative models developed and applied with fairness, accountability and human-centric values. This chapter therefore intends to equip learners and practitioners for critical engagement and for thoughtful application of generative AI in research, industry and more generally.

REFERENCES

Aggarwal, A. (2025). Evolution of recommendation systems in the age of Generative AI. *International Journal of Science and Research Archive*, *14*(1), 485–492. DOI: 10.30574/ijsra.2025.14.1.0061

Batchu, C., & Satya, V. V. (2024). *Generative AI: Evolution and its Future*. International Journal For Multidisciplinary Research., DOI: 10.36948/ijfmr.2024.v06i01.12046

Bisen, W. H., & Agrawal, A. J. (2022). review on natural language generation. *International Journal of Health Sciences*, 10365–10376. Advance online publication. DOI: 10.53730/ijhs.v6nS1.7489

Bishop, C. M. (2006). *Pattern recognition and machine learning*. Springer.

Blei, D. M., Kucukelbir, A., & McAuliffe, J. D. (2017). Variational inference: A review for statisticians. *Journal of the American Statistical Association*, *112*(518), 859–877. DOI: 10.1080/01621459.2017.1285773

Bommasani, R., Hudson, D. A., Adeli, E., Altman, R., Arora, S., von Arx, S., . . . Liang, P. (2021). On the opportunities and risks of foundation models. arXiv preprint arXiv:2108.07258. /arXiv.2108.07258DOI: 10.48550

Brown, T. B., Mann, B., Ryder, N., Subbiah, M., Kaplan, J., Dhariwal, P., . . . Amodei, D. (2020). Language models are few-shot learners. arXiv preprint arXiv:2005.14165. /arXiv.2005.14165DOI: 10.48550

Chataut, S., Bhatta, S., Dahal, B., Ojha, G., Subedi, B., & Bastakoti, B. (2024). Advancements and Applications of Generative AI in Healthcare. *European Journal of Theoretical and Applied Sciences*, *2*(6), 873–895. DOI: 10.59324/ejtas.2024.2(6).77

Doersch, C. (2016). Tutorial on variational autoencoders. arXiv preprint arXiv:1606.05908

Floridi, L., & Cowls, J. (2019). A unified framework of five principles for AI in society. *Harvard Data Science Review*, *1*(1). Advance online publication. DOI: 10.1162/99608f92.8cd550d1

Goodfellow, I., Bengio, Y., & Courville, A. (2016). *Deep learning*. MIT Press.

Goodfellow, I., Pouget-Abadie, J., Mirza, M., Xu, B., Warde-Farley, D., Ozair, S., . . . Bengio, Y. (2014). Generative adversarial nets. In *Advances in Neural Information Processing Systems* (pp. 2672–2680). https://proceedings.neurips.cc/paper_files/paper/2014/file/5ca3e9b122f61f8f06494c97b1afccf3-Paper.pdf

Goodfellow, I. J., Pouget-Abadie, J., Mirza, M., Xu, B., Warde-Farley, D., Ozair, S., & Bengio, Y. (2014). Generative adversarial nets. *Advances in Neural Information Processing Systems*, 27.

Gregor, K., Danihelka, I., Graves, A., Rezende, D. J., & Wierstra, D. (2015). DRAW: A recurrent neural network for image generation. In *Proceedings of the 32nd International Conference on Machine Learning* (pp. 1462–1471). https://proceedings.mlr.press/v37/gregor15.html

Jumper, J., Evans, R., Pritzel, A., Green, T., Figurnov, M., Ronneberger, O., Tunyasuvunakool, K., Bates, R., Žídek, A., Potapenko, A., Bridgland, A., Meyer, C., Kohl, S. A. A., Ballard, A. J., Cowie, A., Romera-Paredes, B., Nikolov, S., Jain, R., Adler, J., & Hassabis, D. (2021). Highly accurate protein structure prediction with AlphaFold. *Nature*, *596*(7873), 583–589. DOI: 10.1038/s41586-021-03819-2 PMID: 34265844

Kingma, D. P., & Welling, M. (2014). Auto-encoding variational Bayes. arXiv preprint arXiv:1312.6114. /arXiv.1312.6114DOI: 10.48550

Lasker, A. (2024). Exploring ethical considerations in generative ai. *International Journal of Advanced Research*, *12*(4), 531–535. Advance online publication. DOI: 10.21474/IJAR01/18578

Maestre, M., Martínez-Murillo, I., Martin, T., Navarro-Colorado, B., Ferrández, A., Suárez Cueto, A., & Lloret, E. (2024). Beyond Generative Artificial Intelligence: Roadmap for Natural Language Generation. /arxiv.2407.10554DOI: 10.48550

Maity, S., & Deroy, A. (2024). Generative AI and Its Impact on Personalized Intelligent Tutoring Systems. /arxiv.2410.10650DOI: 10.48550

Mittal, U., Sai, S., Chamola, V., & Sangwan, D. (2024). A Comprehensive Review on Generative AI for Education. *IEEE Access : Practical Innovations, Open Solutions*, *1*, 142733–142759. Advance online publication. DOI: 10.1109/ACCESS.2024.3468368

Miyato, T., Kataoka, T., Koyama, M., & Yoshida, Y. (2018). Spectral normalization for generative adversarial networks. In *International Conference on Learning Representations (ICLR).* https://openreview.net/forum?id=B1QRgziT-

Murphy, K. P. (2012). *Machine learning: A probabilistic perspective.* MIT Press.

Musalamadugu, T. S., & Hemachandran, K. (2023). *Generative AI for medical imaging analysis and applications*. Future Medicine AI. DOI: 10.2217/fmai-2023-0004

Nag, B. (2024). Navigating Ethical Dilemmas in Generative AI. *Advances in Computational Intelligence and Robotics Book Series*, 189–222. DOI: 10.4018/979-8-3693-9173-0.ch007

Ng, A. Y., & Jordan, M. I. (2002). On discriminative vs. generative classifiers: A comparison of logistic regression and naive Bayes. In *Advances in Neural Information Processing Systems* (pp. 841–848). https://proceedings.neurips.cc/paper/2001/file/1f71e393b3809197ed63b7091b20f0e4-Paper.pdf

Papamakarios, G., Nalisnick, E., Rezende, D. J., Mohamed, S., & Lakshminarayanan, B. (2019). Normalizing flows for probabilistic modeling and inference. arXiv preprint arXiv:1912.02762. /arXiv.1912.02762DOI: 10.48550

Radford, A., Metz, L., & Chintala, S. (2016). Unsupervised representation learning with deep convolutional generative adversarial networks. arXiv preprint arXiv:1511.06434

Radford, A., Narasimhan, K., Salimans, T., & Sutskever, I. (2018). *Improving language understanding by generative pre-training*. OpenAI. https://cdn.openai.com/research-covers/language-unsupervised/language_understanding_paper.pdf

Ramesh, A., Pavlov, M., Goh, G., Gray, S., Voss, C., Radford, A., . . . Sutskever, I. (2021). Zero-shot text-to-image generation. arXiv preprint arXiv:2102.12092. /arXiv.2102.12092DOI: 10.48550

Salimans, T., Goodfellow, I., Zaremba, W., Cheung, V., Radford, A., & Chen, X. (2016). Improved techniques for training GANs. In *Advances in Neural Information Processing Systems* (pp. 2234–2242). https://proceedings.neurips.cc/paper_files/paper/2016/file/8a3363abe7921c41837d905f2b29721f-Paper.pdf

Song, Y., & Ermon, S. (2019). Generative modeling by estimating gradients of the data distribution. arXiv preprint arXiv:1907.05600. /arXiv.1907.05600DOI: 10.48550

Vaswani, A., Shazeer, N., Parmar, N., Uszkoreit, J., Jones, L., Gomez, A. N., . . . Polosukhin, I. (2017). Attention is all you need. In *Advances in Neural Information Processing Systems* (pp. 5998–6008). https://proceedings.neurips.cc/paper_files/paper/2017/file/3f5ee243547dee91fbd053c1c4a845aa-Paper.pdf

Yadav, P., Rathwad, G., & Jain, J. (2024). *Generative AI: Shaping the Future While Disrupting the Present*. International Journal For Multidisciplinary Research. DOI: 10.36948/ijfmr.2024.v06i05.28085

Zhang, D., Malkin, N., Liu, Z., Volokhova, A., Courville, A., & Bengio, Y. (2022, June). Generative flow networks for discrete probabilistic modeling. In *International Conference on Machine Learning* (pp. 26412-26428). PMLR.

Zhao, S., Song, J., & Ermon, S. (2017). *Towards deeper understanding of variational autoencoding models*. arXiv preprint arXiv:1702.08658

Chapter 9
Agriculture 6.0:
Advancing Intelligent and Sustainable Agriculture Through the Convergence of IoT and Machine Learning

Madiha Munawar
https://orcid.org/0009-0001-7211-9022
Stanley College of Engineering and Technology for Women, India

T. Monika Singh
https://orcid.org/0009-0006-8534-9130
Annamalai University, India

Shugufta Fatima
https://orcid.org/0009-0005-4396-4947
Annamalai University, India

R. Madana Mohana
https://orcid.org/0000-0002-7652-6266
Stanley College of Engineering and Technology for Women, India

Anindya Nag
https://orcid.org/0000-0001-6518-8233
Northern University of Business and Technology, Khulna, Bangladesh

ABSTRACT

This chapter explores the convergence of IoT, Machine Learning (ML), and Generative AI as foundational pillars of Agriculture 6.0. It highlights how IoT enables real-time data collection from agricultural environments, while ML analyzes and optimizes

DOI: 10.4018/979-8-3373-4207-8.ch009

this data for intelligent, autonomous decision-making. Generative AI further enhances capabilities such as crop modeling, yield prediction, and synthetic data generation for precision farming. This integration supports sustainable, efficient, and resilient agricultural systems. The chapter also extends the discussion to broader applications in smart cities, healthcare, environmental monitoring, and industrial automation. Innovations like smart grids, predictive maintenance, and traffic optimization illustrate their wider societal impact. Challenges related to data privacy, ethics, and regulatory needs are addressed, along with future trends such as 5G, edge computing, and federated learning as critical enablers of a more connected and intelligent world.

1. INTRODUCTION

1.1 The Role of IoT and Machine Learning in Smart and Sustainable Development

The IoT and ML synergism offer societies an unprecedented paradigm toward sustainable development. The Internet of Things provides connectivity for devices, sensors, and systems for real-time data collection and communication, whereas machine learning acts on this data to provide insights, make decisions, and optimize processes. Both technologies thus permit optimum resource usage, minimal waste, and intelligent solutions for sustainability(Jaber, 2023). Whether in energy management, health care, or urban planning, IoT and ML drive innovation toward a smarter tomorrow.Energy efficiency is one of the major areas in which IoT and ML contribute to sustainability. In smart grids, IoT sensors monitor energy consumption in real-time, while ML algorithms predict patterns of demand and optimize energy distribution. This reduces energy waste, allowing for better integration of renewable energy sources into the power grid, thereby promoting a sustainable and resilient power infrastructure(Revathi et al., 2024).

In the environmental space, IoT-based sensors are tracing air and water quality, deforestation, and climate changes, over time obtaining valuable data that will act as a lever for conservation(Hiran et al., 2021). Pollution trends become decipherable as these data are analyzed with ML models, while environmental hazards will be forecasted and landfilling processes optimized. Another important area is smart agriculture, where an IoT-based monitoring system is used to determine soil health, automate irrigation, and improve crop yield, which helps secure food and mitigate environmental threats.

Figure 1. IoT and ML architecture for smart connectivity

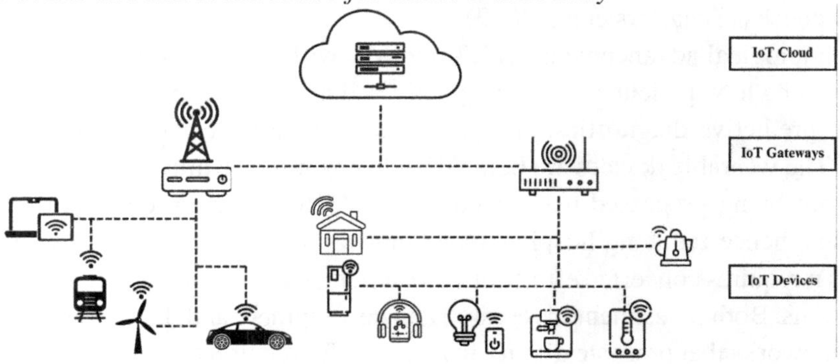

Figure 1 represents the architecture of the Internet of Things (IoT) system and Machine Learning (ML). It also highlights the interaction among IoT devices, gateways, and the cloud. At the base of the IoT architecture are various IoT devices-smart home appliances, connected vehicles, wearables, and industrial sensors-which communicate through wireless networks. Such a device will transmit data to the IoT gateway which will act as an intermediary to process and filter the data before transferring it into the cloud. The IoT clouds would serve as the central hubs for the data storage, analytics, and decision-making purposes. This architecture can provide a very clear communication between all smart devices and can enhance automation, remote monitoring, and real-time data processing of various applications such as smart homes, transportation, healthcare, and energy management.

1.2 Importance of Innovation for a Smarter Future

Innovative capability is the key to building smarter, more effective, and more sustainable systems. As the population expands and urbanization multiplies at an unprecedented rate, the demand for intelligent solutions continues to develop, considering the most important aspects of resource allocation efficiency, productivity maximization, and environmental degradation minimization(Von Stamm&Trifilova, 2009). The latest technologies such as the Internet of Things (IoT), machine learning (ML), artificial intelligence (AI), blockchain, and even 5G technologies are bound to change the way industries run, enhance public services, and thereby contribute to the making of smart cities and sustainable infrastructure.One very important facet of innovation for a smarter future is the automation and digitalization of industries. Smart factories depend on IoT-instrumentation sensors and AI-driven robotics to optimize production lines, reduce defective products in their production, and increase quality checks (Schaffers et al., 2012).

Technological advancements in IoT and ML within the healthcare industry are transforming how patient care is being delivered in enabling real-time health monitoring, predictive diagnostics, and personalized treatment plans(Han &Hawken, 2018). The wearable devices are being used to capture real-time physiological data which are being processed using algorithmic AI for anomaly detection and dire warnings, hence reducing hospital stays. Innovation, however, carries significant risks as it requires concerted efforts of government, business, and scientific research institutions. Both investment in research and development and the formation of policy frameworks that promote dignity in AI as the foundation for making technology work for the larger betterment are extremely important.

1.3 Challenges in Achieving Sustainability Through Technology

Even though IoT and ML have the potential to reshape the world, it is still very challenging to make technology sustainable. Foremost questions popped in the minds of everyone and especially in the engineering field concerning the privacy and security of data(Bambam&Gajrani, 2023). The connecting point between IoT devices is the fact that it has extensive personal and operational data flowing in different channels and stored continuously. Hence, their security against cyber threats is core since they expose sensitive information and vulnerability in critical infrastructures as a result of breach attacks(Jayachandran et al., 2022). Therefore, possible solutions to mitigating these risks include putting up strong encryption methods, developing an AI-powered cybersecurity solution, and formulating robust data governance policies.

Additional challenges are large-scale energy consumption related to IoT deployments and AI processing. Although they conserve energy through their optimal consumption, these advanced technologies also demand a lot of power for computation, especially during model training. Specific examples, except for oil companies and automobiles, are data centers that consume most of the electricity for all other AI systems. Developing energy-efficient algorithms, using edge computing for processing instead of relying on centralized cloud services, and using renewable energy sources in AI infrastructure would go a long way toward ensuring that such advancements do not worsen the environmental dilemma(Izatt et al., 2014).Developing countries often don't have either the financial or technological infrastructure to deploy IoT and AI solutions on a grand scale. This can only be overcome through global cooperation, investment in smart infrastructure, and knowledge-sharing initiatives. If not, excluded underprivileged regions will only see the burdens of sustainability-focused innovations left in the remit of advanced economies(Abdoh, 2024). The development and deployment of these technologies must also, in the last instance, address the ethical and regulatory issues for responsible development. Biases created in AI approaches, transparency in automated solutions, and the societal implications of workforce automation are some problems that need urgent considerations.

2. FUNDAMENTALS OF IOT AND MACHINE LEARNING

2.1 Overview of Internet of Things (IoT)

The Internet of Things (IoT) can be defined as a huge network of interlacing devices, software, and sensors for real-time data collection and exchange(Abdul-Qawy et al., 2015). The devices include appliances from smart homes and industries, and they automate, enhance efficiencies, and provide decisions through remote monitoring and control. IoT systems embody wireless communications, cloud computing, and edge processing managing seamless transmission of data(Gupta &Quamara, 2020). With the increasing demand for IoT, healthcare and agriculture sectors, as well as cities with smart facilities, experience and profit from optimal management of resources, improved operational efficiency, and foresight enabled through predictive analytical.

Real-time monitoring and predictive maintenance are two major benefits that the use of IoT can bring about. The presence of sensors embedded in equipment makes it possible to report to the maintenance department potential defects in machines before occurrence(SwathiSree et al., 2025). Such facilities available in smart cities daily collect information, such as data on energy use, movement of traffic, and other waste management systems, for optimized use of resources(Dudhe et al., 2017). Precision farming that uses IoT provides information to farmers, helping them learn about the moisture content of soil, weather conditions, and crop health. This is a potential agriculture sustainer and booster to yields(Villamil et al., 2020).The Internet of Things faces its own array of challenges, along with its advantages: these challenges include data security, interoperability, and energy consumption(Rangarajan et al., 2025). Since IoT devices continuously exchange sensitive information, protecting it as well as immunity from threats becomes paramount. There is also the hindrance that the presence of different communication protocols does not develop a seamless integration into the different IoT ecosystems. These would include strong cybersecurity mechanisms, regulatory framework enhancement, and development of energy-efficient architectures for IoT(Albreem et al., 2017). While the technology progresses, overcoming these challenges becomes vital for the role IoT will play in driving smart and sustainable development.

2.2 Machine Learning: Key Concepts and Techniques

Machine Learning (ML) is a subset of artificial intelligence that allows computers to learn from data and gradually improve their performance without being explicitly programmed. It uses statistical techniques and algorithms for discovering patterns, predicting behaviors, and optimizing decision-making(Zaki et al., 2020). In general, ML techniques can be classified into these three primary categories: supervised learning in which models learn from labeled data, unsupervised learning which finds hidden structure in unlabeled data, and reinforcement learning where agents learn how to operate in an environment to maximize their rewards. These techniques are being actively applied in various domains, such as healthcare, finance, and autonomous systems.

The ML industry has enhanced automation and intelligence across sectors. ML-based diagnostic engines use medical imaging plus patient records to detect diseases early and suggest personalized treatment options(Vieira et al., 2020). Fraud detection models in financial services spot suspicious transactions by analyzing patterns of user behavior. In cities, ML-based traffic management systems work to optimize signal timings, predict congestion, and improve urban mobility. Therefore, ML algorithms perform better when working with really large datasets to produce data-driven decisions, enhancing efficiency and accuracy.

The issues surrounding ML, however, include bias in training data, interpretability, and computational burden(Jung, 2022). Unfair outcomes may arise from the training data being biased toward certain groups or against others, particularly in critical applications including hiring, lending, and criminal justice. Besides, a lot of ML models are said to work as "black boxes", which generates difficulty in understanding the decisions made. A computationally heavy model would raise concerns over energy consumption in the first place, as many complex models entail their training through heavy processing power.

Machine Learning concepts and techniques are presented in this Table 1, categorized according to type of learning, algorithms used, and areas of application.

Table 1. Machine learning key concepts and techniques

Concept	Type of Learning	Algorithms	Application Area
Supervised Learning	Supervised	Decision Trees, SVM, Linear Regression	Classification, Prediction
Unsupervised Learning	Unsupervised	K-means Clustering, PCA	Pattern Recognition, Segmentation
Reinforcement Learning	Reinforcement	Q-Learning, Deep Q Networks	Robotics, Game AI

continued on following page

Table 1. Continued

Concept	Type of Learning	Algorithms	Application Area
Deep Learning	Supervised/ Unsupervised	Neural Networks, CNN, RNN	Image Recognition, NLP
Semi-supervised Learning	Hybrid	Self-training, Co- training	Data Labeling, Anomaly Detection
Transfer Learning	Transfer	Pre-trained Networks	Medical Imaging, Speech Recognition

2.3 Synergies Between IoT and Machine Learning

Owing to the architecture of the Internet of Things combined with the revolutionary aspects of Machine Learning, automation has become a more intelligent and smarter decision-making machine- industry-wise. Real time data generated by interconnected devices through their sensing elements goes a mile toward enhancing productivity, effectiveness, efficiency, speed, resource consumption, and cost by intelligent processing, analyzing, and drawing out of meaningful insight from that data by ML techniques(Shinde et al., 2023).

Smart cities leverage the combined power of IoT and ML for traffic management, urban planning, and environment monitoring. For instance, IoT-enabled sensors would gather air quality information, and pollution levels would be analyzed through ML models to alert the inhabitants when some action would be most effective(Mendhurwar& Mishra, 2018). Wise transportation systems similarly employ IoT to track the movement of vehicles while using ML algorithms to predict congestion flow patterns and eliminate them. These have thus become a part of all improved efficiencies and have also contributed towards the construction of sustainable energy-efficient models that emit less.

There are several inherent challenges attached to IoT and ML integration; they include data scalability, security risks, and computational constraints. IoT networks have been found to produce data in enormous amounts thereby creating a need for effective storage processing and analytics solutions. Data security becomes more essential as the higher the interconnection among ecosystems, the greater inconvenience and grievous consequences that may probably follow from cyberattacks to IoT(Singh, T. M et al., 2024). Real-time processing requires high computing of advanced intelligence; thus, edge computing turns out to be the best solution. As IoT and ML develop further, these challenges will certainly have to be addressed before realizing the full potential of smart and sustainable development.

3. SMART CITIES AND INFRASTRUCTURE

3.1 IoT-Enabled Smart Cities: Applications and Benefits

The smart city, enabled through IoT, enhances the urban lifestyle with connected sensor applications, real-time data handling, and intelligent automation. Through the integration of IoT technologies into public infrastructure, the governments can boost the efficiency of energy use, waste management, transport, and public safety(Bellini et al., 2022). Smart grids, for instance, monitor electricity usage and distribute it effectively to optimize overall consumption. In parallel, an IoT-based waste management system has sensors installed to recognize the fill levels in the bins for optimizing collection routes, thereby improving efficiency and minimizing environmental impact.

Figure 2. IoT enabled smart city ecosystem

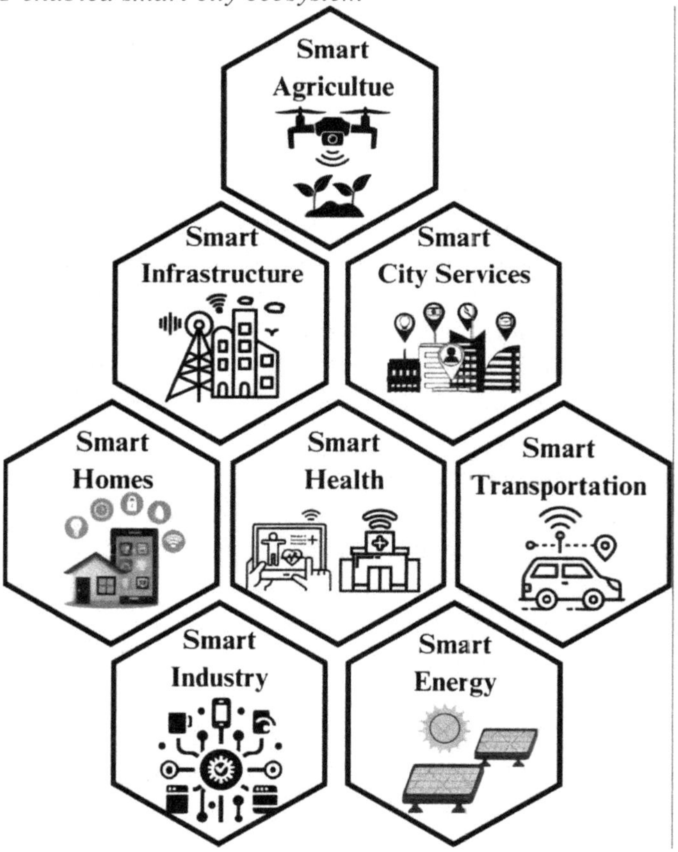

Figure 2 shows the important aspects of the smart city ecosystem which connects all other major sectors that use high technology to optimize the performance and sustainability of the ecosystem. Eight major domains consist of Smart Agriculture, Smart Infrastructure, Smart City Services, Smart Homes, Smart Health, Smart Transportation, Smart Industry, and Smart Energy. From these areas, the new technologies, trends, and processes are brought into urban living. Smart Agriculture entails precision farming, Smart Infrastructure and City Services promote urban mobility and governance, whereas Smart Homes employ IoT-enabled device automation, and Smart Health throws e-health solutions into the fray for present-day treatment protocols. Smart Transportation refers to intelligent mobility systems, while Smart Industry brings automation into the manufacturing sphere. Smart Energy then gives way to renewables integrated into sustainable energy management.

Another major found benefit of IoT-enabled smart cities is providing better urban mobility. Key components being intelligent traffic lights, parking sensors, and real-time public transport tracking systems. All of these effectively help in keeping traffic flowing, while decreasing traffic congestion(Belli et al., 2020). The live traffic data are made available to the citizens via mobile applications, which gives them the choice of taking optimal routes and thereby minimizing delays. Also, smart city projects install connected surveillance systems improving security, using AI-enabled analytics to recognize anomalies and serve to decrease emergency response times. Yet, building up the IoT for smart cities also creates challenges, including expensive infrastructure costs, privacy of data, and interoperability(Gupta & Singh, 2023). Heavy investments are implied for deploying on a large scale, while security in the transmission of data amongst various IoT networks still remains a worry. The compatibility issues caused by integrating systems from different vendors might be another factor of consideration. All of these challenges will require an approach to IoT development of standardization, a level of strong cybersecurity, and government policy to attain digital transformation without limits.

3.2 Machine Learning for Traffic Management and Optimization

Machine learning is indispensable in improving traffic management by analyzing large data sets and optimizing real-time decision-making. Traffic flow data from sensors, cameras, and GPS devices are fed through ML algorithms for predicting the congestion pattern and adjusting the signal timings accordingly. Incorporating machine learning models from historical and real-time traffic data create a smarter city that reduces travel times. ML-powered adaptive traffic control systems avoid stop-and-go traffic whenever possible to save on fuel usage and reduce emissions(Patil, 2022).Furthermore, in terms of predictive analysis, ML supports route optimization for various types of public and private modes of transportation. ML, for instance, is instrumental in accurately forecasting demand, adjusting pricing accordingly, and optimally allocating drivers for ride-sharing services. Public-transit providers monitor commuter behavior to change schedules based on service efficiencies. In such approaches, data-based interventions aim to enhance transport efficiency, save fuel, and boost passenger satisfaction.

The successful deployment of ML for traffic optimization will face several challenges included unreliable data, integration complexity, or concerns like ethics. Traffic data can be sourced by more than one site, where inaccuracies can lead to unsound predictions(Almukhalfi et al., 2024). Furthermore, integration ML models into existing infrastructure will necessitate significant investment and expertise. Similarly, AI decision-making concerning traffic systems will need understanding of the veins of ethics to decide without discrimination for optimization benefit to any of traffic users. It is absolutely necessary that these difficulties are addressed if the successful deployment of ML is to be achieved in traffic management.

3.3 Energy-Efficient Urban Planning and Resource Management

Energy-efficient urban planning is important for sustainable development; the IoT and ML converge much in optimizing the management of resources. Smart grids use IoT sensors to capture electricity demand and adapt supply dynamically, thereby eliminating the unnecessary waste of energy. These grids also facilitate real-time communication between consumers and utility providers who promote efficient consumption of energy via demand-response programs. Automated building services that are compatible with smart meters handle heating, ventilation, and lighting mostly based on occupancy and time, which saves a lot of energy costs.

IoT-based optimization has its application in water and waste management systems. Inside those smart water meters furnished with IoT, humans can tell when and how the water is being used; where leaks exist; and in real time, the customers can monitor and implement conservation effectively, keeping in mind usage patterns. Similar to these, AI intelligence leads to improved analytics of detecting the peak point disposal within waste collection so that it can properly nor streamlined collection scheduling. These technologies can help avoid wastage of resources, reduce the negative impact on the environment, and enhance the efficiencies of service delivery to the public.There are many factors that come into play regarding IoT and ML, such as limitations in data security, scaling, cost, etc. Large investments are expected to be needed for smart infrastructure to be sustainably taken in large numbers. Cybercrime risks with respect to the inter-connected systems have to be mitigated as well. Smart city solutions are to be available on the lines of different socio-economic groups to avoid the potentiality of marginalizing totally by digital means. Governments can therefore be urged to set in partnership the collaborative working between themselves and private sector actors in drawing decent policies and frameworks to push forward on driving closure in smart city-making.

4. SUSTAINABLE ENERGY AND ENVIRONMENTAL MONITORING

4.1 Smart Grids and Renewable Energy Integration

Smart grids in modern energy distribution certainly play a very important role in integrating renewable energy sources and improving energy efficiency. The smart grid employs IoT-enabled sensors and AI analytics in the monitoring of energy consumption, prediction of the fluctuations in demand, and regulation of the power supply. Bidirectional energy flow is facilitated: unlike the traditional power grid, the smart grid allows consumers to send hazardous levels of solar or wind energy back to the system. The decentralized energy distribution mechanism optimizes energy availability, reduces pressure on fossil fuel supplies, and thus, strengthens aggregation.

Machine learning further optimizes smart grid operations by forecasting energy demand and thereby optimizing power generation(Speer et al., 2015). Predictive models handle dynamic energy distribution with minimal wastage by exploiting historical and real-time data. Additionally, smart grids have automated demand-response schemes whereby electricity prices would vary depending on usage data to induce consumers to become flexible in their energy consumption, shifting it to non-peak hours. This aids in energy conservation, resulting in lower carbon footprints and promoting sustainability(Kataray et al., 2023).However, even with application, renewable energy generation with smart grid integration brings issues of energy storage limitation and grid stability. Solar and wind resources are intermittent, thus requiring efficient storage solutions. Besides, ensuring grid stability in the wake of decentralized power generation demands enhanced controls. Accelerating smart grids for sustainable energy management will require both governmental and industrial investments in battery storage technologies, regulatory frameworks, and AI-based optimization strategies.

Figure 3. Virtual power plant (VPP) integrating traditional and renewable energy sources

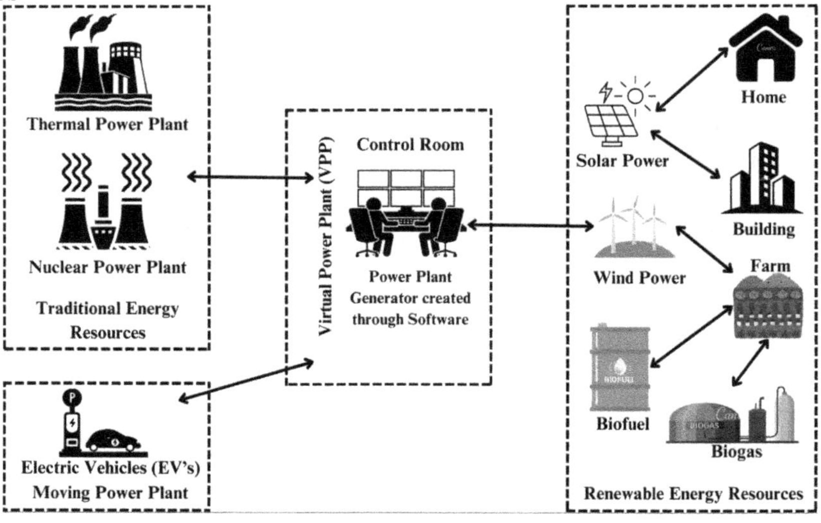

The Concept of Virtual Power Plant is depicted in Figure 3. This refers to integrating energy sources from both conventional and renewable resources along with mobile power plants into a VPP for distributional efficiency. The VPP control room can be defined as an automated central management in terms of software hardware used for generation and distribution of energy. Energy into this virtual plant is fed through conventional power plants, including thermal and nuclear, hydropower plants, as well as renewable sources such as renewables, wind, biogas and biofuel feed sustainable energy. Moreover, the electric vehicles also play their role by acting as mobile power storage units that eventually would supply energy when required. The virtual power plant ensures an optimal energy handling system, routing energy flow across various sectors such as homes, buildings, and farms, while improving energy sustainability and efficiency.

4.2 IoT-Based Climate and Pollution Monitoring Systems

IoT-based climate and pollution monitoring systems provide immediate data for the analysis of the environment for proactive measures to deal with pollution and climate change. These IoT sensors collect the data with regard to air quality, temperature, humidity, and pollutant levels through their deployment in urban and industrial areas. The data that is collected is stored in cloud-based platforms where the trends have been analyzed, anomaly detection, and insight generation against decision-makers by AI algorithms. This data-driven approach allows authorities to formulate policies that target reducing environment hazards.It monitors air quality but also looks into the internal dimensions of the water body, being managed by the IOT. Wastewater quality sensors will detect the kinds of contaminants that are found in lakes, rivers, and reservoirs. In this manner, it will give an early warning of pollution events. Likewise, IoT-based soil monitoring systems ensure that the soil contains balanced moisture levels, pH balance, and nutrients. This will eventually promote sustainable agricultural practices among farmers. These technologies would help Governments and organizations define an environmental degradation strategy and resources conservation plan.

Data accuracy, network reliability, and cybersecurity are some of the challenges limiting the wide adoption of IoT. On the other hand, advanced network infrastructure should be installed to transmit large environmental data from multiple sources. This can be attributed to the fact that hackers can manipulate the environmental data, resulting in misinformation or failure of operational normals. As such, regular calibration and maintenance of the IoT sensor must be performed to ensure accurate measurements. In addressing these challenges: through standardized protocols, encryption measures, and AI-led anomaly detection, IoT-embedded environmental monitoring would be effective.

Table 2 shows components of IoT-based climate and pollution monitoring systems and their applications in environmental monitoring.

Table 2. IoT-Based climate and pollution monitoring systems

Component	Description	Application	Sensors Used
IoT Sensors	Devices to collect data on air quality, temperature, humidity, etc.	Monitoring climate and pollution levels	Air quality sensors, CO2 sensors, Humidity sensors
Wireless Network	Communication infrastructure for transmitting data from sensors	Remote data transmission to cloud	4G/5G, Wi-Fi, LoRaWAN
Cloud Data Processing	Cloud-based platforms to analyze and store data	Real-time analysis and visualization	AWS, Google Cloud, Microsoft Azure
Data Analytics and AI	Machine Learning algorithms for predicting pollution trends	Predicting air quality, climate changes	Regression models, Decision Trees
User Interface (UI)	Dashboard for monitoring pollution levels	Displaying real-time data to users	Web/mobile interfaces

4.3 AI-Driven Optimization for Sustainable Resource Utilization

Resource optimization in sustainability is a cornerstone of artificial intelligence implementation across different sectors for waste reduction and efficiency maximization. In the agricultural sector, AI is utilitarian via its systems to analyze weather, soil condition, and crop health to maximize irrigation and fertilizer use, therefore minimizing wastage of water and chemicals due to overuse(Alharithi&Alzahrani, 2024). In a similar manner, predictive maintenance powered by AI in industries helps to avert the breakdown of machines, hence preventing losses in operational time and energy consumption. In general, these developments lead toward sustainable use of resources.

AI also enhances waste management via intelligent sorting and recycling automation. Machine-learning algorithms identify and classify recyclable materials for optimal waste segregation. Smart waste bins equipped with AI sensors track disposal patterns and schedule efficient collection routes, reducing fuel consumption and operational costs. These innovations support a circular economy, thereby keeping resources in use by reuse and repurposing prior to disposal.AI-sustainability is perceived as a viable extension but is constrained by various challenges, for instance, high implementation costs and data dependency. The processes that characterize the development of an AI model, in itself, are quite computationally intensive and require quality datasets that may not always be readily available. Invarious sectors, integrating AI with legacy systems will come at an additional investment cost while requiring a skilled workforce. Policymakers and organizations must pool their resources to make AI-sustainability initiatives more widespread and scalable, thereby encouraging long-term environmental benefits.

5. INTELLIGENT TRANSPORTATION AND MOBILITY

5.1 IoT in Connected and Autonomous Vehicles

With the incorporation of an Internet of Things element into transportation, it has changed how vehicles operate into intelligent, connected items. Vehicles can now virtually speak to each other, infrastructure, and central control systems through V2X communication. They now share data in real time through various sensors like GPS and through embedded systems built within them to enhance performance, road conditions, and traffic patterns used to improve road safety and navigate the broader network. The whole system enhances the efficiency of buggy driving, mitigates congestion, and pushes the construction of smart transportation networks.

Most of the Autonomous Vehicles (AVs) are greatly reliant upon the usage of Internet of Things (IoT)-based sensors and artificial intelligence to steer without the intercession of a human being. These devices, including LiDAR, radar, and cameras, work in tandem to offer a three-dimensional vision of the environment around the car to enable the identification of any obstacles, pedestrians, and other vehicles. The real-time processing of this data by the IoT platforms allows the instant decision-making by AVs to guarantee safe and smooth transportation. As it reduces the most critical factor of human error, in most cases, responsible for road accidents, IoT-based automation will significantly improve road safety. However, the potential for IoT to be widely accepted into autonomous transportation faces numerous challenges such as cybersecurity threats as well as issues with data privacy. An example is the connection in vehicles because they usually operate on wireless communication, making them very vulnerable to hacking, which can result in accidents or unauthorized control. Besides, there must be legal and ethical implications on issues of liability in case of system failures. It establishes the basis for secure deployment of IoT-driven transportation technologies concerning regulatory frameworks and advancement in cybersecurity.

5.2 Machine Learning for Traffic Prediction and Route Optimization

Machine learning (ML) is the new hype in availing better traffic prediction and travel route optimization in urban mobility. Analyzing historical and real-time traffic data enables ML algorithms to predict the different congestion patterns, recognize alternate routes, and suggest users with the fastest paths(Cohen &Dalyot, 2020). AI models in traffic management systems measure several parameters that include public events, construction activities, and weather conditions to change the timing of traffic lights dynamically to minimize delays. Such intelligent systems improve the flow of traffic into major metropolitan cities while also minimizing the consumption of fuel.

Apart from making cities smarter, routes of public transportation are also optimized by machine learning for better operations. AI algorithms analyze the passenger demand and optimally schedule the bus, train timetables accordingly in a bid to reduce waiting times while improving commuter experience. Ride-hauling services employ machine learning efficiently match two people together and completing the empty trip to reduce traffic congestion in urban. Smart parking also utilizes the same intelligent technology to navigate roads into possible parking lots where one can park without spending a whole lot of searching for it, which saves on emissions. However, machine learning applied in traffic management depends on good-quality data and advanced computing. Inaccuracy or incompleteness of information at any point results in wrong predictions of traffic phenomena, which would, in turn, affect route planning. Furthermore, the establishment of an AI-harnessed traffic optimizer generally requires massive capital investment in infrastructure, not possible in many cities. Public and private stakeholders should come together to make reliable data collection, financial investments in AI research, and developing cost-efficient solutions in using ML technology in urban transportation systems.

Table 3 illustrates the applications of machine learning algorithms in traffic prediction and route optimization.

Table 3. Machine learning for traffic prediction and route optimization

Application Area	Algorithm/ Technique	Purpose	Benefits
Traffic Flow Prediction	Neural Networks, Time Series Forecasting	Predict future traffic flow based on historical data	Improved accuracy in predicting traffic
Congestion Prediction	Decision Trees, SVM	Detect and predict traffic congestion	Reduced congestion and delay
Dynamic Route Optimization	Genetic Algorithm, Reinforcement Learning	Optimize routes based on real-time traffic conditions	Reduced travel time and fuel consumption
Accident Prediction	Random Forest, XGBoost	Predict potential accident hotspots	Enhanced safety, proactive planning
Public Transport Optimization	K-means Clustering, Reinforcement Learning	Route and schedule optimization for buses and trains	Improved service reliability and efficiency

5.3 Reducing Carbon Footprint with Smart Mobility Solutions

The transportation sector contributes heavily to global carbon emissions, thereby necessitating sustainable mobility solutions to combat climate change. Smart mobility is about IoT and AI intervening to facilitate energy-efficient modes of transportation such as electric vehicles (EVs), shared mobility, and intelligent public transit systems(Liu et al., 2023). The adoption of EVs has been trending upward by virtue of breakthroughs in battery technology and charging infrastructure; meanwhile, AI is enhancing charging and battery management for EVs, thus prolonging life and boosting efficiency. Sustainability is then backed by the intelligently distributed and redistributed energy on the grid from the use of AI-enabled electric vehicle charging stations. It is carpooling as well as micro-mobility e-scooter and bike-sharing services that address alternatives to single-occupancy vehicles and contribute to carbon emissions reduction. AI platforms analyze commuter preferences with live traffic conditions and recommend environment-friendly travel options. Meanwhile, smart city initiatives are promoting walkable neighborhoods with integrated multimodal transport systems that encourage citizens to opt for sustainable travel modes over private cars.

Notwithstanding these strides, the transitioning into sustainable mobility will still grapple with many impediments, such as poor infrastructure state and social barrier to the acceptance of a new mobility scheme. Heavy investment is needed in expanding EV charging networks, while shared mobility services are plagued with challenges too, like vehicle availability and regulatory constraints.

6. SMART HEALTHCARE AND WELL-BEING

6.1 IoT in Remote Patient Monitoring and Wearable Devices

It is the Internet of Things (IoT) that has changed the face of monitoring vital signs and health states. In fact, it has ushered in real-time monitoring. RPM systems collect data using these sensors and use intelligent activity tracking devices, which can include anything that measures blood glucose levels, heart rates level of oxygen in the blood and blood pressure. This information gets transmitted to healthcare providers, who can track the condition of the patients continuously, thus being able to intervene whenever necessary. Such technology has been subsequently useful in the management of chronic conditions, removing the need for hospital visits and thus improving patient outcomes, for instance, in diabetes and hypertension management.

Wearable devices like smartwatches, fitness bands, and ECG monitors give users personalized insights into their health while assisting doctors with making decisions based on data. Such devices detect anomalies like abnormal heart rhythms or sudden drops in oxygen saturation, which otherwise demand potential medical assistance on a priority basis. Apart from that, IoT in wearables fosters elderly care through movement pattern tracking that facilitates fall detection and medication adherence, thereby extending the safety and independence of senior citizens.However, challenges related to security of data, interoperability, and regulatory compliance confront IoT health applications. Strong countermeasures for cybersecurity are required in this situation to protect medical devices collecting sensitive patient information from unauthorized access. Interoperability and standardization of IoT protocols across various healthcare systems are also needed to ensure the smooth exchange of data between them. These hindrances can be solved by advanced encryption, robust data management policies, and compliance with global healthcare regulations for a safe and effective implementation of IoT-enabled healthcare solutions.

6.2 Machine Learning for Disease Prediction and Diagnosis

ML is transforming disease prediction and accuracy into diagnosis through early condition detection at a much higher accuracy than earlier. It examines huge datasets of medical records to find patterns that indicate the presence of diseases or enhanced disease progression. For instance, deep learning algorithms detect early stage cancerous tumors by interpreting medical images like X-rays, MRIs, and CT scans. AI-enabled diagnostic tools complement radiologists and pathologists in improving accuracy, reducing human error, and accelerating treatment planning. Apart from imaging, machine learning can now assess different genetic, lifestyle, and environmental factors in the likelihood of developing a disease and enhance predictive analytics in health care. AI-based risk assessors can help physicians identify risk-prone individuals for future cardiovascular diseases, Alzheimer's, and diabetes before these develop symptoms. It can then be possible to have tailored medications that ultimately may reduce or delay disease onset while at the same time improving patient results and cutting out on healthcare expenses.

Nonetheless, while implementing ML in healthcare, there is a need for well-distributed, good-quality datasets, which will guarantee unbiased and accurate predictions. Most AI models are characterized by biases mostly arising from non-inclusiveness of certain masses in their training data. Disparities thus arise in diagnoses as well as suggestions for treatment. Additionally, there would always be regulatory approvals and ethical issues surrounding AI decision-making that must be addressed to ensure transparency and accountability. Continuous advancement in AI ethics, model validation, and collaboration between healthcare providers and AI researchers will drive the necessary trustworthy machine learning solutions in medical diagnostics.

6.3 AI-Powered Personalized Healthcare Solutions

As such, according to a precise treatment, individuals differ from one another with different medical histories, genetic setups, and lifestyle choices. Thus, under the aspects of precision medicine, AI takes genetic data as target therapy recommendation for the occasion of the disease such as cancer should effectively treat and leave fewer side effects(Rahman, 2023). AI can therefore predict the feasibility of potential compounds in drug discovery, assuring faster rollout of new possible drugs.

Virtual health assistants and AI chatbots are meant to provide patients with personalized support, such as reminders for medication, dietary suggestions, and symptom tracking. By utilizing these applications with AI capabilities, routine queries can be taken care of by these assistants, providing real-time support for patients with chronic illness, hence relieving healthcare professionals to a great extent. AI-based tools designed for mental health monitor emotional well-being using natural language processing (NLP) techniques, thereby offering tailored coping skills and treatment suggestions. Healthcare provision using AI is still facing hurdles based on ethical considerations and data privacy. AI must ensure transparency in decision-making, particularly with regard to biases of any type, in order to provide equitable access to healthcare. Also, encrypted health information and compliance with health regulation principles such as HIPAA and GDPR will go a long way in maintaining trust in AI-based medical solutions. In this way, an ever-evolving AI can improve patient outcomes while protecting the ethical and secure implementation of personalized healthcare systems.

Figure 4. AI-Driven personalized healthcare process

As shown in figure 4, an artificial intelligence-based healthcare process is necessary in patient diagnosis and therapy. The process is divided into five parts: patient data collection involves compiling medical history and health parameters; data processing with AI algorithms means analyzing the patient data using machine learning techniques; personalized diagnosis and treatment plan recommended by AI; real-time monitoring means continuous health condition monitoring using wearable devices and IoT sensors; and feedback loop and improvement, learning lessons from monitoring and improving future therapy. All of these enhance health care efficiency in terms of precision, time, and flexibility in real-time services for patients.

7. INDUSTRIAL AUTOMATION AND SMART MANUFACTURING

7.1 IoT-Enabled Predictive Maintenance in Industries

With the use of IoT technologies, predictive maintenance is having a profound impact on industrial operations through the real-time monitoring of equipment and machinery. Conventional maintenance approaches rely on the regularity of inspections or reactive repairs that may lead to untimely downtimes or greater costs. With an array of sensors, machine learning, smart sensors, and smart analytics, industries are able to monitor machine performance continuously with respect to temperature, vibration levels, and energy consumption(Compare et al., 2019). By analyzing such data, predictive maintenance schemes can forecast potential failures even before they happen, thus reducing production disruptions further and enhancing the service life of the equipment.

The primary benefit, in this case, is enhanced optimization of resource utilization and reducing operational costs. AI algorithms sift through massive amounts of sensor data in order to detect anomalies and patterns signifying wear and tear(Kalusivalingam et al., 2020). Maintenance activities have been protecting the production lines in manufacturing plants, such that failures in robotic arms or conveyor belts are detected by predictive maintenance before causing considerable damage. The energy sector similarly uses IoT sensors such that turbines and generators detect possible onset of mechanical stress to avert expensive failures as well as enhance operational efficiencies.Predictive maintenance implementation, however, brings a big price tag with it in terms of investment for creating IoT infrastructure, a data management system, and training the workforce. Across industries, there is difficulty bringing together IoT and other legacy systems, thus requiring seamless connectivity. It is also crucial to ensure the cybersecurity of IoT-enabled devices to prevent them from data breaches and other cyber threats. However, by tackling the above challenges through very strong security for data and scalable IoT architecture, predictive maintenance will become a means of driving efficiency, sustainability, and cost savings across industries (Puttanapura, J et al., 2024).

7.2 AI in Process Optimization and Quality Control

The Artificial Intelligence is revolutionizing the optimization of industrial processes, increasing efficiency when reducing waste and improving control over quality. Traditional manufacturing processes are beset by inefficiencies due to human errors, equipment variability, and the possibility of disruptions in the supply chain. Using automation driven by machine learning based on the skimmed data of the past and the data of the present moment, the workflows are optimized in such a way that the factories can fine-tune the parameters like speed of production, usage of raw materials, and consumption of power to achieve maximum efficiency.

Quality control in manufacturing has greatly improved, thanks to computer vision-based AI that, with great accuracy, examines products for any defects. For instance, advanced image recognition algorithms will run real-time examinations of surface defects, dimensions, and assembly correctness to make sure that only products of excellent quality are on the market. For example, in the automotive industry, AI-driven quality control detects minute imperfections in car parts from going into the market. In drug formulation and packing, AI does well to ensure the accuracy and safety of the patient.Yet, there are hurdles to adopting AI for the same industry improvements, namely skill adoption, high implementation costs, and issues of data privacy and security. Many employees need upskilling for effective interaction with AI-driven systems. In addition, it is essential to ensure that AI models are trained with unbiased and diverse datasets so that inaccuracies in production optimization may be prevented. It is foreseen that for AI-driven manufacturing to be further embraced by industries, further advances in efficient and ethical team structures between engineers, data scientists, and policymakers will be required.

Table 4 discusses AI employed for industrial process optimization and quality control, showing techniques and their results.

Table 4. AI in process optimization and quality control

Application Area	AI Technique	Use Case	Outcome / Benefits
Predictive Maintenance	Machine Learning, Predictive Analytics	Predict equipment failure before it occurs	Reduced downtime, cost savings
Quality Inspection	Computer Vision, Deep Learning	Automating visual inspection of products	Enhanced precision, reduced human error
Supply Chain Optimization	Reinforcement Learning, AI Algorithms	Optimizing supply chain logistics	Cost reduction, increased efficiency
Energy Consumption Monitoring	Machine Learning, Neural Networks	Analyzing and optimizing energy use in manufacturing processes	Reduced energy costs, sustainability
Manufacturing Process Control	AI-based Control Systems	Adaptive control of manufacturing processes	Increased process stability and yield

7.3 Enhancing Workplace Safety and Efficiency

Industrial automation is largely needed for making work environments safe and efficient by minimizing the human touches in hazardous places. Safety systems in the IoT bases smart sensors and wearable devices for continuous health information transmission as they recognize hazards imposed on the worker to prevent accidents at the workplace. For instance, smart helmets and vests with IoT sensors measure various environmental parameters from toxic gas levels, temperature, and the condition of structure all at construction sites or chemical plants. These devices send real-time alerts that help workers take precautionary measures before an accident sets in.

Robotic automation has been shown to have a much safer working environment because there is no involvement of human beings in those dangerous job functions. These robots use AI to perform welding, lifting, and working in extreme high or low temperatures; thus eliminating direct contact between the employees and the employees which are potentially dangerous due to heavy manual labor. Autonomous mobile robots (AMRs) carry out the transportation of heavy loads in warehouses and logistics centers, putting less physical strain on workers and making the supply chain run efficiently. AI ergonomics assessment tools measure feedback about the posture and movements of employees, allowing one to intervene before injuries occur.Despite these advances, ensuring that workplace automation is ethical remains a challenge. The transition to more robotic automation runs the risk of job displacement, which means industries will be pressured to develop workforce retraining and job reallocation strategies.

8. CHALLENGES AND ETHICAL CONSIDERATIONS

8.1 Data Security and Privacy Concerns in IoT and AI Systems

The application of AI and IoT in different industries has presented serious challenges concerning the security and privacy of data. There is a constant flow of sensitive data in the form of personal details, health records, and information about industrial operations being siphoned off by IoT devices. In the absence of effective security mechanisms allowing unauthorized access, such IoT devices could be on the receiving end of cyberattacks leading to data breaches, identity theft, and operational attacks. Further, high risk is posed to systems of unauthorized data access and misuse due to AI systems being trained on and making decisions with massive data sets.

The improper encryption and protection of IoT devices pose a risk of all-mentioned data leaks. Many IoT networks are operated using poor authentication mechanisms that can be easily compromised by hackers. Cybercriminals may exploit those vulnerabilities to tamper with AI-controlled systems that could turn smart grids into infrastructure chaos or interfere with the safe operation of autonomous vehicles. In other occurrences, the use of cloud-based AI services raises the question of sensitive data access to third parties, prompting industries to develop strict access control and data governance policies to restrict third-party access to sensitive data. To some extent, we can defuse these hazards through advanced cybersecurity frameworks, with end-to-end encryption, multi-factor authentication, and blockchain-based data management applications in place. Regulators must help with data protection laws like the General Data Protection Regulation (GDPR) and industry-wise compliance ends. Such organizations willing to invest in AI and IoT should base themselves on enhancing user consent, developing transparent data usage policies, and ensuring continuous security updates to protect user trust and privacy.

8.2 Ethical Implications of AI Decision-Making

In the age of decision-making with AI, industries are being radically affixed by automation, predictive analytics, and personalization. Since there are ethical concerns associated with the AI-driven system, these concern bias, fairness, and accountability(Osasona et al., 2024). A machine learning model uses historical data that contains biases potentially impacting the decisions made by AI, thus causing unfair treatment in services such as healthcare, hiring, and law enforcement. These could create mental blocks and ethical dilemmas in automated decision-making if AI systems are not designed carefully.

Another major ethical dilemma regarding AI decision-making is the opacity of these systems, the so-called "black-box" problem. Many AI model implementations, particularly deep learning algorithms, predict but fail to elucidate their reasoning (Giuffrida, 2019). With this lack of transparency, it becomes extremely difficult for both users and regulators to assess the fairness and reliability of the AI-led decisions. AI systems for loan approvals might deny an application without a concrete reasoning made available to the applicants, thereby further easing the process for an individual in challenging the unfair outcome. It is thus crucial to explain AI in fostering trust and accountability (Rithika, A et al., 2025).To combat these ethical problems, organizations should adopt responsible AI practices, including bias detection methods, algorithmic transparency, and human-in-the-loop decision-making. Ethical AI frameworks should ensure that automated systems conform to human values, give explanations for their decisions, and afford opportunity for disputation. Governments and technology leaders must work hand-in-hand to establish ethical guidelines that will ensure an equilibrium between AI innovation and societal well-being to minimize harm while maximizing AI's potential for the common good.

Figure 5. Ethical implications of AI decision-making

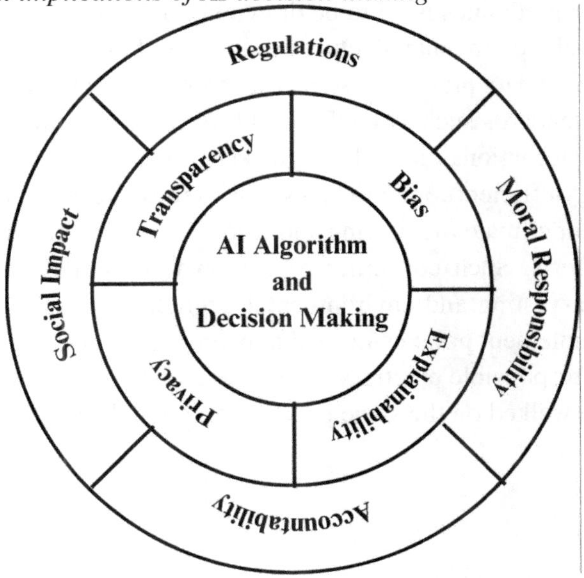

Figure 5 expounds upon vital ethical factors dictating AI algorithms and decision-making. At the center lie AI Algorithm and Decision Making and surrounding attributes that ensure responsible AI deployment: Transparency, Privacy, Accountability, Explainability, and Bias. The outer layer signifies greater societal concerns that govern AI, including Regulations, Moral Responsibility, and Social Impact. These in turn are instrumental in nourishing ethical AI practice, ensuring fairness and trust while meeting legal and societal expectations.

8.3 Regulatory and Policy Considerations for Sustainable Innovation

The rapid advancement of AI and IoT technologies calls for the establishment of wide-ranging regulatory frameworks to ensure sustainable and responsible innovations. In this respect, governments and policymakers will be vital in laying down rules for ethical AI application, data privacy, and environmental sustainability. On the flip side, unregulated development and implementation of AI and IoT could be disastrous, creating avenues for surveillance concerns, monopolistic behaviors, and environmental degradation.

One focal area of regulations is the definition of legal accountability for AI-considered decisions. In autonomous operations, such as self-driving tech and AI diagnostics, there are questions around liability arising in the event of an error or mishap. Clarified guidelines need to be drawn up to identify whether the liability falls with the developer, the manufacturer, or the user. Further, regulatory agencies must endorse strong data protection standards preventing misuse of personal data, with respect to standards such as GDPR and California Consumer Protection Act (CCPA) on the international arena.Policymakers must advance sustainable AI by embedding energy-efficient computing, responsible e-waste management, and green AI research. AI for climate monitoring and resource optimization may help further sustainability efforts. Such consumer protection frameworks must also prevent unethical data harvesting and environmental degradation. The government must formulate and implement policies in such a manner that encourages the growth of technology with responsible practices so that the growth of technology and welfare of society can be walked on the same path to a smart and sustainable future.

9. FUTURE TRENDS AND EMERGING TECHNOLOGIES

9.1 Advances in AI and IoT for Next-Generation Smart Systems

The advances of AI and IoT will provide great prospects for changing industries and building smarter systems. AI is likely to become the backbone of IoT, which will allow real-time decision-making and autonomous operations spanning hospitals, manufacturing, transportation, and so on. The greater cooperation expected between AI and IoT with edge computing is expected to foster greater opportunities with the faster processing of data close to the source, thereby minimizing latency and maximizing efficient functioning of systems. This will enable better performance of the smart devices with higher autonomy, providing useful insight at the moment of engagement.

Another noteworthy trend relates to improvements in algorithms for AI, particularly in the areas of natural language processing and computer vision, thus allowing the ability of IOT devices to understand human interactions in a more human-like way. AI predictive analytics abilities will be enhanced and will allow the system to anticipate needs while optimizing processes in real-time and minimizing energy consumption. This level of sophistication will foster smart homes, cities, and workplaces that autonomously control everything from heating to traffic management. Furthermore, collaboration and interoperability will become immensely important as these technologies progress. The standardization of IoT devices and AI algorithms will be vital for seamless integration across the various platforms and industries.

9.2 The Role of 5G and Edge Computing in Sustainable Innovation

The proper deployment of 5G technology will bring a remarkable change in the way IoT and AI systems are evolving. With increased data speed, lower latency, and higher bandwidth, 5G will allow real-time transmission of data collected from billions of IoT devices to cloud-based AI systems. It is this improvement in connectivity that will enable a variety of smart devices to be deployed in some industries, including healthcare, transportation, and manufacturing, for immediate decision-making and autonomous system management. As a counterpart to 5G, edge computing processes data closer to where it is generated, thereby lessening reliance on central cloud infrastructure and enhancing response time. This transition to edge computing will also curb energy consumption and grant better scalability to smart systems. In smart cities, for instance, edge computing might optimize traffic flow by processing traffic data locally, without needing to send all information to a remote data center. This localized processing also ensures privacy-sensitive data, such as health data, remains within secure networks and hence, ensures data security.

By working in tandem, 5G technology and edge computing encourage a new wave of innovation, one that is far more efficient and sustainable than ever before. They facilitate fast, uninterrupted communication between devices and systems, propelling the co-evolution of smart, green cities, transportation systems, and energy-efficient networks. What is more, these developments will open avenues for many enterprises that wish to create AI-powered solutions in resource-efficient ways, incurring less waste and reducing environmental footprints.

9.3 Towards a Smarter and More Sustainable Global Future

The investment would be in global economies of resource and waste towards renewability-and efficiency-the future hubs that would define the sustainability and efficiency of global economies. It is here that the promise of these technologies will touch serious environmental management, optimizing the use of resource, waste reduction, and integration of renewable energies. In energy, AI provides context for a smarter approach toward management of smart grids in the balancing of demands and supplies while working with renewable sources, such as wind and solar energy. The same analytic approach of AI would yield predictions for equipment failure in industries, thereby reducing unnecessary energy consumption as well as expensive repairs. Such technologies will be increasingly focused on the movement of autonomous vehicle fleets using IoT and AI throughout the next generation of transportation, which will drive today deeper into the future of a carbon-free environment. The systems will improve air quality and better fuel economy as the routes and traffic patterns they would adopt would reduce fuel consumption. The creation of such technologies will also lead to the use of shared mobility solutions that employ vehicles in a more efficient way and downsize the larger state of vehicles on the roads. These innovations shall turn toward downward development-in a circular economy that reuses resources and minimizes waste.

10. CONCLUSION

AI and IoT integration may change the ways to address some of the major global problems in environmental monitoring, sustainability, and intelligent development. What we can see through this document-the merger of two cutting-edge technologies-means better resource management, improved data analysis, and tracking real-time environmental changes with the most unhindered accuracy possible.

From intelligent transportation systems and energy-efficient urban planning to smart health care applications, AI and IoT innovation lives at the core of industry transformations pushing the frontier for sustainable solutions for a better and smarter tomorrow.In environmental monitoring, the combination of vast-source data processing by AI in real-time, with IoT gathering environmental data from several sources, is transforming how we understand and address environmental issues. AI analytics help to predict environmental trends, assess potential threats, and provide adaptive strategies for climate change mitigation. Likewise, IoT sensors in discrete locations can help monitor air quality, water levels, and pollution-all this data will immensely inform policymakers and raise awareness among the public. These innovations are, therefore, critical to the solutions for environmental challenges from climate change mitigation to sustainable conservation of our ecosystems.

This allegiance of AI and IoT in smart cities and infrastructure is already being augmented with new opportunities for energy optimization, waste management, and transportation amelioration. IoT-and AI-powered- smart cities can dynamically self-regulating systems, in which resource allocation is managed in real time depending on demand for energy conservation and maximizing efficiency. AI will greatly minimize carbon footprint through traffic optimization, autonomous vehicles, and energy-efficient buildings. Furthermore, these cities can integrate renewable energy into smart grids, aiming for a future in which energy is harnessed more efficiently, minimizing reliance on non-renewable resources and reducing greenhouse gas emissions.

The capacity that these technologies inherently possess is enormous, but many challenges stand in the way of the successful implementation and wide acceptance of these technologies. Among the foremost concerns are data security and privacy, with increased interconnectivity of devices and huge amounts of data generation putting these systems at great risk of cyber threats. Stronger cybersecurity systems and well-designed regulations to protect personal data will become essential to develop trust in these systems. The ethical implications related to AI making decisions will require further investigation, particularly with respect to areas such as healthcare or environmental monitoring, where impacts are highly sensitive. Ultimately, making AI systems transparent, accountable, and fair will be crucial for ending bias and ensuring equitable access to the benefits of such technologies.

In summary, the AI and IoT future is promising for sustainable smart solutions to help fix the world's issues. From optimizing urban environments to reversing environmental degradation to providing a pathway to healthy-liveable cities, these technologies stand on the foremost front of innovation. The full potential of these technologies will depend, however, on sustained coordination and collaboration between governments, businesses, and research institutions that are dedicated to promulgating standards, ethical frameworks, and security. In accepting these technologies in a responsible and innovative approach, a pathway for a smarter sustainable world for all people and the planet will be opening.

REFERENCES

Abdoh, S. A. (2024). Art and sustainability: Can digital technologies achieve sustainability? *Journal of Cultural Heritage Management and Sustainable Development*, *14*(4), 555–562. DOI: 10.1108/JCHMSD-03-2022-0038

Abdul-Qawy, A. S., Pramod, P. J., Magesh, E., & Srinivasulu, T. (2015). The internet of things (iot): An overview. *International Journal of Engineering Research and Applications*, *5*(12), 71–82.

Albreem, M. A., El-Saleh, A. A., Isa, M., Salah, W., Jusoh, M., Azizan, M. M., & Ali, A. (2017, November). Green internet of things (IoT): An overview. In 2017 IEEE 4th International Conference on Smart Instrumentation, Measurement and Application (ICSIMA) (pp. 1-6).IEEE.

Alharithi, F. S., & Alzahrani, A. A. (2024). Enhancing environmental sustainability with federated LSTM models for AI-driven optimization. *Alexandria Engineering Journal*, *108*, 640–653. DOI: 10.1016/j.aej.2024.09.058

Almukhalfi, H., Noor, A., & Noor, T. H. (2024). Traffic management approaches using machine learning and deep learning techniques: A survey. *Engineering Applications of Artificial Intelligence*, *133*, 108147. DOI: 10.1016/j.engappai.2024.108147

Bambam, A. K., & Gajrani, K. K. (2023). Challenges in achieving sustainability during manufacturing. In *Sustainable Materials and Manufacturing Technologies* (pp. 108–124). CRC Press. DOI: 10.1201/9781003291961-9

Belli, L., Cilfone, A., Davoli, L., Ferrari, G., Adorni, P., Di Nocera, F., Dall'Olio, A., Pellegrini, C., Mordacci, M., & Bertolotti, E. (2020). IoT-enabled smart sustainable cities: Challenges and approaches. *Smart Cities*, *3*(3), 1039–1071. DOI: 10.3390/smartcities3030052

Bellini, P., Nesi, P., & Pantaleo, G. (2022). IoT-enabled smart cities: A review of concepts, frameworks and key technologies. *Applied Sciences (Basel, Switzerland)*, *12*(3), 1607. DOI: 10.3390/app12031607

Cohen, A., & Dalyot, S. (2020). Machine-learning prediction models for pedestrian traffic flow levels: Towards optimizing walking routes for blind pedestrians. *Transactions in GIS*, *24*(5), 1264–1279. DOI: 10.1111/tgis.12674

Compare, M., Baraldi, P., & Zio, E. (2019). Challenges to IoT-enabled predictive maintenance for industry 4.0. *IEEE Internet of Things Journal*, *7*(5), 4585–4597. DOI: 10.1109/JIOT.2019.2957029

Dudhe, P. V., Kadam, N. V., Hushangabade, R. M., & Deshmukh, M. S. (2017, August). Internet of Things (IOT): An overview and its applications. In 2017 International conference on energy, communication, data analytics and soft computing (ICECDS) (pp. 2650-2653).IEEE.

Giuffrida, I. (2019). Liability for AI decision-making: Some legal and ethical considerations. *Fordham Law Review*, *88*, 439.

Gupta, B. B., & Quamara, M. (2020). An overview of Internet of Things (IoT): Architectural aspects, challenges, and protocols. *Concurrency and Computation*, *32*(21), e4946. DOI: 10.1002/cpe.4946

Gupta, S. K., & Singh, N. (2023).AI and IoT-Enabled Technologies and Applications for Smart City.In Artificial Intelligence in Cyber-Physical Systems (pp. 175-193). CRC Press. DOI: 10.1201/9781003248750-10

Han, H., & Hawken, S. (2018). Introduction: Innovation and identity in next-generation smart cities. *City, Culture and Society*, *12*, 1–4. DOI: 10.1016/j.ccs.2017.12.003

Hiran, K. K., Khazanchi, D., Vyas, A. K., & Padmanaban, S. (Eds.). (2021). *Machine learning for sustainable development* (Vol. 9). Walter de Gruyter GmbH & Co KG. DOI: 10.1515/9783110702514

Izatt, R. M., Izatt, S. R., Bruening, R. L., Izatt, N. E., & Moyer, B. A. (2014). Challenges to achievement of metal sustainability in our high-tech society. *Chemical Society Reviews*, *43*(8), 2451–2475. DOI: 10.1039/C3CS60440C PMID: 24464346

Jaber, M. (2023). IoT and machine learning for enabling sustainable development goals. *Frontiers in Communications and Networks*, *4*, 1219047. DOI: 10.3389/frcmn.2023.1219047

Jayachandran, M., Gatla, R. K., Rao, K. P., Rao, G. S., Mohammed, S., Milyani, A. H., Azhari, A. A., Kalaiarasy, C., & Geetha, S. (2022). Challenges in achieving sustainable development goal 7: Affordable and clean energy in light of nascent technologies. *Sustainable Energy Technologies and Assessments*, *53*, 102692. DOI: 10.1016/j.seta.2022.102692

Jung, A. (2022). *Machine learning: the basics*. Springer Nature. DOI: 10.1007/978-981-16-8193-6

Kalusivalingam, A. K., Sharma, A., Patel, N., & Singh, V. (2020). Enhancing Predictive Maintenance in Manufacturing Using Machine Learning Algorithms and IoT-Driven Data Analytics. *International Journal of Artificial Intelligence and Machine Learning*, *1*(3).

Kataray, T., Nitesh, B., Yarram, B., Sinha, S., Cuce, E., Shaik, S., Vigneshwaran, P., & Roy, A. (2023). Integration of smart grid with renewable energy sources: Opportunities and challenges–A comprehensive review. *Sustainable Energy Technologies and Assessments*, *58*, 103363. DOI: 10.1016/j.seta.2023.103363

Liu, J., Li, J., Chen, Y., Lian, S., Zeng, J., Geng, M., Zheng, S., Dong, Y., He, Y., Huang, P., Zhao, Z., Yan, X., Hu, Q., Wang, L., Yang, D., Zhu, Z., Sun, Y., Shang, W., Wang, D., & Chen, X. M. (2023). Multi-scale urban passenger transportation CO2 emission calculation platform for smart mobility management. *Applied Energy*, *331*, 120407. DOI: 10.1016/j.apenergy.2022.120407

Mendhurwar, S., & Mishra, R. (2018). Emerging synergies between Internet of Things and social technologies. *Journal of Global Information Technology Management*, *21*(2), 75–80. DOI: 10.1080/1097198X.2018.1462918

Osasona, F., Amoo, O. O., Atadoga, A., Abrahams, T. O., Farayola, O. A., & Ayinla, B. S. (2024). Reviewing the ethical implications of AI in decision making processes. *International Journal of Management & Entrepreneurship Research*, *6*(2), 322–335. DOI: 10.51594/ijmer.v6i2.773

Patil, P. (2022). Applications of deep learning in traffic management: A review. *International Journal of Business Intelligence and Big Data Analytics*, *5*(1), 16–23.

Puttanapura, J., Singh, T. M., Reddy, C. K. K., & Doss, S. (2024). Improvised prediction of rainfall using random forest classifier over Taiwan. In *Proceedings of the 2024 4th International Conference on Soft Computing for Security Applications (ICSCSA)* (pp. 152–156). IEEE. DOI: 10.1109/ICSCSA64454.2024.00031

Rahman, I. (2023). AI-powered Personalized Treatment Recommendation Framework for Improved Healthcare Outcomes. *Journal of Computational Social Dynamics*, *8*(11), 42–51.

Rangarajan, D., Rangarajan, A., Reddy, C. K. K., & Doss, S. (2025). Exploring the next-gen transformations in healthcare through the impact of AI and IoT. In *Intelligent Systems and IoT Applications in Clinical Health* (pp. 73–98). IGI Global.

Revathi, S., Ansari, A., Susmi, S. J., Madhavi, M., Gunavathie, M. A., & Sudhakar, M. (2024).Integrating Machine Learning-IoT Technologies Integration for Building Sustainable Digital Ecosystems.In Multidisciplinary Applications of Extended Reality for Human Experience (pp. 259-291).IGI Global. DOI: 10.4018/979-8-3693-2432-5.ch012

Rithika, A., Singh, T. M., Reddy, C. K. K., & Doss, S. (2025). Leveraging generative AI for designing next-generation haptic systems: Advancements, challenges, and applications. In Mamodiya, U., Kalam, A., Bossoufi, B., Awad, M., & Somwanshi, D. (Eds.), *Integrating AI with haptic systems for smarter healthcare solutions* (pp. 327–350). IGI Global., DOI: 10.4018/979-8-3373-2307-7.ch015

Schaffers, H., Komninos, N., &Pallot, M. (2012).Smart cities as innovation ecosystems sustained by the future internet.

Shinde, N. K., Seth, A., &Kadam, P. (2023).Exploring the synergies: a comprehensive survey of blockchain integration with artificial intelligence, machine learning, and iot for diverse applications. Machine Learning and Optimization for Engineering Design, 85-119.

Singh, T. M., Reddy, C. K. K., & Lippert, K. (2024). The revolution and future of blockchain technology in cybersecurity: A comprehensive analysis. In *Artificial intelligence for blockchain and cybersecurity powered IoT applications* (1st ed., p. 28). CRC Press., DOI: 10.1201/9781003497585-5

Speer, B., Miller, M., Schaffer, W., Gueran, L., Reuter, A., Jang, B., & Widegren, K. (2015). *Role of smart grids in integrating renewable energy (No.NREL/TP-6A20 -63919). National Renewable Energy Lab.* NREL. DOI: 10.2172/1215177

SwathiSree. M., Reddy, C. K. K., Vaishnavi, K., &Harika, V. (2025). Informatics and Internet of Things uses in clinical medicine. In Intelligent systems and IoT applications in clinical health (pp. 99-126).IGI Global.

Vieira, S., Pinaya, W. H. L., & Mechelli, A. (2020). Main concepts in machine learning. In *Machine learning* (pp. 21–44). Academic Press. DOI: 10.1016/B978-0-12-815739-8.00002-X

Villamil, S., Hernández, C., & Tarazona, G. (2020). An overview of internet of things. *Telkomnika, 18*(5), 2320–2327. DOI: 10.12928/telkomnika.v18i5.15911

Von Stamm, B., & Trifilova, A. (Eds.). (2009). *The future of innovation.* Gower Publishing, Ltd.

Zaki, M. J. MeiraJr, W., &Meira, W. (2020). Data mining and machine learning: Fundamental concepts and algorithms. Cambridge University Press.

Chapter 10
Revolutionizing Agriculture and Industries With Smart Solutions:
Case Studies on the Transformative Impact of IoT and Machine Learning

Laya Billinty Varra
https://orcid.org/0009-0001-1205-3789
Stanley College of Engineering and Technology for Women, India

T. Monika Singh
https://orcid.org/0009-0006-8534-9130
Annamalai University, India

Shugufta Fatima
https://orcid.org/0009-0005-4396-4947
Annamalai University, India

B. V. Ramana Murthy
Stanley College of Engineering and Technology for Women, India

Hitesh Manapragada
California State University, USA

DOI: 10.4018/979-8-3373-4207-8.ch010

ABSTRACT

This chapter explores how the convergence of the Internet of Things (IoT) and Machine Learning (ML) is revolutionizing agriculture and a range of industries, including healthcare, smart cities, manufacturing, retail, finance, education, and smart homes. Through real-life case studies, it demonstrates how these technologies are enabling intelligent automation, operational efficiency, real-time monitoring, and predictive decision-making. Special attention is given to agricultural innovations such as precision farming, smart irrigation, and crop health analytics, alongside comparable breakthroughs in other sectors. The chapter also critically examines challenges related to data security, privacy, ethical considerations, and system integration. Furthermore, it addresses the future trajectory of IoT-ML ecosystems, including the role of emerging technologies like edge computing, 5G, and federated learning. Readers will gain valuable insight into how IoT and ML are driving a paradigm shift towards smarter, more connected systems, transforming the way industries operate and evolve.

1. INTRODUCTION

The fast pace growth of the Internet of Things (IoT) and Machine Learning (ML) has reshaped various different industries, resulting in groundbreaking solutions that boost efficiency, productivity, and user experience. The combination of the two brings together smart solutions using real-time data processing, automation, and smart decision-making, creating a more interlinked and intelligent world (Kumar, K., et al., 2023). This chapter delves into real-life case studies where IoT and ML have been implemented effectively to solve problems and design intelligent solutions in various industries such as healthcare, smart cities, industry, agriculture, retail, finance, consumer applications, and education.

This chapter delves into practical, real-life case studies where the integration of IoT and ML has been successfully implemented to tackle complex challenges and design intelligent, scalable solutions across diverse sectors. In healthcare, for example, wearable sensors combined with ML models are revolutionizing patient monitoring, enabling early detection of anomalies such as arrhythmias, glucose level fluctuations, and respiratory issues. In smart cities, IoT-based infrastructure, augmented by ML algorithms, is optimizing traffic management, energy usage, and waste collection, thereby improving the quality of urban living.

In industrial settings, predictive maintenance powered by IoT-enabled equipment and ML algorithms is reducing downtime and extending machinery lifespan. In agriculture, smart farming solutions leverage IoT devices to monitor soil moisture, temperature, and crop health, while ML models analyze this data to optimize irrigation, fertilization, and pest control strategies. Similarly, in the retail sector, IoT and ML technologies enhance inventory management, customer experience personalization, and dynamic pricing strategies by analyzing real-time consumer behavior and purchasing trends.

The financial industry benefits from this convergence through advanced fraud detection systems, personalized financial advising, and automated trading platforms. In consumer applications, smart homes equipped with intelligent assistants adapt to user preferences by learning from interaction data. The education sector is also witnessing a transformation, with smart classrooms and learning management systems utilizing IoT for attendance tracking and environmental control, and ML for personalized learning and student performance analytics.

By presenting these varied use cases, this chapter highlights how the synergy between IoT and ML is not only solving existing problems but also paving the way for a future of intelligent, connected ecosystems. It underscores the importance of cross-disciplinary approaches, robust data governance, and scalable architectures in realizing the full potential of these technologies in a responsible and sustainable manner.

1.1 Role of IoT and ML in Industry Transformation

IoT and ML are now vital in driving digital transformation by enabling smart, data-driven solutions across industries. Sensor and internet-enabled IoT devices collate vast amounts of data in real-time, which is processed and analyzed by ML algorithms in order to extract meaningful intelligence. Yet, in unison, these technologies provide the ability to automate quickly, analyze predictive data, and build on-the-spot judgments, facilitating more effective service delivery. From running production processes smoothly to customizing customer experiences, IoT and ML are the engines propelling the digital transformation across a barrage of sectors (Visconti, P., et al., 2024). They also contribute to operational resilience by enabling systems to self-correct, adapt, and improve based on historical trends and anomalies. Moreover, by streamlining supply chains and reducing human error, businesses can achieve higher efficiency and cost savings. As industries continue to evolve, the integration of IoT and ML becomes not just an advantage but a necessity for sustainable innovation and competitiveness.

1.2 Research Trends and Industry Adoption

IoT and ML research is constantly growing, fuelling innovation and mass adoption in various industries. Federated learning and Edge AI are gaining momentum to make decentralized processing of data possible for increased efficiency and privacy. 5G network implementation is enhancing the performance of IoT applications through faster and more reliable connectivity, facilitating real-time decision-making. AI-driven automation is revolutionizing industries through automated processes, optimized functions, and restricted human intervention. Green technology and sustainability are becoming the principal areas, with smart solutions for reducing energy consumption, waste preservation, and green practices. Cybersecurity solutions are also being made more effective to address growing concerns of data security and privacy with AI-driven threat detection and anomaly detection to keep things secure. While startups and industry players continue to fund these technologies, the convergence of IoT and ML is essentially changing business operations in such a way as to support greater efficiency, cost reduction, and enhanced delivery of services. (Naidoo, P., et al., 2024)

2. HEALTHCARE: SMART AND CONNECTED MEDICAL SYSTEMS

Figure 1. IoT and machine learning integrated healthcare architecture

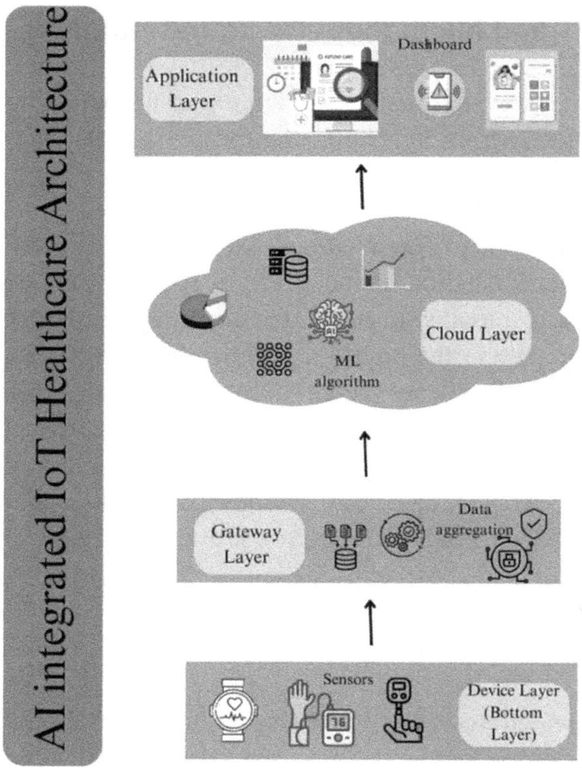

In Figure 1, a layered diagram explains the architecture of an IoT and ML integrated healthcare system. It contains four important layers-a Device Layer, a Gateway Layer, a Cloud Layer, and an Application Layer. The device layer encompasses a range of different medical IoT devices, including wearables, sensors, and implantable devices, used for live collection of patient data. The Gateway Layer aggregates the data transmitted over the device layer, pre-processes it, and securely transmits it further for edge computing. The Cloud Layer contains a data store with advanced ML algorithms therein for the purpose of predictive analytics and anomaly detection. Finally, the Application Layer provides healthcare practitioners with real-time insights through patient portal interfaces and alert systems. The architecture guarantees secure, interoperable, and quality-driven data management for applications such as real-time or remote patient monitoring, predictive analysis, and personalized medicine.

2.1 Case Study: AI-Powered Remote Patient Monitoring

Remote Patient Monitoring (RPM) with artificial intelligence enhances healthcare via continuous monitoring of vital signs, medication adherence, and patient status by wearable technologies and smart sensors. AI identifies early health risk through real-time data analysis and predictive analytics and thus enables early interventions and individualized treatment strategies. This prevention boosts patient results, reduces readmissions to the hospital, and lowers healthcare costs. Besides, AI-based RPM also promotes patient engagement through real-time feedback. Data privacy, integration, and technology investment are some of the challenges. Even with these, AI-based RPM is revolutionizing healthcare with improved, cost-efficient, and personalized care options. (Nigar, N., 2024)

Example: How Johns Hopkins Hospital uses AI for ICU monitoring: Johns Hopkins Hospital has deployed AI-based solutions to augment ICU monitoring, with improved patient care through early warning and real-time intervention. A prime example is the application of AI-enabled headsets to decrease delirium among ICU patients by monitoring eye movements and providing audio cues to preserve patient orientation, with a possible long-term benefit to cognitive function. Also, the Targeted Real-Time Early Warning System uses AI to scan patient data and identify sepsis almost six hours sooner than conventional methods, greatly enhancing survival rates. Another innovation is non-invasive intracranial pressure monitoring, in which AI algorithms translate routine ICU vital signs to precisely estimate brain pressure with minimal use of invasive procedures. These AI tools enable more precise monitoring, earlier risk detection, and individualized interventions, putting Johns Hopkins at the forefront of AI-based critical care.(Seth, A., 2023; Best Practice AI, n.d)

2.2 Case Study: AI-Driven Early Disease Detection

Early disease detection using AI relies on ongoing gathering, analysis, and interpretation of patient health information from wearable sensors, medical imaging, and electronic health records. Machine learning algorithms in these systems recognize patterns and deviation from normal health parameters, and early detection of probable diseases even before the emergence of symptoms is made possible (Khan,A.O.R., et al., 2024). For example, AI-powered wearables track heart rate, oxygen saturation, blood pressure, and glucose levels against real-time measurements compared to customized baselines and detect possible dangers such as arrhythmias, diabetic swings, or respiratory impairment. If there is an anomaly, the system alerts healthcare practitioners or patients in a timely manner to prevent possible complications (Zaman,Q., 2024)

Aside from wearables, AI for diagnostics improves disease diagnosis by scrutinizing medical imaging, lab tests, and patient history to detect early disease signals. Deep learning algorithms developed on large databases find early-stage cancer, cardiovascular conditions, and neurological diseases at high accuracy, often better than conventional diagnostic techniques (Zaman, 2024). These AI algorithms also forecast disease development and suggest customized treatment plans, allowing for proactive and data-based healthcare (Khan,A.O.R., et al., 2024). Through the relocation of focus from treatment to early prevention, AI-based detection is enhancing patient outcomes and clinical decision-making.

Example: Google's DeepMind and AI for early-stage cancer detection: Google's DeepMind AI made major breakthroughs in detecting early-stage cancer, specifically breast and lung cancer, using deep learning models trained on huge medical imaging databases. The AI program reads mammograms and CT scans and detects cancer early, even before it is detectable by human radiologists. In a trial, DeepMind's computer program outperformed human experts by reducing false positives by 5.7% in the US and 1.2% in the UK while reducing false negatives, avoiding additional cases of cancer going undetected (McKinney,S.M., et al., 2020). The AI functions by matching new scans to thousands of previous examples, learning from labeled images to become more accurate over time. It not only highlights possible cancerous lesions but also gives risk scores and diagnostic confidence, assisting radiologists in prioritizing high-risk cases. DeepMind's success demonstrates the way AI can enhance early cancer diagnosis, reduce diagnostic errors, and accelerate treatment choices, eventually translating to improved survival rates among patients. With improved algorithms for AI, it has the capability to detect other types of cancer and advance precision medicine approaches globally.

2.3 IoT-Integrated EMS (Emergency Medical Services)

When IoT is embedded in EMS it has revolutionized the field by optimizing real-time monitoring, communication, and overall efficiency of an emergency response. It is used in automatic warning devices on ambulances, where Wi-Fi or RFID technology is utilized to alert the drivers and other traffic users of the approaching ambulance, thus preventing traffic jams and optimizing patient care (Kumar Stephen, V., et al., 2024).

In addition, the real-time vital signs of the patient, such as blood pressure, oxygen saturation, and heart rate, were collected and communicated through IoT-based patient monitoring systems so that physicians could make an informed decision within a reasonable time frame before the patient's admission to the hospital. All these advancements allow for a better-coordinated and improved EMS network due to better collaboration between the patient, the hospitals, and emergency responders.

3. SMART CITIES: ENHANCING URBAN INFRASTRUCTURE

Case Study: AI-Based Traffic Management in Barcelona: Barcelona's traffic management has evolved from the TRYS system (1996) to an AI and IoT-driven model, significantly improving congestion control. The TRYS system relied on local and coordinator agents, 52 VMS panels, adaptive traffic lights, and 300+ loop detectors to optimize traffic flow through real-time signal adjustments. Although efficient, it was not predictive and adaptive enough to support growing urban mobility needs. By contrast, the new AI system takes advantage of 527 road segments with smart sensors, infrared cameras, and loop detectors, combining AI models such as XGBoost and ARIMA to make real-time traffic forecasts and adaptive signal control. As opposed to TRYS, the new system is based on data-driven decision-making and predictive congestion management, improving urban mobility. Moreover, synchronized traffic lights and VMS panels dynamically adapt using real-time data, making a future-proof smart city infrastructure. This transition from rule-based to AI-driven traffic management has brought down congestion levels considerably, enhanced air quality, and maximized urban transport in Barcelona. (Cuena, J., et al., 1996; Garcia, E., et al, 2024). This AI and IoT-based system successfully modernized Barcelona's traffic management, reducing congestion and improving urban mobility.

3.1 Case Study: IoT in Waste Management: The Smart Bin Project

Application of Internet of Things (IoT) technology in the management of wastes over the recent years saw the use of intelligent bins that ensure sustainability and efficiency. A notable example is the Smart Bin initiative that makes use of IoT sensors to keep a real-time account of wastes present. Using an array of sensors, these trash cans measure the amount of garbage put into them and send it to a main unit. Through this system, waste collection agencies are able to plan collections beforehand, lowering costs of operation and promoting environmental tidiness (Rai, S., et al., 2020).

Another breakthrough in this field is the creation and deployment of an IoT-enabled intelligent bin that has the capability of sorting waste autonomously. It applies machine learning algorithms and sensors to differentiate between organic and inorganic waste, channeling them into specific compartments. Also, the bin tracks its levels and notifies collection staff when it is full to a certain point, guaranteeing timely waste removal. In a move to improve sustainability, the bin has an electricity-generating capacity through solar power, hence less dependence on traditional forms of energy (Antora, N. A., et al., 2022).

Example: Amsterdam's IoT-powered waste collection system: Amsterdam's smart waste collection network is based on a high concentration of underground containers with fill sensors that send live data to create optimized collection routes, resulting in huge cost reductions each year (Nijhuis, S., et al., 2024). AI solutions improve cleanliness further through the detection and mapping of litter on streets to allow for prompt removal by council services. This efficient integration of IoT and AI earned Amsterdam recognition with the GO SMART Award, showcasing its innovative approach to sustainable urban waste management.

4. INDUSTRY 4.0: AI-DRIVEN MANUFACTURING AND AUTOMATION

Industry 4.0 or the Fourth Industrial Revolution is transforming the manufacturing sector through Artificial Intelligence (AI), IoT (Internet of Things), robotics, and digital twins. These technologies enable predictive maintenance, real-time decision-making, and smart automation, making industries efficient, cost-effective, and resilient.

4.1 Case Study: Predictive Maintenance in BMW Factories

BMW ambitiously attempted to overcome the challenge of machine reliability and the expense of unplanned downtimes in automobile manufacturing by outfitting its Electric Monorail System (EMS) with a number of very affordable IoT sensors including accelerometers to measure vibration, current sensors to monitor electrical anomalies, and temperature sensors to detect overheating (Strauß et al., 2018). The EMS produces approximately 1.2GB of data per hour and assesses 2.4 million messages in real-time for condition monitoring. Machine learning models, both Isolation Forest (89% F1-score) and Random Forest (98.1% F1-score) successfully detected anomalies and predicted failures, and the production lines were digitally twinned in order to simulate, preemptively modify production, and plan maintenance to reduce inefficiencies or damages caused by faulty machinery (Asimiyu, Z., 2024). The AI-enabled solution resulted in unplanned downtimes that were curtailed by nearly one hour per failure, an overall increase of 25% production efficiency, and a 30% decrease in the rate of defects that improved quality control. Overall, the adoption of Industry 4.0 technologies—IoT, big data analytics, and AI—produced a 40% reduction in downtime, better forecasting across the supply chain, and integrated predictive maintenance to be incorporated into the overall existing manufacturing processes smoothly (Asimiyu, Z., 2024; Strauß et al., 2018).

The implementation of IoT-based predictive maintenance solutions in manufacturing industries is summarized in Table 1, showing their technologies, attributes, and impact. AI-powered IoT sensors perform real-time observation and prediction to ensure downtime is economically hampered within the automobile industry. The aerospace industry uses digital twin technology combined with IoT analytics for predictive maintenance on jet engines to prevent unexpected failures and improve fuel efficiency. In steel, an AI vibration analysis is fused with IoT sensors for detecting anomaly in real-time, thus saving costs on repairs and improving machine reliability. In food processing, smart, IoT-enabled equipment improves predictive maintenance in packaging and processing units while increasing equipment uptime and production efficiency. Notably, the textile industry has adopted IoT-integrated loom machines with smart sensors to auto-alert when attention is needed to repair wear-and-tear. Such improvements have reduced material wastage and improved machine performance. These implementations show where IoT-based predictive maintenance optimizes efficiency, minimizes failures, and improves resource utilization over broad manufacturing sectors.

Table 1. Predictive maintenance: IoT-Based solutions in manufacturing

Industry	IoT Technology Used	Key Features	Impact on Maintenance Efficiency
Automobile (BMW, Germany)	AI-Powered IoT Sensors	Real-time condition monitoring, failure prediction using ML algorithms	Reduced downtime, extended equipment lifespan
Aerospace (GE Aviation, USA)	Digital Twin & IoT Analytics	Predictive analytics for jet engine maintenance, real-time performance tracking	Fewer unexpected failures, optimized fuel efficiency
Steel Manufacturing (Tata Steel, India)	AI-Based Vibration Analysis	IoT sensors for real-time vibration monitoring, automated anomaly detection	Cost savings in maintenance, improved machine reliability
Food Processing (Nestlé, Switzerland)	Smart IoT-Enabled Equipment	AI-based predictive maintenance in packaging and processing units	Increased machine uptime, enhanced production efficiency
Textile Industry (China)	IoT-Integrated Loom Machines	Smart sensors for wear-and-tear detection, automated maintenance alerts	Reduction in material wastage, improved machine performance

The table 2 compares Digital Twins and Conventional manufacturing models along certain parameters like data usage, predictive maintenance, product design, efficiency, cost, complexity, and scalability. Digital Twins utilize real-time data analytics and AI-driven insights, while classical models depend on periodic collection and manual analysis. For predictive maintenance, Digital Twins use AI to prevent failures, while the classical systems depend on reactive maintenance, leading to some unfortunate breakdowns. In the area of product design, Digital Twins perform virtual simulations and reduce material costs, while classical models depend on physical prototyping which requires more inputs to come up with a finished product (Rithika, A. et al., 2025). Operational efficiencies of Digital Twins are achieved through the loosening up of workflows, contrary to classical models that are tied into fixed procedures. It comes to reduce expenses more conventionally over a longer span, although installation is very difficult as it requires IoT, AI, and cloud onboarding. Digital Twins are scalable because they expand easily through a cloud-based model; traditional manufacturing, on the other hand, warrants significant infrastructural changes to scale up.

Table 2. Comparison of digital twins and traditional manufacturing models

Aspect	Digital Twins	Traditional Manufacturing Models
Data Utilization	Real-time data analytics, AI-driven insights	Periodic data collection, manual analysis
Predictive Maintenance	AI-powered predictive models, reduced downtime	Reactive maintenance, higher risk of unexpected failures
Product Design & Testing	Virtual simulations, digital prototyping	Physical prototyping, higher material costs
Operational Efficiency	Optimized workflows, dynamic process adaptation	Fixed processes, less flexibility
Cost Efficiency	Long-term cost savings through optimization	Higher maintenance and production costs
Implementation Complexity	Requires IoT, AI, and cloud integration	Simpler setup but limited adaptability
Scalability	Easily scalable with cloud-based models	Limited scalability without major infrastructure changes

5. AGRICULTURE: AI AND IOT FOR SMART FARMING

Figure 2. IoT and machine learning integrated healthcare architecture

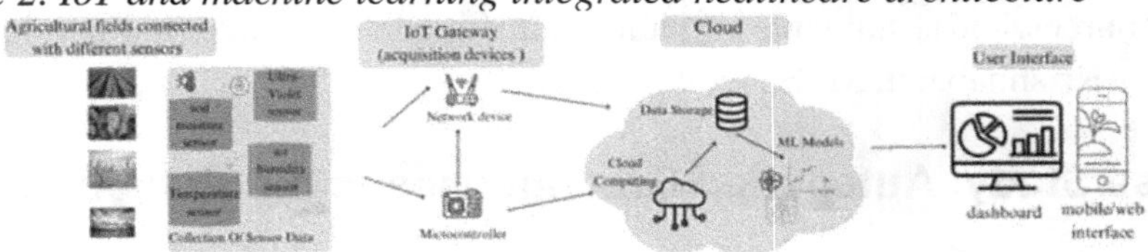

Figure 2 presents the architecture of Smart Agriculture solution using IoTs and machine learning (ML) to augment farm efficiency and sustainability. Various sensors including soil moisture, temperature, humidity, light, CO_2, rainfall sensors, and pest detection cameras are used for the continuous collection of real-time environmental data to inform IoT gateways and microcontrollers before uploading it to a cloud data management application. On the cloud platform, this data is collected and analyzed through ML algorithms that include decision trees, support vector machines, and neural networks, thereby generating predictive insights into irrigation scheduling, nutrient management, crop health monitoring, yield predictions, and crop rotation planning. This information is then made available for farmers through web dashboards and mobile apps, allowing them to monitor their farms, receive alerts, and have farm management automated in real-time. The applications of IoTs and ML in agriculture run from smart precision irrigation to AI-enabled pest and disease detection, automated farm operations, and predictive analytics related to yield forecasting and crop planning. This smart agricultural framework thus offers great promise in transforming modern farming into an advanced, efficient, and sustainable system through resource optimization, cost saving, improved sustainability, and data-driven decision-making.

5.1 Case Study: IoT-Based Precision Farming in India:

IoT and AI are revolutionizing Indian agriculture using precision farming methods. IoT-enabled devices such as soil moisture smart sensors, temperature, humidity, and nutrient level smart sensors provide real-time monitoring and data-based insights, maximizing irrigation, pest control, and resource allocation. Machine learning models and predictive analytics from AI technologies maximize productivity further through crop yield forecasts, soil condition monitoring, and automation of drudgery.

Effective case studies such as Microsoft's Project FarmBeats and Tata's mKRISHI platform prove the prowess of IoT-enabled precision farming in India, which increases crop output by up to 30% and supports sustainability. With barriers such as hefty prices, scalability, and concerns about privacy over data, government policies and joint efforts with tech firms are spurring demand. Future developments combining AI, IoT, and genomics hold the promise of more durable crops, effective use of resources, and sustainable agriculture, making India a leader in smart farming solutions. (Krishnababu, M. E., et al., 2024)

5.2 Case Study: Automated Greenhouses in the Netherlands

The Netherlands has led the way in AI-managed greenhouse automation, significantly increasing crop yields and resource utilization. Advanced networks of sensors and machine learning manage critical parameters such as temperature, humidity, CO_2 levels, and light to provide optimal growing conditions. Real-time greenhouse settings are controlled and climate models predicted by AI-based models adapt to environmental changes, optimizing plant growth while reducing energy use. Dutch greenhouses employ completely closed structures with robotic ventilation and CO_2 enrichment systems, maximizing photosynthesis efficiency. Research indicates that using AI-based climate control has resulted in yield gains of as much as 78 kg/m^2 for vegetables such as tomatoes and peppers, relative to conventional methods. Precision climate management cuts crop loss from capricious weather conditions and enhances general agricultural sustainability.

Drought is one of the biggest threats to contemporary farming, and automation in Dutch greenhouses overcomes this by incorporating IoT-based smart irrigation systems. These systems take advantage of live soil moisture monitoring sensors, auto-drip irrigation, and fertigation technologies for maximizing water supply according to plants' requirements. The AI analytics combined with IoT assure various practices of conserving water along with a reduced consumption rate. The smart irrigation systems monitor electrical conductivity, pH, and levels of nutrient uptake in the soil, assuring optimum utilization of resources. Rainwater harvesting and recycling systems in Dutch greenhouses add to considerable water conservation and thus sustainability and eco-friendliness. (Luna., et al., 2019)

5.3 AI-Powered Pest and Disease Detection

Artificial Intelligence (AI) integration in agriculture has remarkably enhanced the detection of pests and diseases, which has allowed early intervention, minimization of losses in crops, and increased productivity in farms. AI-based solutions utilize machine learning, sensor systems, image analysis, and decision support systems to detect and suppress possible threats in real-time. AI-based smart farming combines inputs from soil sensors, climate data bases, and high-resolution images to identify abnormalities that impact the health of crops. Machine learning algorithms process agricultural data patterns and forecast pest infestations early on, and computer vision and IoT sensors automate plant disease identification with high precision. These AI-based decision support systems provide actionable insights to farmers to maximize interventions, reducing the use of chemicals and improving sustainability.

Empirical evidence from agricultural engineers shows that AI adoption and efficient pest management are highly correlated, with scalability and flexibility proving to be crucial success factors for AI. Practical applications, in the form of mobile apps for small farmers, allow users to leverage AI to determine plant diseases, while research institutions have developed AI software that uses satellite images to detect plant diseases in crops, with a potential to save significant portions of crop yields. These examples show how AI-assisted disease and pest detection is transforming modern-day agriculture by encouraging more data-driven, sustainable, and resilient agriculture.(Hashem, T. N., et al., 2024)

6. RETAIL AND E-COMMERCE: ENHANCING CUSTOMER EXPERIENCE

Figure 3. Applications of AI integrated IoT in retail and e-commerce

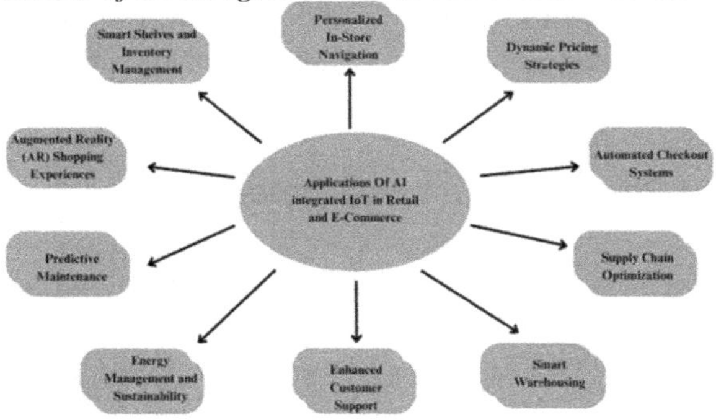

The various applications of AI-integrated IoT in retail and e-commerce sectors are shown in Figure 3. At the center lies the core theme, "Applications Of AI Integrated IoT In Retail and E-Commerce". Outlining the keynote applications include smart shelves and inventory management equipped with IoT sensors for real-time stock management; personalized in-store navigation that gives AI-based guidance for shoppers; dynamic pricing techniques, according to real time demand and market situations. Essentially, they include automated checkouts that cut on wait lines, optimization of supply chains for better logistics, logistics-based smart warehouse/inventorization, enhanced operational capabilities for customer support systems via AI chatbots, energy conservation through improved energy management systems, predictive maintenance curbing equipment failures, as well as AR-based shopping experiences that enhance customer interaction with virtual interfaces. The whole figure illustrates the way AI and IoT work together in enhancing efficiency, customer experiences, and streamlining retail and e-commerce operations.

6.1 Case Study: Amazon Go: AI-Powered Cashierless Stores

Amazon Go transformed retail with AI-fueled cashierless stores, doing away with the conventional checkout lines. By applying computer vision, deep learning, and machine learning (ML) algorithms together, the system monitors customer movement, product choice, and buying habits in real time. Scores of AI-powered cameras and sensors detect when a customer picks up or puts back an item, making sure that there is an accurate virtual cart. As the shopper leaves, payment is automatically collected by the Amazon Go app when they leave, providing a hassle-free shopping experience that saves time and increases convenience. Other shoppers, following on from the success of Amazon Go, have embraced similar "Just Walk Out" technology, moving the frontiers of AI-automated innovation in retail forward.

Amazon Go uses IoT sensors and AI-based inventory management systems to hold real-time inventory levels. Weight sensors on the shelves track product movement, and RFID tags help improve accuracy in inventory tracking and loss prevention. The backend system of the store interprets customer buying habits, maximizing restocking and minimizing supply chain inefficiencies. Through the use of IoT and AI-based data analytics, Amazon Go facilitates faster inventory turnover, lowers operational expenses, and makes products available according to real-time demand. This frustration-free automation has established a new benchmark in smart retail, whose effects on future developments in cashierless shopping globally are noticeable. (Szabó-Szentgróti, E., et al., 2023)

6.2 Case Study: AI-Based Personalized Recommendations in E-commerce

Example: How Netflix and Amazon use AI for content and product recommendations:

(Habil, S., et al., 2023) Netflix and Amazon employ AI-powered recommendation engines to improve user interaction and retention via user behavior, history, and preference data. Amazon's AI-powered recommendation engine deploys collaborative filtering and deep models to suggest items on the basis of purchase history, browsing history, and interaction, boosting cross-selling and upselling by an enormous percentage. Likewise, Netflix uses deep neural networks and reinforcement learning to make personalized content suggestions based on watch history, genre, and user rating. This type of personalization increases customer experience, leading to increased engagement and satisfaction.

7. FINANCIAL SERVICES AND AI-DRIVEN BANKING

7.1 Case Study: AI in Fraud Detection: How PayPal Uses Machine Learning

PayPal, a global digital payment system, is increasingly under complex fraud threats that necessitate real-time adaptive fraud detection systems. PayPal has implemented various machine learning algorithms (supervised, unsupervised, and reinforcement models), to analyze large amounts of transaction history in order to detect suspicious activity at very high levels of precision (Ramadugu & Doddipatla,

2022; PayPal n.d.). Supervised models detect known fraud patterns while unsupervised algorithms detect anomalies, using graph database models and studying the over 1.5 billion edges to expose hidden relationships between accounts (H2O.ai, 2020). In addition to machine learning algorithms PayPal uses H2O.ai's Driverless AI as a feature engineering automation system that accelerates the detection model development and deployment process. The AI system displays significant advantages over traditional rule-based methodologies—by reducing the high volume of false positives the automated systems have produced and the speed/accuracy of fraud detection iteratively improve. Therefore, systemic fraud detection will create safer transaction experiences for millions of people around the world. In the future, biometric authentication processes (using biosensors, such as fingerprints and facial recognition) supported by a predictor indicator for propensity to commit fraud—could add an additional layer to prevent fraud and increase customer confidence (Ramadugu & Doddipatla, 2022).

Table 3 provides various case studies of financial institutions that use artificial intelligence in order to prevent, detect, and investigate the fraudster's actions. PayPal uses algorithms, machine-learning and pattern-recognition approaches to detect suspicious transactions, improving accuracy and reducing false positives. Utilizing AI-powered transaction monitoring, JP Morgan Chase is capable of real-time fraud detection and preventing financial loss. In conjunction with Mastercard, AI biometric authentication enhances security and reduces identity fraud. To this, HSBC responds by deploying NLP for phishing and fraudulent messages, with protection for its clients. American Express utilizes deep learning to analyze spending patterns and allows faster event detection of frauds and minimal unauthorized transactions.

Table 3. AI in fraud detection: Case studies in financial services

Financial Institution	AI Technology Used	Fraud Detection Method	Impact on Fraud Prevention
PayPal	Machine Learning & Pattern Recognition	Identifies suspicious transactions based on behavioral anomalies	Reduced false positives and improved fraud detection accuracy
JPMorgan Chase	AI-Powered Transaction Monitoring	Real-time analysis of transaction data to detect fraud patterns	Faster fraud identification, minimizing financial losses
Mastercard	AI-Driven Biometric Authentication	Uses facial recognition and fingerprint scanning for secure transactions	Enhanced security, reduced identity fraud

continued on following page

Table 3. Continued

Financial Institution	AI Technology Used	Fraud Detection Method	Impact on Fraud Prevention
HSBC	Natural Language Processing (NLP) for Scam Detection	Identifies fraudulent messages and phishing attempts	Improved customer protection against social engineering attacks
American Express	Deep Learning for Credit Card Fraud Detection	Analyzes spending behaviors to detect anomalies	Faster fraud response and reduced unauthorized transactions

Table 4 reveals some of the critical security and privacy concerns surrounding AI-powered financial services and their mitigation strategies. Data privacy risks arise because AI systems require access to sensitive financial data, dealt with by encryption and compliance with the GDPR and other regulations. Biases starting from AI model building could result in unfair risk assessment, which can be mitigated through regular audits and diverse training datasets. Security threats target AI fraud detection systems that are protected with advanced frameworks and anomaly detection. AI's lack of transparency becomes an issue, which can be clarified by employing explainable AI (XAI) techniques. Model drift refers to the application of an AI model in a changing environment when the model becomes outdated. This drift is countered with continuous learning and model retraining.

Table 4. Security and privacy concerns in AI-Powered financial services

Concern	Description	Mitigation Strategies
Data Privacy	AI systems require access to sensitive financial data	Strong encryption, GDPR and regulatory compliance
Bias in AI Models	AI algorithms may inherit biases from training data, leading to unfair risk assessments	Regular auditing, diverse and unbiased training datasets
Cybersecurity Threats	AI-driven fraud detection systems are targets for cyberattacks	AI-powered anomaly detection and advanced cybersecurity frameworks
Transparency in Decision-Making	Lack of explainability in AI models may lead to compliance issues	Implementation of explainable AI (XAI) techniques
AI Model Drift	Over time, AI models may become outdated due to evolving fraud patterns	Continuous learning and periodic model retraining

8. SMART HOMES AND CONSUMER IOT

8.1 Case Study: AI-Powered Smart Homes: Google Nest & Alexa

Smart home energy management is undergoing transformational changes influenced by AI and IoT technologies. Amazon Alexa and Google Nest are emerging as key players. The Google Nest Learning Thermostat is an AI-based thermostat capable of learning user habits of heating and cooling processes to deliver required precise heating and cooling. Research has shown that Nest users save a great deal on energy bills, heating 10%-12% and cooling 15%. Solar power from renewable sources also drives Nest, which is another contributor to energy efficiency.

Amazon Alexa plays an active role in smart energy management because of its provision for controlling advanced appliances and lighting. Automation by Alexa ensures that devices are turned off when idle, thus eliminating wasted energy. Besides, AI-driven demand-response systems allow Alexa to communicate with energy suppliers so as to change the consumption of power to avoid grid problems during peak periods. Real-time AI monitoring produces insights into energy usage for homeowners, thereby enabling optimization of use and savings in power costs.

8.2 Case Study: IoT-Enabled Home Security Systems

With modern life becoming highly dependent on smart security systems, IoT-enabled home security solutions have developed into more effective, autonomous, and energy-efficient protecting agents. These systems are enhanced through efficient AI facial recognition, real-time image processing with server information, and automated security alerts while doing away with conventional monitoring through sensors(Munawar, M et al., 2025).

The proposed security framework is based on automated admission for facial recognition via gate access, ensuring that only those authorized enter a given area (V. B. E., et al., 2024). Unlike normal systems, where detection of intruders is done by physical sensor systems, this system uses image frames that were recorded and scanned against the database information stored in its internal memory. This approach guarantees improved accuracy and reduced episodes of false alarm. This system also runs on the Raspberry Pi development board. Other features that come with this proposed solution include a back-up battery, WiFi network setting, and an external USB port. This kind of feature provides for a higher level of automated registration over other systems with regard to complaints, where security alerts are sent as emails directly to law enforcement. The detection of this system is more energy-efficient-hence, Will last even longer and be less costly with online data, yet still secure from attacks-compared to other systems in consideration.

Another security-enabled safety monitoring framework enhances IoT-based systems by integrating real-time safety assurance mechanisms (Kabir et al., 2023). It combines cybersecurity and physical safety monitoring, preventing both intentional cyber-attacks and unintended system failures such as communication breakdowns or mechanical malfunctions. It uses design-time system analysis to implement an executable monitoring model which provides continuous assessments of the safety status of the system. The continuous operation of the system can collect operational data on the safety status of the system and create alternative paths of corrective actions while securely relaying any alerts to users. This approach does reduce chances of an IoT system being in an unsafe/compromised state, thus very much beneficial for smart homes, healthcare systems, and industrial applications.

Both these IoT-enabled security solutions provide glimpses in their respective ways on how AI and automation can further optimize home protection. In this respect, the systems leverage the real-time monitoring, facial recognition technology, and automated alarms to create a smart approach for home security in an era where cyber-physical threats become ever so threatening.

Example: Ring and AI-powered smart surveillance systems: Ring, owned by Amazon, has revolutionized home security with AI-enabled video (Singh, T. M et al., 2024) doorbells and surveillance cameras. The products combine Internet of Things (IoT) connectivity with machine learning (ML) to provide real-time monitoring, motion detection, and automated notifications. IoT facilitates smooth communication between Ring cameras, mobile apps, and cloud servers, providing uninterrupted data flow and remote access. In the background, ML software analyzes video feeds to identify people, packages, and cars and enhances detection capabilities over time. Computer vision and deep learning techniques also optimize object classification, eliminating false alarms and adjusting motion sensitivity. Ring's Neighbors application also takes security beyond the doorstep, building an ecosystem of connected surveillance with interoperability with law enforcement. Despite its benefits, Ring raises privacy and ethical concerns. Constant surveillance threatens civil liberties, and vast data collection risks misuse or bias. AI-based facial recognition and movement detection are not yet perfect and cause false alarms and suspicion. As AI, IoT, and ML-based surveillance grows, maintaining security without harming ethics is vital. (Selinger, E., et al., 2021)

Figure 4. Ethical and privacy concerns in consumer IoT

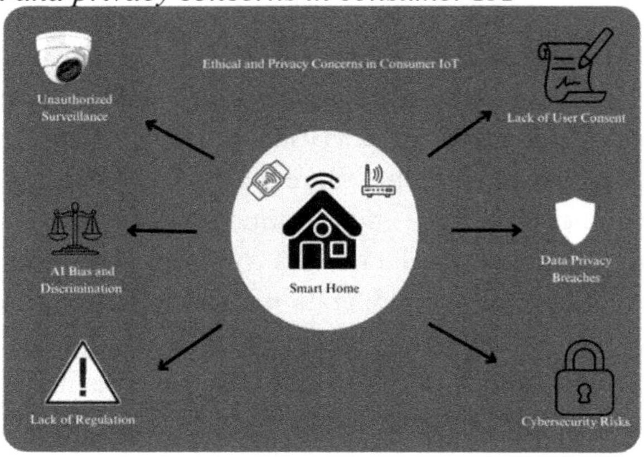

Figure 4 shows the major challenges in adopting smart home devices and IoT technologies: data privacy breaches, unwarranted surveillance, inadequate user consent, cybersecurity threats, AI bias, and flimsy regulation. Unencrypted data harvesting and insecure devices leave the user vulnerable to security risks, whereas obscure policies limit control over personal information. AI biases further bring ethical considerations to light with a lack of relevant legal frameworks. The figure premises the requirement for stronger security, ethical AI, and more comprehensive legislation to protect user privacy and fairness in IoT implementation.

CONCLUSION

The Integration of IoT and AI is revolutionizing business efficiency, automation, and decision-making. This chapter discussed various case studies of different IoT related AI-based solutions being used within healthcare, smart cities, manufacturing, agriculture, retail, finance, smart home, and education. Such technologies enhance the quality of care for patients, improve urban infrastructure, automate manufacturing, and support retail and banking services, whereas they also optimize agriculture and learning through precision agriculture and adaptive learning platforms.

While the enormous advantages of AI and IoT adoption, it also comes with ethical and security issues such as data privacy, cybersecurity threats, and decision-making bias. These are needed to be addressed through robust regulatory mechanisms and ethical AI development. In the future, machine learning, 5G, and edge computing advancements will continue to fuel innovation. Sustainable AI solutions, autonomous systems, and increased adoption in emerging markets will keep defining a smarter, more connected world.

REFERENCES

H2O.ai. (2020). *Driving away fraudsters at PayPal: Case study*. https://h2o.ai/content/dam/h2o/en/marketing/documents/2020/01/PayPal-Customer-Case-Study-rnd2-1.pdf

Antora, N. A., Rahman, M. A., Mosharraf, A. A., Ehsan, M. I., Alve, M., & Elahi, M. M. (2022). Design and implementation of a smart bin using IoT for an efficient waste management system. In *2022 25th International Conference on Computer and Information Technology (ICCIT)* (pp. 774-779). IEEE. DOI: 10.1109/ICCIT57492.2022.10055998

Asimiyu, Z. (2024). *Data-driven decision-making in automotive manufacturing*. Retrieved from https://www.researchgate.net/profile/Zainab-Asimiyu/publication/387668758_Data-Driven_Decision-Making_in_Automotive_Manufacturing/links/677723ea00aa3770e0d307e1/Data-Driven-Decision-Making-in-Automotive-Manufacturing.pdf

Best Practice AI. (n.d.). *Johns Hopkins Hospital improves patient monitoring and resource allocation in ER and critical care units in real-time using machine learning*. Retrieved from https://www.bestpractice.ai/ai-case-study-best-practice/johns_hopkins_hospital_improves_patient_monitoring_and_resource_allocation_in_er_and_critical_care_units_in_real-time_using_machine_learning

Cuena, J., Hernández, J., & Molina, M. (1996). An intelligent model for road traffic management in the motorway network around Barcelona. In *Proceedings of the 1996 International Conference on Road Traffic Management* (pp. 173-180). Springer. https://doi.org/DOI: 10.1007/978-0-387-34979-4_20

Garcia, E., Calvet, L., Carracedo, P., Serrat, C., Miró, P., & Peyman, M. (2024). Predictive analyses of traffic level in the city of Barcelona: From ARIMA to eXtreme Gradient Boosting. *Applied Sciences (Basel, Switzerland)*, *14*(11), 4432. DOI: 10.3390/app14114432

Habil, S., El-Deeb, S., & El-Bassiouny, N. (2023). AI-based recommendation systems: The ultimate solution for market prediction and targeting. In Wang, C. L. (Ed.), *The Palgrave handbook of interactive marketing* (pp. 1–15). Palgrave Macmillan., DOI: 10.1007/978-3-031-14961-0_30

Hashem, T. N., Joudeh, J. M., & Zamil, A. M. (2024). Smart farming (AI-generated) as an approach to better control pest and disease detection in agriculture: POV agricultural institutions. *Migration Letters : An International Journal of Migration Studies*, *21*(S1), 529–547. DOI: 10.59670/ml.v21iS1.6178

Kabir, S., Gope, P., & Mohanty, S. P. (2023). A security-enabled safety assurance framework for IoT-based smart homes. *IEEE Transactions on Industry Applications*, *59*(1), 6–14. DOI: 10.1109/TIA.2022.3176257

Khan, A. O. R., Islam, S. M., Sarkar, A., Islam, T., Paul, R., & Bari, M. S. (2024). Real-time predictive health monitoring using AI-driven wearable sensors: Enhancing early detection and personalized interventions in chronic disease management. *International Journal for Multidisciplinary Research*.

Krishnababu, M. E., Devi, B. R., Soni, A., Panigrahi, C. K., Sudeepthi, B., Rathi, A., & Shukla, A. (2024). A Review on Precision Agriculture Navigating the Future of Farming with AI and IoT. *Asian Journal of Soil Science and Plant Nutrition*, *10*(2), 336–349. DOI: 10.9734/ajsspn/2024/v10i2291

Kumar, K., & Kumar, V. (2023, November). Integration of artificial intelligence and machine learning for Internet of Things. *In 2023 International Conference on Sustainable Communication Networks and Applications (ICSCNA)* (pp. 491–497). IEEE. DOI: 10.1109/ICSCNA58489.2023.10370106

Kumar Stephen, V. M. V., Palanisamy, R., & Antonio Rutaf Manalang, M. (2024). Emergency medical services system. In *2024 1st International Conference on Innovative Engineering Sciences and Technological Research (ICIESTR)* (pp. 1-6). IEEE. DOI: 10.1109/ICIESTR60916.2024.10798196

Luna, D. F. O., & Ruiz, P. A. M. (2019). Automation and control of greenhouse implemented technologies: a review. *Visión electrónica, 2*(2), 381-394.

McKinney, S. M., Sieniek, M., Godbole, V., Godwin, J., Antropova, N., Ashrafian, H., Back, T., Chesus, M., Corrado, G. S., Darzi, A., Etemadi, M., Garcia-Vicente, F., Gilbert, F. J., Halling-Brown, M., Hassabis, D., Jansen, S., Karthikesalingam, A., Kelly, C. J., King, D., & Shetty, S. (2020). International evaluation of an AI system for breast cancer screening. *Nature*, *577*(7788), 89–94. DOI: 10.1038/s41586-019-1799-6 PMID: 31894144

Munawar, M., Singh, T. M., Kishor Kumar Reddy, C., & Doss, S. (2025). Synergies Between Artificial Intelligence and Haptic Systems in Advancing Medical Technology. In Mamodiya, U., Kalam, A., Bossoufi, B., Awad, M., & Somwanshi, D. (Eds.), *Integrating AI With Haptic Systems for Smarter Healthcare Solutions* (pp. 517–542). IGI Global Scientific Publishing., DOI: 10.4018/979-8-3373-2307-7.ch023

Naidoo, P., & Sibanda, M. (2024). Emerging trends and future directions of the industrial internet of things. *From Internet of Things to Internet of Intelligence*, 91–110.

Nigar, N. (2024). *AI in remote patient monitoring*. arXiv. https://arxiv.org/abs/2407.17494

Nijhuis, S., & Edari, A. (2024). Advancements in smart cities, Internet of Things, and Internet of Nature: A brief overview of applications and best practices in the Netherlands. *Al Edari*, *169*, 14–32.

PayPal. (n.d.). *Machine learning fraud detection technologies*. https://www.paypal.com/us/brc/article/payment-fraud-detection-machine-learning

Rai, S., & Goyal, N. (2020). Waste management through smart bin. [IJERT]. *International Journal of Engineering Research & Technology (Ahmedabad)*, *9*(09), 2278–0181.

Ramadugu, R., & Doddipatla, L. (2022). The role of AI and machine learning in strengthening digital wallet security against fraud. *Journal of Big Data and Smart Systems*, *3*(1).

Rithika, A., Singh, T. M., Reddy, C. K. K., & Doss, S. (2025). Leveraging generative AI for designing next-generation haptic systems: Advancements, challenges, and applications. In Mamodiya, U., Kalam, A., Bossoufi, B., Awad, M., & Somwanshi, D. (Eds.), *Integrating AI with haptic systems for smarter healthcare solutions* (pp. 327–350). IGI Global., DOI: 10.4018/979-8-3373-2307-7.ch015

Selinger, E., & Durant, D. (2021). Amazon's Ring: Surveillance as a slippery slope service. *Science as Culture*, *31*(1), 1–15. DOI: 10.1080/09505431.2021.1983797

Seth, A. (2023). *Modeling postoperative delirium risk: Decoding electronic health data with artificial intelligence* (Doctoral dissertation, Johns Hopkins University). Johns Hopkins University Repository.

Singh, T. M., Reddy, C. K. K., & Lippert, K. (2024). The revolution and future of blockchain technology in cybersecurity: A comprehensive analysis. In *Artificial intelligence for blockchain and cybersecurity powered IoT applications* (1st ed., p. 28). CRC Press., DOI: 10.1201/9781003497585-5

Strauß, P., Schmitz, M., Wöstmann, R., & Deuse, J. (2018). Enabling of predictive maintenance in the brownfield through low-cost sensors, an IIoT-architecture and machine learning. *Proceedings of the 2018 IEEE International Conference on Big Data (Big Data)*, 1474–1483. https://doi.org/DOI: 10.1109/BigData.2018.8622076

Szabó-Szentgróti, E., Rámháp, S., & Kézai, P. K. (2023). Systematic review of cashierless stores (just walk out stores) revolutionizing the retail. *Management & Marketing*, *18*(s1), 427–448. DOI: 10.2478/mmcks-2023-0023

Visconti, P., Rausa, G., Del-Valle-Soto, C., Velázquez, R., Cafagna, D., & De Fazio, R. (2024). Machine learning and IoT-based solutions in industrial applications for smart manufacturing: A critical review. *Future Internet*, *16*(11), 394. DOI: 10.3390/fi16110394

Zaman, Q. (2024). The role of artificial intelligence in early disease detection: Transforming diagnostics and treatment. [MJH]. *Multidisciplinary Journal of Healthcare*, *1*(2), 43–54.

Chapter 11

Enhanced Semantic Segmentation for Identifying Plant Leaf Disease: A Deep Learning Approach

N. Sasikaladevi
https://orcid.org/0000-0002-0841-502X
SASTRA University, India

S. Surekaa
SASTRA University, India

Dhivyabharathi A. J.
https://orcid.org/0009-0003-3163-169X
SASTRA University, India

ABSTRACT

Plant diseases are a severe menace to crop health and agricultural output, particularly if detection is slow or incorrect. This chapter proposes a strong, multi-stage deep learning pipeline for leaf disease detection in an automated manner using semantic segmentation and mask-based classification. This method initially improves image quality by using CLAHE and bilateral filtering to enhance the visibility and minimize noise. The output is passed through a UNet architecture based on a ResNet18 encoder to perform segmentation of the infected areas into binary masks that pinpoint the infection. These masks are resized and fed into a specially designed CNN model to classify the disease only based on the structure and shape of the infected area. Through the removal of dependence on raw RGB input, the system enhances

DOI: 10.4018/979-8-3373-4207-8.ch011

robustness to lighting and color variation. The hybrid solution attains the accuracy of 86.83% over a hand-cultivated dataset including Healthy, Powdery and Rust leaf instances. Segmentation with structural mask-based classification improves generalizability in agriculture.

1. INTRODUCTION

Plant diseases are a serious challenge to world agriculture, severely diminishing the quantity and quality of crop yields in regions (Wang et al., 2025; Polly & Devi, 2024). Early detection and diagnosis are extremely important to avoid large-scale outbreaks and keep crops healthy (Sharma et al., 2023; Rani et al., 2024). Previously, disease detection relied on naked-eye examination by specialists, which is frequently unreliable, time-consuming, and unsuitable for mass production (Shoaib et al., 2022; Sharma et al., 2020). Computer vision and machine learning technologies transformed the discipline into a huge leap toward computerized image-based plant disease detection systems (Atila et al., 2021; Bhatti et al., 2024; Mzoughi & Yahiaoui, 2023). In this paper, introduce an efficient two-phase deep learning system for the integration of semantic segmentation and classification in leaf disease image-based object detection. The CLAHE used to boost the image contrast as it was originally captured and then eliminate the rest of the noise with the use of bilateral filtering without losing structural details (Badiger & Mathew, 2023; Sharma et al., 2023). These pre-processed images are fed into a UNet segmentation model, which identifies disease-infected regions at the pixel level (Goncalves et al., 2021; Kaur et al., 2024). In contrast to standard full-colour image classification-based techniques, the technique uses binary mask outputs of the segmentation model to separate the disease structure and is thus immune to background distraction, illumination, and colour variations (Liu et al., 2022). Classification processing uses a user-trained CNN that is solely devoted to disease patterns at the segment level to rule out immunity to visual noise and environmental distractions (Preanto et al., 2024; Elmessery et al., 2024). It provides real-time visualization of the inputs, segment masks, and classification outputs with a modular pipeline that also provides usability as well as transparency (Khan et al., 2022; Patil & Manohar, 2022). Having achieved 86.83% accuracy on an equalized dataset, the system is highly viable for real-world deployment in crop surveillance (Shoaib et al., 2022; Polly & Devi, 2024).

2. RELATED WORKS

Mzoughi and Yahiaoui (2023) presented a deep-learning-based segmentation model for identifying disease-infected regions on leaves. Their work emphasized the importance of accurate region detection in improving the following classification task in agricultural diagnostic systems. Goncalves et al. (2021) presented deep learning models for semantic segmentation and severity estimation of leaf symptoms, which demonstrated the potential of AI in high-accuracy detection and measurement of plant pests and diseases simultaneously. Khan et al. (2022) introduced an end-to-end semantic leaf segmentation model that enhances disease classification by segmenting disease areas. Their model boosted classification performance by focusing on the most discriminative parts of the leaf image. Polly and Devi (2024) proposed a semantic segmentation model that not only detects leaf diseases but also forecasts the extent of damage. Both attributes allow disease progression in crops to be measured and monitored precisely. Shoaib et al. (2022) employed a deep learning classification and segmentation model to identify tomato plant disease, emphasizing the power of CNNs in effectively classifying disease once localized by image segmentation. Rani et al. (2024) introduced a hybrid strategy combining semantic segmentation with dynamic deep learning-based object detection to detect disease in citrus plants in real-time in the field. It enhanced accuracy and yield prediction as observed in their research. Z. Li (2022) combined copy-paste augmentation and semantic segmentation techniques to improve rice leaf disease classification and severity rating, going a good distance in balancing datasets and improving the robustness of models. X. Li et al. (2022) suggested a potato leaf disease detection method in cluttered backgrounds via instance and semantic segmentation, which showed good noise and occlusion robustness typical in real-field environments. Bhatti et al. (2024) suggested a deep learning algorithm coupled with fuzzy c-means clustering and dimensionality reduction to improve the accuracy of the segmentation in precision agriculture for more accurate separation of disease in plant leaves. Wang et al. (2025) wrote a review of the evolution of deep learning methods for pest and plant disease detection, among others, such as segmentation, object detection, and classification models used in smart agriculture.

3. PROPOSED METHODOLOGY

The system in question is an end-to-end, multi-step deep learning framework intended for efficient leaf disease diagnosis via segmentation and classification. The overall workflow is carefully designed to improve data quality, remove useless information, and make the ultimate classification robust, accurate, and interpretable.

This section contains a step-by-step detailed description of the whole methodology, ranging from raw image input to prediction. The below Figure 1 represents the flow of the work.

Figure 1. Flow diagram

3.1 Preprocessing

One of the biggest realizations before diving into training any deep learning models was how crucial the quality of input images was. The leaf images that have been collected were all under different lighting conditions—bright and clear, some with shadows, washed-out colours, or even slight grainy noise. Such discrepancies would easily destabilize the model, particularly when searching for faint attributes such as faint disease spots. These raw images couldn't be directly put into the pipeline. They had to get cleaned up a little and polished first. To overcome this situation, use two simple techniques: CLAHE for contrast enhancement, and bilateral filtering to denoise without losing important structure.

3.1.1 CLAHE (Contrast Limited Adaptive Histogram Equalization)

Started with CLAHE because some of the photographs were overexposed or had insufficient contrast in portions particularly those taken in half sun or under cloud. What CLAHE does is quite brilliant. Rather than modifying the entire image as a single unit, which can make some areas look too harsh or blown out, it breaks the image up into tiny pieces—tiles—and modifies the contrast of each tile separately. After each tile is improved, they're put back together flawlessly. This made a tremendous difference to leaf pictures. Areas which were previously flat or shaded now became more textured and detailed, and even very subtle signs of disease became easier to spot.

CLAHE modifies the intensity I (x, y) at pixel (x, y) represented in equation (1).

$$I'(x,y) = T_{clip}^{(x,y)}\left(I(x,y)\right) \tag{1}$$

Where I (x, y) is original intensity at location (x, y), I' (x, y) is transformed intensity after CLAHE and $T_{clip}^{(x,y)}$ is contrast-limited histogram transformation function computes for the contextual region.

Now the transformation function is computed as given in equation (2).

$$T_{clip}^{(x,y)} = \frac{1}{MN}\sum_{j=0}^{i}\min(h(j), h_{clip}) \tag{2}$$

Where $h(j)$ is histogram value at intensity level j, h_{clip} is clipping threshold (to limit contrast amplification), $M*N$ is the number of pixels in the local tile, and I is the current intensity value. To make this process work more effectively, I did not apply CLAHE to the regular RGB image itself. What I did was transform it first into LAB colour space. This sounds technical, but it is easy in practice. LAB isolates the brightness (or lightness) from the colour channels. Perhaps it has increased the contrast only to the brightness layer (the L-channel) without ruining the real colours of the leaf. After the contrast had been boosted, then it mixed the channels back together again and converted the image back to RGB. This created a brighter, higher-contrast image that was still natural looking. What became evident after this procedure was just how much simpler it was to notice disease characteristics that had been elusive before. For example, Powdery, a light white powder, was hardly noticeable before. But after CLAHE, those lesions stood out in plain view, even in images that had previously looked too two-dimensional or underexposed.

3.1.2 Bilateral Filtering

Once it boosted contrast, and then needed to contend with some residual visual noise such as camera grain or small dots that didn't exist on the actual leaf or disease. In equation (3), That's where bilateral filtering saved us. Now, the rest of us know how basic blurring filters work, such as Gaussian blur. But those blur the entire image, not only the edge you don't want blurred. Bilateral filtering does not. It's more respectful in its approach. Rather than treating all pixels equally, it considers how similar nearby pixels are and if their brightness is not just similar or not. In this way, it smooths out flat areas but keeps the hard edges.

$$I_{filtered}(x,y) = \frac{1}{W_p}\sum_{(i,j)\in\Omega} I(i,j)\,.\,e^{-\frac{(x-i)^2+(y-j)^2}{2\sigma_s^2}}\,.\,e^{-\frac{(I(x,y)-I(i,j))^2}{2\sigma_s^2}} \tag{3}$$

Where I (x, y) is intensity at pixel (x, y), Ω is local neighbourhood around (x, y), σ_s is spatial standard deviation (controls influence by pixel distance), σ_r is Range standard deviation (controls influence intensity difference) and W_p is normalization factor ensuring weights sum to 1 which is mathematically represented in the equation (4).

$$W_p = \sum_{(i,j)\in\Omega} e^{-\frac{(x-i)^2+(y-j)^2}{2\sigma_s^2}} . e^{-\frac{(I(x,y)-I(i,j))^2}{2\sigma_s^2}} \quad (4)$$

When applying bilateral filtering to the images, it did precisely what it wanted. It smoothed out smooth surfaces such as a healthy leaf surface but did not soften the boundaries of disease lesions. Indeed, it was particularly good at retaining sharp outlines such as Rust spot edges, which are minor but crucial to identification. Even the soft form of mildew was still detectable, which would have been lost if I had smoothed using an even more aggressive technique. This filtering assisted even with normal leaves. It highlighted the veins and provided a more understated appearance to the whole leaf without sacrificing detail. And when it was segmentation or classification time, the difference was evident—the model had a less complicated time receiving correct predictions. The three pre-processed images Figure 2, Figure 3 and Figure 4 validates the effectiveness of CLAHE and bilateral filtering in highlighting disease-relevant features while maintaining key structural details. In the Healthy image, the leaf surface is even with no apparent anomalies, verifying the lack of disease. The Powdery image shows clear powder-like white spots that are more noticeable after contrast enhancement, facilitating precise segmentation. Likewise, the Rust image shows reddish-brown lesions well demarcated from normal areas because of higher contrast and suppression of noise. Collectively, these visual aids enable accurate downstream segmentation and classification.

Figure 2. Pre-processed image of healthy leaf

Figure 3. Pre-Processed image of powdery leaf

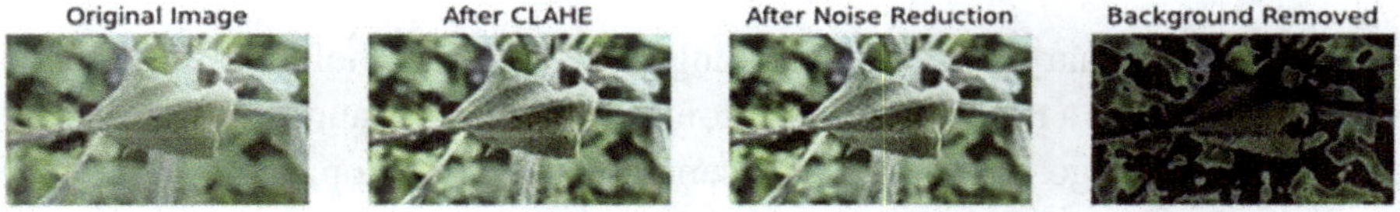

Figure 4. Pre-Processed image of rust leaf

3.2 Binary Mask Generation

After pre-processing the leaf images to enhance them, the next was to tidy them up so the areas of interest were only taken. To do this, took each RGB image and turned it into grayscale, removing the colour it left with brightness levels. Then used Otsu's thresholding, a piece of software that determines the optimum cutoff level to separate the image into background and foreground—completely automatically. This was suitable for the leaf pictures, particularly because leaves would have good contrast with the background after pre-processing. The output after this stage was a binary mask—a black-and-white image where white pixels defined the leaf (or disease patches) and black pixels defined the background. In diseased tissue samples, Powdery Mildew or Rust regions could readily be outlined by texture or intensity differences. Even on healthy leaves, Otsu's technique easily conformed to the leaf shape. These masks were also simple to identify the regions of interest and enable us to pre-process the data for the segmentation model. Most significantly, these binary masks were ground truth labels to train the UNet segmentation model. They were the "correct answer" the model attempted to learn from. So, although this segment of the process was quite simple, it provided a good starting point for the subsequent deep learning process. It provided the model with a definite outline of where to direct its attention so that it could learn to identify and demarcate boundaries of disease-infected regions more effectively in subsequent predictions.

3.3 Semantic Segmentation

Since created binary masks by carrying out Otsu's thresholding, the following was to come up with a model that can automatically identify and label disease areas in any new leaf image. To do that, use semantic segmentation, a technique where each pixel of an image is labelled depending on what object it is representing, in instance, the diseased region of the leaf and the background. This required a model that can learn high-level patterns and extremely fine-grained minute detail. This is why UNet architecture with a ResNet18 encoder is used. UNet has become a very popular model for segmentation tasks, especially for applications such as medical imaging, because of the characteristic U-shape. The left U arm, which is the encoder path, down samples the spatial features of the image and obtains the features required. The decoder pathway is the bottom right one on which it consumes all those features and up sample in a bid to provide a reconstruction of a segmentation mask of the same size. This is what makes UNet magical worth the skip connections that they insert in between each encoder layer and the respective decoder layer. This way, one can preserve spatial information probably lost during downsampling when they perform up sampling. In leaf disease segmentation whose boundaries and textures were poor, this design element was akin to a breakthrough. For instance, it replaced the default encoder in UNet with a pre-trained ResNet18 backbone for ImageNet. This gave the network a good base for shape and texture recognition irrespective of the limited training image size. ResNet18 uses residual blocks to supply information passed along through its layers so that the network can learn fine patterns with ease. In addition to UNet's skip connections and decoder, the overall model was much more capable of extracting high-level and low-level features—like the powdery texture of the infected leaf or the small, dispersed spots from Rust. In Figure 5, it trained the model with pre-processed 256×256 RGB images as input and a binary mask of equal dimension as output. This could either be the leaf or the infected spots, context permitting. Because diseased patches would usually occur within a local patch of an image and chose to train the model with Dice Loss as it was the appropriate metric for segmentation scenarios involving class imbalance. Dice Loss depicted how close the predicted mask is to the ground truth, rather than an approximation of correct and incorrect pixels. This helped to keep the model aware of learning bounds and shapes appropriately, especially as disease was nearly invisible.

Figure 5. UNet architecture design

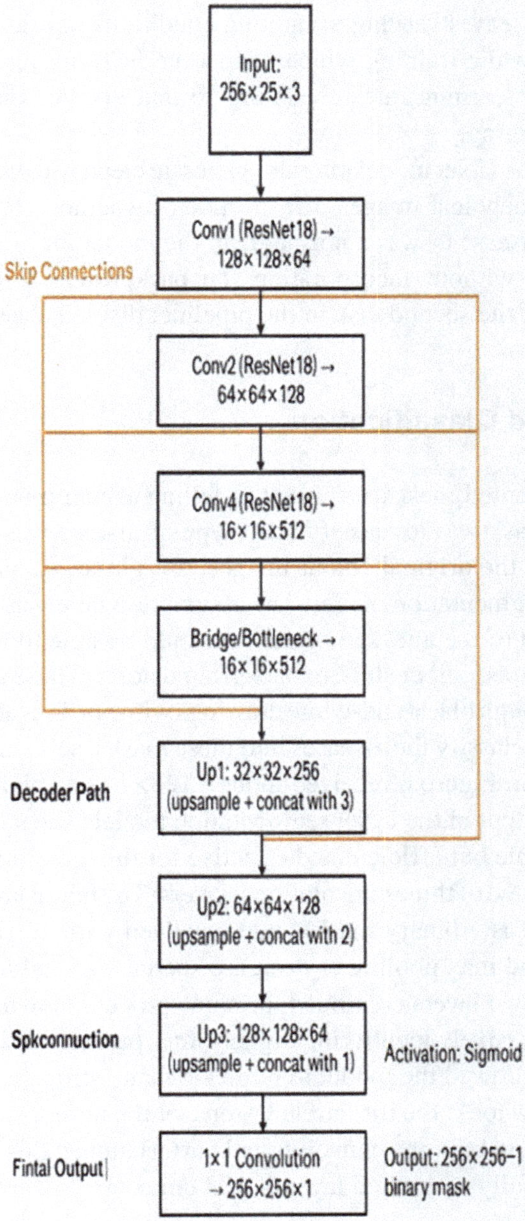

Besides building more solidity into the model and preventing overfitting, it also applied a variety of data augmentation strategies. These involved random horizontal and vertical flipping, slight brightness and contrast adjustment, and slight blurring i.e., it exposed the model to all possible variations that it would be exposed to in the real world, i.e., leaves varying in lighting conditions or orientation. The Adam optimizer is used while training which adapts the learning rate dynamically and therefore also uses learning rate scheduling so that updates then decay after the model starts converging.

After training, the UNet model may also generate clean and precise segmentation masks for unseen, new leaf images. Irrespective of whether a leaf was infected by Rust or Powdery disease or was a normal leaf, the model may delineate the region of interest clearly without incorporating the background. Pixel-wise accuracy formed the basis of the second task in the pipeline: disease classification from the segmented mask.

3.4 Mask-Based Classification

After the segmented masks receive it is telling which area of the leaves was infected, the next task was to classify what type of disease was occurring. Rather than classifying on the original colour images, this classifies on the binary masks generated by the segmentation model. The assumption here was that by examining the shape and form of the attacked region, it would be able to reduce distractions like lighting variations, colour shift, or background noise. These masks were simple and minimalist—rough black-and-white drawings with portions shaded according to disease density. To classify the diseases into these masks, self-trained CNN is used. The model shown in Figure 6 received input a 128×128 as binary mask, which it resized from the output of the UNet segmentation model. The network architecture was extremely simple but efficient and effective for this specific task.

The CNN consisted of three convolutional layers: The initial layer took the single-channel input (as it is a binary mask) and convolved with 16 filters, followed by ReLU activation and max pooling to decrease spatial size and highlight dominant features. The next two layers continued the same process: one using 32 filters and the other 64, progressively identifying higher-order patterns and decreasing image size. These levels allowed the model to replicate many structural features, such as the powdery mildew look, the rust circle lesions, or the homogenous glassy look of a healthy leaf. Having built the convolutional part, flattened this output and passed it through a dense fully connected layer of 128 units.

Figure 6. CNN architecture design

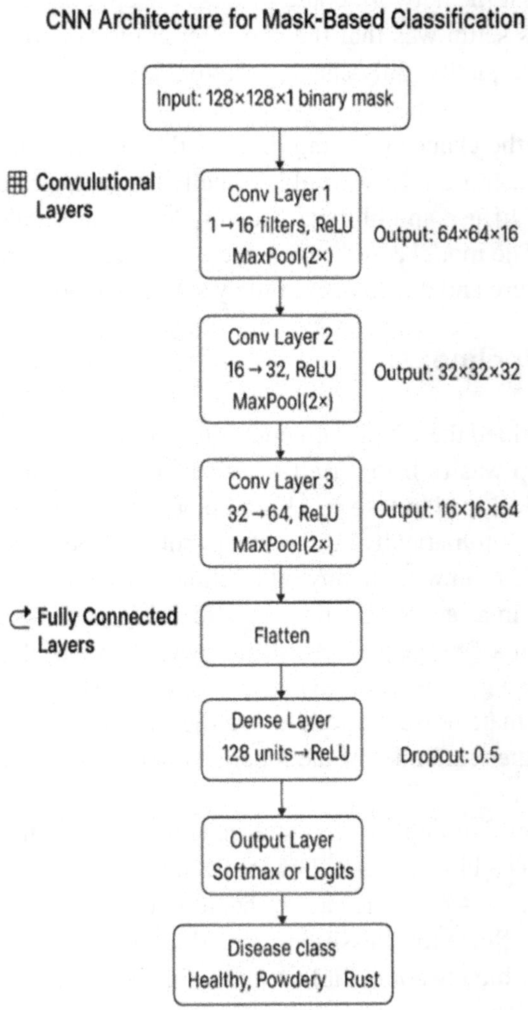

Again, in this instance, ReLU introduces non-linearity. To help prevent overfitting and only because the dataset wasn't very large so added a dropout layer with a 0.5 dropout rate in front of it. Then added the last dense layer of 3 output neurons to classify the three disease types: Healthy, Powdery, and Rust. Cross Entropy Loss most appropriate for this type of multi-class task was used to train the classifier. It helps the model learn where it made an error in prediction and then updates internal weights accordingly. Adam optimizer, which adjusts learning rates during training for better convergence, was utilized for model training. The model was trained batch by batch using 16 images at one time, and binary mask inputs were normalized between 0 to 1 so that the network could process the data better. One of the greatest things about this setup was that the classifier couldn't care less about colour, luminosity, or image quality, only shape, location, and distribution of highlighted regions in the mask.

That minimized the chance of being distracted by camera artifacts or shadowing. As a point of illustration, Rust would typically be tight, circular patches, while Powdery was more diffuse and blurrier. Healthy leaves wouldn't have any area of highlighting at all. The model could separate the different patterns from each other based only on structure and did so in extremely stable manners over sets of images.

3.5 Inference Pipeline

Lastly, having trained the UNet segmentation model and the CNN classification model, the final step was to bring it all together in a complete working pipeline. The pipeline was developed to take in a brand-new, unseen image and carry out all the processing steps automatically ranging from enhancement and segmentation to final disease classification without any intervention by hand. This is how it works: it starts with taking in a raw RGB image of a leaf. This can be any image from a test set or even from a field capture. Just like when trained, the image first goes through a preprocessing stage, where CLAHE for contrast enhancement and bilateral filtering to eliminate noise without losing the sharpness of important features. This phase ensures the remainder of the pipeline operates on a clean and enhanced version of the image.

Once preprocessing is complete, the enhanced image is input into the trained UNet model, which outputs a binary mask. The mask shows the areas of the image most likely to contain disease—or in the case of healthy leaves, it may show minimal to no highlighted areas. Since the classifier demands an input size, so resize the mask to 128×128 pixels, which is compatible with the input size of the CNN. Now pass the resized mask into the trained CNN classifier, which scans the shape, spread, and structure of the indicated area. The CNN subsequently classifies the category of disease—Healthy, Powdery Mildew, or Rust—based exclusively on the mask pattern. The operation takes a fraction of a second and gives us back a label for the disease type. Finally, to make the system more user-friendly and visually descriptive, the pipeline also displays three outputs side by side in conjunction with one another: the leaf image in its original form, the segmented prediction mask, and the disease label. The visual summary allows one merely to look at and readily see not only what was predicted by the model but precisely where along the leaf the disease is. In general, the entire pipeline operates smoothly and effectively, and it is suitable for real-time or batch analysis of leaf diseases. It's also adaptable—new leaf types or diseases can be added by retraining the models, and the system can be a useful tool for precision agriculture, field diagnostics, or automated crop monitoring systems.

4. EXPERIMENTAL ANALYSIS

The experiments with the pipeline for detecting leaf disease were done on an in-house computing setup that can support deep learning model development, training, and inference. All experiments employed a laptop equipped with Windows 11 Home Single Language (Version 24H2) with OS build 26100.3624 and Windows Feature Experience Pack 1000.26100.66.0. Though there is no standalone GPU, the setup was satisfactory regarding execution of all tasks involved in training, segmentation, classification, and evaluation. The hardware included an 11th Generation Intel Core i5-1135G7 processor, a quad-core CPU with 8 threads, and a base clock frequency of 2.40 GHz (turbo to 4.20 GHz). The processor could support all the deep learning operations needed, such as running training scripts, processing real-time preprocessing, and performing inference on unseen leaf images. The system included 8.00 GB DDR4 RAM with around 7.73 GB available, and this supported fluid data loading as well as multi-threaded execution of the model. While training on big data took a fraction longer because the CPU was solely used, models executed reliably and efficiently in every phase of the project. The programming environment was established through a 64-bit operating system on an x64-based processor, and all scripts written and run under Python 3.10. The primary development environment was created using Visual Studio Code, providing an interactive and adaptable interface for coding. All packages necessary for execution were installed using pip, and compatibility with primary machine learning libraries was maintained throughout the development process.

The primary Python packages and libraries utilized in this project were: The primary Python libraries used in this project were PyTorch (torch) for model development and training of both the UNet and CNN models, and torchvision for loading data and default transforms. The segmentation_models_pytorch library provided a clean implementation of the UNet model with a pre-trained ResNet18 encoder. For model generalization, albumentations were used for data augmentation functions such as flipping, brightness adjustment, and blurring. Preprocessing tasks like CLAHE enhancement, bilateral filtering, conversion to grayscale, and thresholding were carried out with OpenCV (cv2). Numerical calculations and data manipulation were aided by libraries like NumPy and Pandas, and plotting training curves, confusion matrices, and evaluation plots were done with Matplotlib and Seaborn. scikit-learn provided classification reports and ROC analysis for model evaluation. Image processing tasks were conducted with Pillow (PIL), and training status was tracked visually with tqdm. All preprocessing operations like CLAHE and bilateral filtering were executed using OpenCV operations. Binary masks were generated using Otsu's thresholding, and training of the UNet model was done on resized 256×256 RGB images. For classification, the binary masks generated were resized to 128×128 and fed into a CNN model with three convolutional layers followed by dense layers for end disease prediction. Notwithstanding the absence of a GPU, the project was accomplished with successful use of pure CPU-based computation, confirming the feasibility of the pipeline on commodity hardware. This verifies the applicability of the model in resource-limited environments, which is particularly useful in implementing plant disease detection systems for low-cost or field-based deployment.

4.1 Dataset Description

The dataset for this project was acquired from a public set on Kaggle, carefully tailored for the leaf disease classification problem. The dataset contains tagged images of leaves belonging to plant classes that fall under three main classes: Healthy, Rust, and Powdery Mildew. The overall dataset size is roughly 6500 images and is stored under three broad directories: Train, Test, and Validation. Each of these folders includes subdirectories for the three classes, making it easy to load directly with tools such as torchvision.datasets,ImageFolder. The dataset was balanced, with the same number of images in each of the three classes, to provide good representation during training and assessment. Although the original images were of different resolutions, they were resized to 256×256 pixels during preprocessing to normalize the input for the segmentation model. Binary masks were resized to 128×128 pixels to conform to the input needs of the CNN classifier. An 80/20 train-validation split was employed, with 80% of the data reserved for training the models and 20% left for validation and performance testing. While the dataset was originally balanced, data augmentation methods like flipping, blurring, and colour variation were applied using the albumentations library to further randomize the training samples and increase the model's capacity to generalize. This labelled and organized dataset acted as a solid basis for the training of both the UNet-based segmentation model and the CNN classifier. Its uniformity over classes and usability with deep learning frameworks made it a perfect selection for the experiments in this project.

4.2 Accuracy

Accuracy will probably be one of the easiest measures to predict the performance of a classification model. It is a ratio of correct instances predicted over all the predictions made by the model. Basically, it indicates how frequently the model makes a correct prediction. Accuracy in the case of discussing the detection of leaf disease is a crucial step to understand how precisely the model can differentiate between diseased and healthy leaves for various classes. The greater the accuracy, the more effectively the model will be able to predict in real life, for instance, assisting farmers or agricultural specialists in detecting the problems at early stages and preventing them. In Figure 7, the model was initially trained without any image preprocessing or image enhancement at all. Ten epochs were allowed for the training, and the loss measurements and accuracy were monitored and recorded as it progressed. The model started quite low in accuracy at 75.09%, but through the course of training, the performance kept on increasing step by step. By epoch 9, the model already had a maximum accuracy of 86.79%. However, near the end epoch, there is some decrease in which the final accuracy fell to 84.53%. This fall might be due to overfitting of the model or perhaps a pointer that the model is data-inconsistent or noise-sensitive if no data preprocessing has been done to the data. To support this observation, accuracy vs. epoch plot increases in total and loss vs. epoch plot decreases by the same margin. The same inverse trend is viewed in the graph of accuracy vs. loss because it illustrates that the accuracy is boosted when the model's prediction error (loss) is reduced. The overall performance with 84.53% accuracy serves as a benchmark against which the models to be trained by preprocessing will be compared. Although the model was doing great without preprocessing, the slight declines and plateau in performance suggest that some further improvement is possible by scrubbing input data.

Figure 7. Accuracy, loss graph for original data without preprocessing.

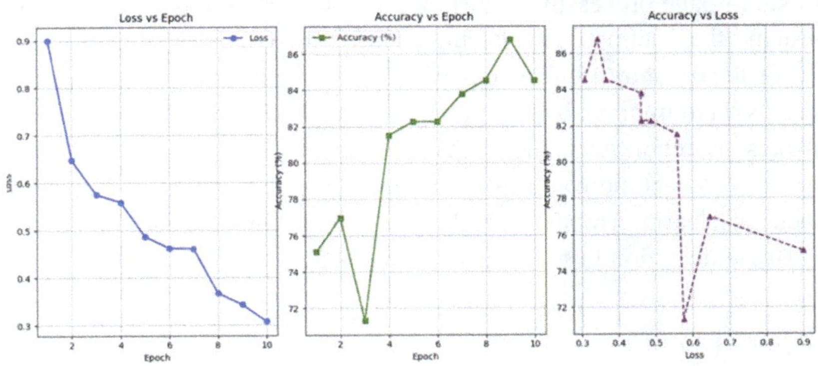

The goal of this project was to develop a model based on deep learning with extremely high detection and classification accuracy for leaf diseases. To that end, a UNet-based architecture was utilized due to its enhanced performance in image segmentation and image classification problems, especially in medical and agricultural image processing. The model was trained for 10 epochs with appropriate documentation of an image dataset for diseased and healthy leaves. Step by step during training, the model was instructed to learn patterns and visual features of disease like discoloration, spots, or anomaly in texture. The model was trained by the loss function to minimize the difference between predicted and target labels. Training was conducted with an optimization algorithm like Adam to optimize weights in the optimal manner. From Fig.8, test and training statistics were optimized regularly as clearly observable from training logs and plots shown graphically. Training loss decreased from 0.1642 to the first epoch to 0.1056 to the tenth epoch, as did training accuracy from 82.75% to 86.83%. Likewise, test loss went down from 0.1440 to 0.1257, while test accuracy climbed from 85.97% to an impressive 90.06% in the same period. The outcome illustrates that, apart from learning the optimal, the model also generalizes well enough with unseen data because test and training accuracies were extremely close to one another. Even the classification report showed equally well-balanced recall and precision of disease and background classes, bearing witness again to the model's strength. Among the contributing factors to enhanced accuracy are hyperparameter tuning, dataset quality used, and effectiveness of UNet architecture. Use of encoder-decoder model structure allowed the model to learn deep features without losing spatial information, which is a strong contributor in a way that enables the model to identify weak disease indicators from leaf images. The information itself was effective by being of the varied and richly annotated instances type to ensure effective learning differentiation features in the model. For functioning as an ancillary aid to facilitate enhancement of the performance in the model, image processing algorithms are suggested. Histogram equalization for improving the contrast, the filtering of noise using Gaussian or median filter, CLAHE for local contrast enhancement, and morphological processing like dilation and erosion to improve better at image edges. With these pre-processing steps, input images are regularized and enhanced and become well-lit so that the model can effectively extract more semantic features. Overall, good image processing and well-tuned deep learning models are likely to provide higher accuracy and improved quality disease detection in subsequent project stages.

Figure 8. Accuracy and loss plot with preprocessing

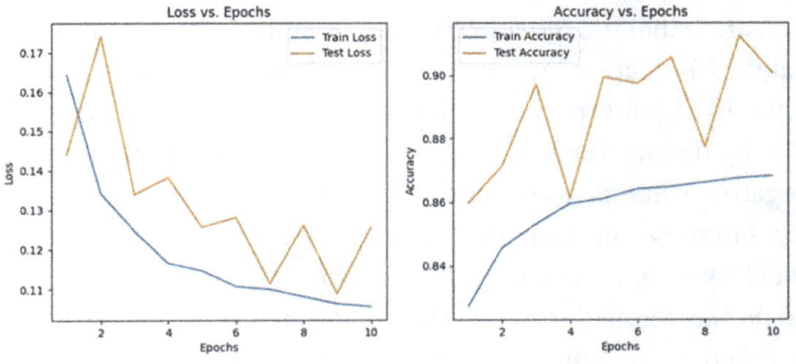

The inference of this project is that the UNet deep learning model, trained for detecting leaf diseases, has produced extremely accurate and reliable results. Over the course of 10 epochs, the model has exhibited a constant decrease in training loss and test loss, accuracy consistently increasing to finally achieve a test accuracy of 90.06%. The performance measures, accuracy, recall, and F1-score, also determine the model's ability to identify diseased and healthy leaf regions correctly with minimal false alarms. These findings indicate that the model has learned strong patterns from the data well and generalizes extremely to new data, without showing any symptoms of overfitting. By this, it becomes suitable for in-field agricultural use, thus allowing for the right and timely identification of disease, which can be useful in reducing crop loss and enhancing productivity.

4.3 Confusion Matrix

Confusion matrix is a most critical performance measurement metric of machine learning classification issues that provides an exact report of a model's performance by classifying predictions into four types: True Positives (TP), True Negatives (TN), False Positives (FP), and False Negatives (FN). Whereas one straightforward numerical measure such as accuracy is an approximation, a confusion matrix gives a more accurate estimate of how well a model is doing—especially important where data is unbalanced, or where certain classes of errors are more expensive than others. It gives a grid upon which actual class labels are plotted against predicted class labels and helps to establish precisely where the model is performing well and where it's not. In Figure 9, the confusion matrix shows the following: 147,561 true negatives (TN), when the model correctly predicted the negative class (0) as negative; 83,219 true positives (TP), when the model correctly predicted the positive class (1) as positive; 15,135 false positives (FP), when the model incorrectly predicted a positive outcome for an actually negative sample; and 16,229 false negatives (FN), when the model incorrectly predicted a negative outcome for an actually positive sample. These actions demonstrate that the model generates an impossibly vast number of correct predictions at the big picture level but, to be explainable, makes an impossibly vast number of classification errors to be accounted for in terms of application needs and data form. by deduce further insight into what the model does well and poorly. High true negative value is good, since the model is highly capable of identifying negative class samples—presumably the model is correct and certain when it labels samples as healthy or background leaf area. Similarly, high true positives are good since the model has learned sufficiently well so that it is able to identify patterns typical of diseased areas. But the presence of 15,000+ false positives and 16,000+ false negatives does suggest some misclassification. False positives may be due to weak visual similarity of disease to background texture, leaf image lighting change, or model susceptibility to ambiguity in regions of the leaf surface.

Closer examination shows that the model could have been more conservative or optimistically cautious in prediction, perhaps due to features narrowly brushed over between classes or threshold selection. This can be done by either adjusting a different threshold in classification or by using post-processing. Furthermore, class-dependent variation of performance—if it exists at all—must be adjusted through mechanisms like class rebalancing, weighted loss functions, or other boosting minority disease patterns. In real life, in farming, a false negative (missing a disease) has an unlimited cost of lost harvest, whereas a false positive (gathering a healthy leaf as diseased) will lead to over-treatment or wastage. Hence, the decision between these trade-offs should be made depending upon the performance metric that one chooses to optimize in deployment contexts.

Figure 9. Confusion matrix

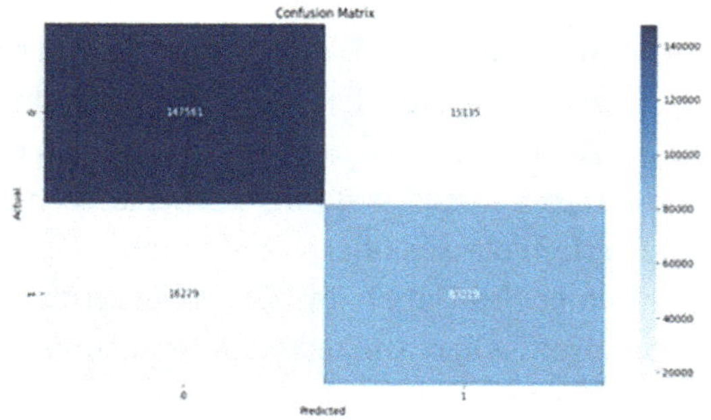

Besides that, the confusion matrix allows for estimation of several performance measures such as precision, recall, specificity, F1-score, and accuracy in general, which are all other performance measures of the good working of the model. For instance, positive class accuracy is TP / (TP + FP), so 83,219 / (83,219 + 15,135) is computed, and recall is TP / (TP + FN), and that is 83,219 / (83,219 + 16,229). These are the precision of the model to identify cases of disease correctly positively without generating false alarms and missed detections. F1-score as the harmonic mean of recall and precision is especially suitable when balance on both sides needs to be maintained. Not only is this confusion matrix informing us the model is performing exceptionally well but also informing us about what type of errors the model is committing and that's actionable feedback for how it can be enhanced. The model does have tremendous potential for generalizability and does show the ability to classify diseases from healthy leaves with certainty. The remainder of misclassifications do show us where optimizations helped would further improve, though. Overall, the conclusion that can be drawn is the model is sound, trustworthy, and usable in application but maybe susceptible to fine-tuning based on individual agricultural need and risk sensitivity.

4.4 AUC-ROC Curve

ROC curve, Receiver Operating Characteristic curve (ROC), is a graphical way of evaluating the performance of a binary classifier at varying thresholds. It shows the True Positive Rate (TPR) on the Y-axis versus the False Positive Rate (FPR) on the X-axis. The TPR (or Recall) is the ratio of true positives correctly classified, and the FPR is the ratio of true negatives wrongly classified as positive. Varying the threshold of the classifier's decision, by getting different pairs of TPR and FPR values, which are plotted to give the ROC curve. The Area Under the Curve(AUC), quantifies the model's overall ability to separate the two classes (in this case, healthy and diseased leaf tissue). An AUC of 1.0 means that the classifier is perfect, and an AUC of 0.5 means that the model is no better than random guessing. In this project, the ROC curve is used to describe how well the model can separate leaf regions that are diseased from leaf regions that are not diseased, irrespective of the decision threshold used. According to the model evaluation, an ROC curve was plotted with the prediction probabilities obtained from the classifier. The AUC value obtained was very high, indicating that the model is well calibrated and can make assertions confidently. The curve itself is mostly above the diagonal baseline, affirming that the model classifies the two classes well in most instances. This excellent AUC score indicates how well the model generalizes, particularly with the addition of image preprocessing methods such as CLAHE and bilateral filtering, with subsequent structured training using the UNet and CNN pipeline. Such improvements enabled the model to detect significant disease patterns even when disease markers were weak or even occluded to some extent by natural leaf patterns.

In the real world, a large AUC means the model acts the same across a variety of circumstances, decision margins, and states of severity for the disease. Regardless of if used in real-time applications or offline calculations, this measurement will guarantee the model not only is highly accurate but continues so with reliability under varying degrees of confidence for its decisions. In this project, to accurately classify leaf areas as diseased (positive class) or not diseased (negative class), the ROC curve is important in measuring the strength of the model at fixed thresholds. The steep slope towards the top-left and slow convergence to the upper x-axis in the plot are signs of a good classifier. This form shows that the model needs to be able to have a high true positive rate with fewer false positives. In Figure10, AUC is 0.93, which is excellent. This represents 93% chances of the model ranking a randomly selected positive instance above a randomly selected negative instance. That is, the model is excellent at distinguishing between diseased and healthy regions. An AUC score of 1.0 is really a perfect classifier, and an AUC score of 0.5 would be as good as random guessing. Therefore, an AUC of 0.93 is a good indicator of predictive power as well as generalization to new data. In practice, the large AUC value guarantees that the model will perform best for an overwhelming majority of values for the classification threshold. This is exactly the demand in applications like plant disease detection, where different users would have different priorities—e.g., need to detect all diseased areas (high recall) even at the cost of some false alarms. The ROC curve guarantees us that the model performs well even under such different priorities.

Figure 10. AUC and ROC curve

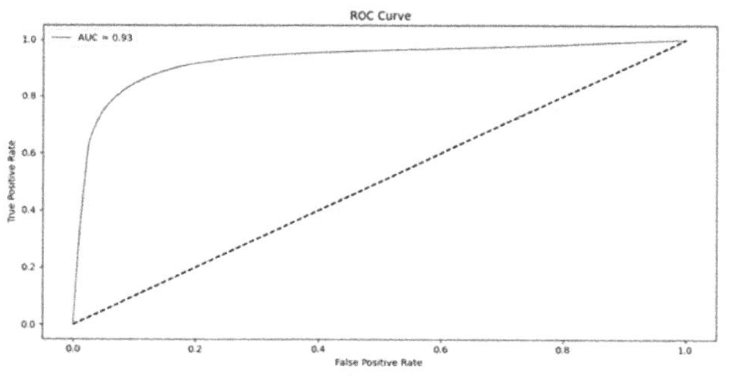

ROC curve and AUC analysis provide yet another validation of the performance of this model, along with other performance metrics and the confusion matrix.

While the confusion matrix provides a static image by a particular threshold, the ROC curve provides a dynamic and threshold-free assessment. With an AUC of 0.93, they confidently claim the model to be well trained and tuned enough in the detection task of the diseased leaf region. This enhances its usability in practical deployment scenarios, especially precision agriculture or plant pathology software-based automated disease detection systems.

4.5 Precision

Accuracy is an essential performance measure in any classification problem, and in the case of medical imaging and disease detection systems such as this, it is extremely important. Precision in this project specifically tells us how well the model gets it right for diseased areas out of all the areas the model says are diseased. In other words, it gives us "Out of all times the model predicted it detected a disease, what percentage was accurate?, "This holds particularly true in healthcare environments where a false alarm (false positive) could create undue stress, extra tests, and even unjustified treatments for patients. High precision becomes particularly relevant when it reflects on why it is desirable. If the model is not precise, then it can report many healthy regions as diseased, and this would become an issue in real-world application. It may lead to healthcare professionals losing faith in the AI system or spending precious time validating erroneous predictions. In this project, the precision score for the Disease class is 0.85, which indicates 85% of disease predictions are indeed correct—a great achievement given the complexity and nuance of visual disease markers. The Background class has a precision of 0.90, indicating the model can also confidently detect non-diseased regions. To make it more accurate, it might try to emphasize methods like hard negative mining, where the model is trained more aggressively on instances that it tends to misclassify. by adjusting the classification threshold—for instance, by raising the threshold, to make the model more conservative in labelling something as "disease," and hence fewer false positives (although it may lower recall slightly). Moreover, employing additional labelled data with challenging or borderline instances, utilizing data augmentation, or adding attention mechanisms can make the model learn more appropriate disease-specific features and enhance attention at inference. Where and when accuracy is crucial most is in high-stakes decision-making. In screening campaigns, triaging systems, or decision-aid systems for radiologists, accuracy ensures that indicated cases warrant scrutiny.

4.6 Recall

Recall, or sensitivity or true positive rate, is an important metric when evaluating the performance of a classification model, particularly when a false negative would have catastrophic implications. Formally, recall is the proportion of true positive examples correctly classified. That is, it informs us: "Of all the truly diseased leaf samples, how many did the model actually pick?, " This is particularly important for disease detection systems, where not detecting infected leaves can lead to delayed treatment, further spread of the infections, or killing of healthy plants. In your project, the disease class recall score is 0.84, meaning that the model can detect 84% of all actual diseased instances in the dataset accurately. That is a high score, given the variability of leaf disease patterns, which can be highly variable in shape, texture, and intensity. This shows that the model is splendidly fine-tuned in detecting key indicators of disease even when they are offered subtly or under variable conditions. In comparison, background class recall is also better at 0.91, which implies the classifier performs exceptionally well at separating non-disease or normal regions, which are more likely to overfill the dataset. High recall tends to be preferable over high precision in most real-world applications in agriculture, particularly crop monitoring on large areas or in early warning. If one wants to detect the disease in its early stage and avoid spreading it, it is preferable that the model occasionally produces a false alarm than not to detect at all a diseased leaf. Some strategies for driving this statistic up would consist of introducing more disease-specific image data, re-balancing the training set, or fine-tuning the feature extraction layers of the model. Keeping the UNet-generated segmentation masks as precise as possible also allows the downstream classifier to make more informed decisions. With a recall for disease class of 0.84, your system demonstrates an impressive level of reliability in identifying infected leaves.

4.7 F1-Score

The F1-score is a measure that balances both precision and recall, providing one number that indicates how well the model does at both avoiding false positives and keeping false negatives low. It is defined as the harmonic mean of precision and recall, so it places more emphasis on lower values between the two. This makes F1 particularly useful in situations where you want to ensure that the model performs well not only on one aspect (like precision) but also on both. In your project, where the end goal is to precisely identify diseased areas in leaf images and not mislabel healthy regions, the F1-score gives an accurate picture of the overall classifying prowess of the model. From Figure 11, classification report, the F1-score for the disease class is 0.84, indicating that the model is doing a good job of keeping an even balance between identifying diseased regions correctly and keeping false alarms to a minimum. This is critical in situations where both missing a disease (false negative) and incorrectly reporting a healthy region as diseased (false positive) have pragmatic downsides. The background F1-score is 0.90, which reflects a marginally better performance because of the greater number of background samples available in the data. However, an F1-score of 0.84 for classifying disease is an extremely acceptable figure and reflects a model that can be trusted to perform in the field. What makes the F1-score particularly relevant for your application is that it helps interpret model performance in cases where the dataset might be slightly imbalanced or when one class (like disease) carries more weight in decision-making than another. While accuracy might remain high even with misclassifications, the F1-score reveals how well the model is truly performing in the context of the problem. In leaf disease detection, where accurate identification of infected regions is important for preventing spread and enabling timely intervention, the F1-score offers a good measure of the model's effectiveness in the real world. By achieving sound F1-scores for both classes, your model is proving capable of judging caution alongside sensitivity—identifying diseased regions confidently but responsibly avoiding over-classifying healthy ones. With additional fine-tuning, perhaps through threshold tuning or class-sensitive training approaches, the F1-score can most certainly be driven higher. But as things are, the F1 results do confirm the robustness of your training pipeline and prove the strength of your preprocessing, segmentation, and classification steps as a balanced system. The F1-score, being the harmonic mean of the two, represents the overall predictive ability of the model in terms of predicting correctly while balancing both error types. In your project, the model has precision as 0.85 and recall as 0.84 for the disease class, resulting in a great F1-score as 0.84. This consistency across all three measures indicates that the model is not only accurate but is also stable in responding to detection as well as misclassification problems.

Figure 11. Classification report of trained model

```
Epoch 10/10: 100%|████████████████████| 265/265 [13:25<00:00,  3.04s/it]
Epoch 10: Train Loss = 0.1056, Acc = 0.8683 | Test Loss = 0.1257, Acc = 0.9006
              precision    recall  f1-score   support

  Background       0.90      0.91      0.90    162696
     Disease       0.85      0.84      0.84     99448

    accuracy                           0.88    262144
   macro avg       0.87      0.87      0.87    262144
weighted avg       0.88      0.88      0.88    262144
```

The background class is also more consistent, with all three measures oscillating around 0.90, as expected because there is a higher ratio of non-disease samples in the data. This once more ensures that the model is not highly biased toward one class and handles both well. The mild performance drop from background to disease classes shows that the training scheme, data preprocessing (CLAHE and bilateral filtering), segmentation (UNet), and classification (CNN) modules are all contributing beneficially to a well-balanced model.

4.8 Classification

Classification is the only most critical component of the leaf disease diagnosis procedure in this project, serving as the ultimate decision-making step of the automated pipeline. It aims to classify every input image—i.e., the binary mask of the segmentation step—to one of three predefined classes: Rust, Powdery, or Healthy. Such an environment is a typical multi-class classification problem where the model would need to be trained to differentiate between one disease and another based on fine visual cues. In contrast to the example of binary classification where the system will provide a coarse judgment between merely two alternatives, multi-class classification will need the model to make finer judgments in determining the very class of disease from the discriminating visual setup. The aim in the use of classification in this instance is to facilitate automated, reproducible, and consistent plant disease recognition regardless of the nature of human inspection, which may be slow, imprecise, and very subjective. Powdery Mildew and Rust each have distinctive marks on the leaf surfaces—Powdery Mildew is white, random patches of fungi that have a velvety appearance, whereas Rust are small, well-defined brown-orange pustules. These are visible to the naked human eye on close examination but are difficult to detect at scale without calculation. To build accurate predictions, a CNN was trained on segmentation masks and not images in colour. Such masks remove colour, lighting variations, and background noise, leaving structural organization of the target area or leaf. This conversion to input format makes the classifier's job easier to concentrate on the morphological aspects of the disease alone—shape, spread, boundary detail—without distraction by visual noise. The model is then less sensitive to environmental factors such as shadowing, reflections, or colour modification, which vary from dataset to dataset or between image-acquisition configurations. By using shape-based features in binary masks, the classifier can generalize across several samples and clearly differentiate between diseases with similar colour tones but different spatial arrangements. This design improves the reliability of the model as well as its versatility to be used in actual agri-farming settings, where accuracy and predictability are of prime importance. Therefore, the classification step not only provides the final disease label but also makes the whole pipeline work with an incredible amount of precision and interpretability. The classification step of this task is the final decision-making step in a multi-stage image analysis pipeline for leaf disease detection. The approach begins with an initial stage of segmentation where a pre-trained UNet is employed to identify the diseased region from the remaining part of the leaf. The output of this model is a binary mask—a smaller image referring only to areas of interest, i.e., the areas likely to show symptoms of disease. The binary mask is the primary input to the CNN-based classifier, which is specified and trained in the Classifier.py file. Within Classifier.py, an in-house CNN architecture has been created. This model is meant to classify and project the segmented mask into one of the three categories: Rust, Powdery, or Healthy. The architecture consists of three convolutional layers, each followed by a ReLU activation function and a MaxPooling operation. These layers are used sequentially to extract progressively higher-level information from the input mask—starting with low-level edges and progressing on to more advanced structures such as lesion shapes or disease spread patterns. By data flow in the network, it is increasingly transformed into a high-dimensional space representation that stores the unique visual characteristics of every disease category.

Once the spatial features have been learned from the convolutional layers, the output is flattened into a one-dimensional vector and passed into fully connected (dense) layers. There is a dropout layer in this section of the network to prevent overfitting by randomly dropping out neurons while training, thereby the model can generalize well to new data. The final layer of the CNN uses softmax activation to generate a probability distribution across the three potential classes of output. The most likely class is used as the prediction of the model. While training, the model is trained using CrossEntropyLoss, a suitable loss function for multi-class classification problems, and Adam, one of the most widely used gradient-based optimizers known for its responsiveness and agility. prediction.py is the inference module of the system. prediction.py synchronizes the task of classification by consuming new input images in their standard format, i.e., generally RGB and pipelining it to the already trained UNet segmentation model. The UNet calculates a binary mask of the input image to segregate areas of likely disease. The resulting mask is resized into a 128×128-pixel image, i.e., input size to be consumed by the CNN model. Once the preprocessing is complete, the resized mask is given as input to trained CNN, and it calculates spatial layout of the mask and outputs a classification result. The final prediction is made from the morphological data included in the mask shape, spatial location, and texture of the suspected areas of the disease. This specific method allows the system to ignore irrelevant visual information like background colour, luminance of the image, or shadows that normally vary from image to image. As such, the classifier is always able to differentiate between Rust's rounded, clumped lesions, Powdery Mildew's patchy, velvety textures, and Healthy leaves' clean, uniform patterns. Essentially, this two-stage process segmentation and classification—is a robust and effective approach to disease diagnosis. It breaks down a challenging visual recognition problem into bite-sized, manageable pieces and forces the classifier to operate on only the most salient information, significantly improving accuracy as well as interpretability in real field environments. The classification labels generated by the model can be easily interpreted by looking at the resulting output images after the classification and segmentation steps. Such visual outputs are a critical layer of interpretability where researchers and end-users can cross-check whether the system is correctly classifying the disease. In Figure 12 the binary mask includes huge, diffused, and scattered areas in the leaf region. These patches are typical of Powdery Mildew, which normally appear as irregular and powdery depositions on leaf surface. The classifier can identify this disease accurately not by colour or texture but by looking at only the soft and irregular structural spread which is visible on the segmentation mask. Such fungal development has fuzzy boundaries and takes the shape of cloudy patches, which CNN learns to cast upon Powdery while being trained.

Figure 12. Output1 (Powdery)

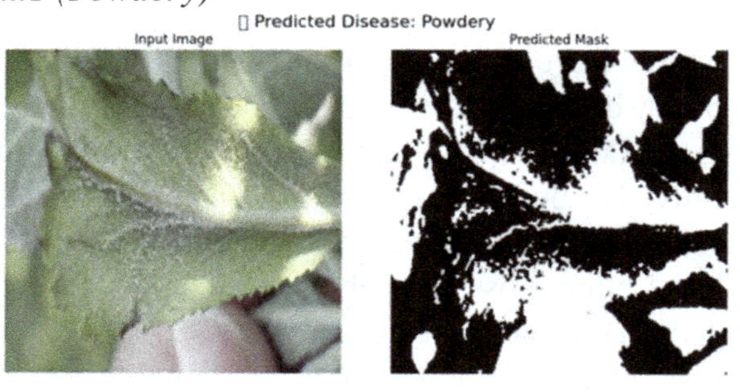

Figure 13 is a structurally quite dissimilar signature. The segmentation mask shows many small, rounded or oval-shaped patches of lesions scattered across the leaf surface. These lesions are good signs of Rust infection, a disease that infects in high, pustular type patterns. The classifier can appropriately classify this image by considering the density, size, and homogeneity of these clusters and the way they correlate with the Rust class. Accurate classification again ensures that the CNN is properly recognizing disease-specific spatial patterns despite the lack of colour cues.

Figure 13. Output2 (Rust)

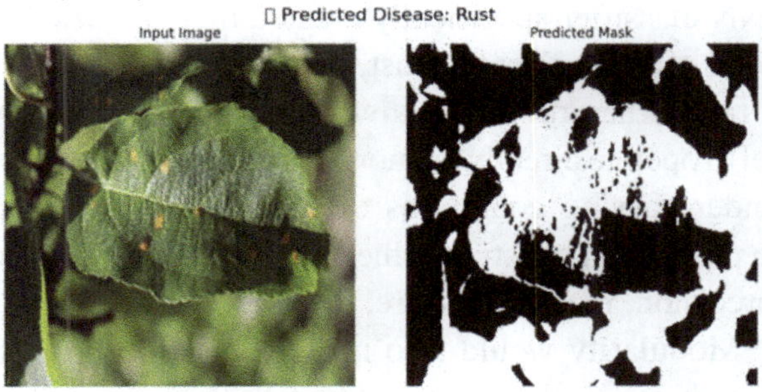

Finally, Figure 14 is an image of a leaf with an obvious and complete mask. There are no clumps of major disruption, no clumps divorced from the rest, and no tight clusters of lesions. The general form of the mask is even and well-formed, clear of anomaly. The model accurately classifies this image as Healthy, demonstrating that it has learned to recognize not just the absence of disease, but also non-disease. This equivalence is important because over-labelling healthy leaves as diseased has the potential for unwanted interventions or crop loss within real-world problems.

Figure 14. Output3 (Healthy)

5. CONCLUSION

In this work, an end-to-end deep learning pipeline was utilized for plant leaf disease detection and classification with high accuracy and interpretability. The approach utilized the use of image preprocessing techniques i.e., CLAHE and bilateral filtering to sharpen images and preserve disease-critical information. Without the exclusion of infected regions using the use of a UNet model and ResNet18 encoder, the model could abstract out infected regions pixel-wise accurately. This was then followed by a CNN classifier specifically created that classified segmented areas as belonging to one of three classes: Rust, Powdery Mildew, or Healthy. The use of binary masks rather than raw images was a significant positive factor since it allowed the model to operate directly on morphological shapes, and this will improve generalization under varying conditions of the image. Performance assessment validated that the pipeline in question achieved 86.83% classification accuracy and valid values of precision, recall, F1-score, and ROC-AUC, which bore witness to method stability. Modularity would also make extension to other crops or types of diseases easy with minimal retooling. Overall, this paper is a sound basis for smart plant disease monitoring. The blend of conventional image processing and deep learning methods provides transparency and accuracy and has the potential to make the system strong, scalable, and deployable to precision agriculture. With additional advancements and field-level verifications, such models can be a great lead to greener and technology-supported farm solutions.

6. FUTURE SCOPE AND CHALLENGES

Although the advanced leaf disease classification model has worked very well under restrictive conditions, there is still a lot of scope for enhancing its extension and robustness for real-world application. Better performance could involve integrating the model to have more coverage of plant species and types of diseases so that it can become more universal for a variety of uses in agriculture. Installation on mobile phones or integration with drones and IoT platforms can bring in-field detection of diseases in real-time, and that can significantly benefit farmers residing in rural and remote areas. But there are some issues that need to be addressed for its effective deployment. These are handling low light illumination, partial occlusion, and overlapping symptoms, and all these perturb segmentation, thus impact classification accuracy. Also, the model's reliance on precise binary masks can be avoided by increased resilience in segmentation as well as by noise-resistant classifiers. Management of class imbalance in bigger data sets and addition of explainability features like visual overlays for the explanation of predictions would increase trust and usability. Deep classification and segmentation will generally be the cornerstones in guaranteeing model reliability and flexibility for effective integration into smart farm systems.

REFERENCES

Atila, Ü., Uçar, M., Akyol, K., & Uçar, E. (2021). Plant leaf disease classification using EfficientNet deep learning model. *Ecological Informatics*, *61*, 101182. DOI: 10.1016/j.ecoinf.2020.101182

Badiger, M., & Mathew, J. A. (2023). Tomato plant leaf disease segmentation and multiclass disease detection using hybrid optimization enabled deep learning. *Journal of Biotechnology*, *374*, 101–113. DOI: 10.1016/j.jbiotec.2023.07.011 PMID: 37543108

Bhatti, M. A., Zeeshan, Z., Bhatti, U. A., Khan, A., Ghadi, Y. Y., Alsenan, S., & Afzal, T. (2024). Advanced plant disease segmentation in precision agriculture using optimal dimensionality reduction with fuzzy c-means clustering and deep learning. *IEEE Journal of Selected Topics in Applied Earth Observations and Remote Sensing*. Advance online publication. DOI: 10.1109/JSTARS.2024.3437469

Elmessery, W. M., Maklakov, D. V., El-Messery, T. M., Baranenko, D. A., Gutiérrez, J., Shams, M. Y., El-Hafeez, T. A., Elsayed, S., Alhag, S. K., Moghanm, F. S., Mulyukin, M. A., Petrova, Y. Y., & Elwakeel, A. E. (2024). Semantic segmentation of microbial alterations based on SegFormer. *Frontiers in Plant Science*, *15*, 1352935. DOI: 10.3389/fpls.2024.1352935 PMID: 38938642

Goncalves, J. P., Pinto, F. A., Queiroz, D. M., Villar, F. M., Barbedo, J. G., & Del Ponte, E. M. (2021). Deep learning architectures for semantic segmentation and automatic estimation of severity of foliar symptoms caused by diseases or pests. *Biosystems Engineering*, *210*, 129–142. DOI: 10.1016/j.biosystemseng.2021.08.011

Kaur, P., Harnal, S., Gautam, V., Singh, M. P., & Singh, S. P. (2024). Performance analysis of segmentation models to detect leaf diseases in tomato plants. *Multimedia Tools and Applications*, *83*(6), 16019–16043. DOI: 10.1007/s11042-023-16238-4

Khan, K., Khan, R. U., Albattah, W., & Qamar, A. M. (2022). End-to-End Semantic Leaf Segmentation Framework for Plants Disease Classification. *Complexity*, *2022*(1), 1168700. DOI: 10.1155/2022/1168700

Li, X., Zhou, Y., Liu, J., Wang, L., Zhang, J., & Fan, X. (2022). The detection method of potato foliage diseases in complex background based on instance segmentation and semantic segmentation. *Frontiers in Plant Science*, *13*, 899754. DOI: 10.3389/fpls.2022.899754 PMID: 35865287

Li, Z., Chen, P., Shuai, L., Wang, M., Zhang, L., Wang, Y., & Mu, J. (2022). A copy paste and semantic segmentation-based approach for the classification and assessment of significant rice diseases. *Plants, 11*(22), 3174. DOI: 10.3390/plants11223174 PMID: 36432903

Liu, E., Gold, K. M., Combs, D., Cadle-Davidson, L., & Jiang, Y. (2022). Deep semantic segmentation for the quantification of grape foliar diseases in the vineyard. *Frontiers in Plant Science, 13*, 978761. DOI: 10.3389/fpls.2022.978761 PMID: 36161031

Mzoughi, O., & Yahiaoui, I. (2023). Deep learning-based segmentation for disease identification. *Ecological Informatics, 75*, 102000. DOI: 10.1016/j.ecoinf.2023.102000

Patil, M. A., & Manohar, M. (2022). Enhanced radial basis function neural network for tomato plant disease leaf image segmentation. *Ecological Informatics, 70*, 101752. DOI: 10.1016/j.ecoinf.2022.101752

Polly, R., & Devi, E. A. (2024). Semantic segmentation for plant leaf disease classification and damage detection: A deep learning approach. *Smart Agricultural Technology, 9*, 100526. DOI: 10.1016/j.atech.2024.100526

Preanto, S. A., Ahad, M. T., Emon, Y. R., Mustofa, S., & Alamin, M. (2024). A Semantic Segmentation Approach on Sweet Orange Leaf Diseases Detection Utilizing YOLO. *arXiv preprint* arXiv:2409.06671.

Rani, N. S., Krishna, A. S., Sunag, M., Sangamesha, M. A., & Pushpa, B. R. (2024). Infield disease detection in citrus plants: Integrating semantic segmentation and dynamic deep learning object detection model for enhanced agricultural yield. *Neural Computing & Applications, 36*(35), 22485–22510. DOI: 10.1007/s00521-024-10451-4

Sharma, M., Kumar, C. J., Singh, T. P., Talukdar, J., Sharma, R. K., & Ganguly, A. (2023). Enhancing disease region segmentation in rice leaves using modified deep learning architectures. *Archiv für Phytopathologie und Pflanzenschutz, 56*(20), 1555–1580. DOI: 10.1080/03235408.2024.2310326

Sharma, P., Berwal, Y. P. S., & Ghai, W. (2020). Performance analysis of deep learning CNN models for disease detection in plants using image segmentation. *Information Processing in Agriculture, 7*(4), 566–574. DOI: 10.1016/j.inpa.2019.11.001

Shoaib, M., Hussain, T., Shah, B., Ullah, I., Shah, S. M., Ali, F., & Park, S. H. (2022). Deep learning-based segmentation and classification of leaf images for detection of tomato plant disease. *Frontiers in Plant Science*, *13*, 1031748. DOI: 10.3389/fpls.2022.1031748 PMID: 36275583

Wang, S., Xu, D., Liang, H., Bai, Y., Li, X., Zhou, J., Su, C., & Wei, W. (2025). Advances in Deep Learning Applications for Plant Disease and Pest Detection: A Review. *Remote Sensing (Basel)*, *17*(4), 698. DOI: 10.3390/rs17040698

Chapter 12
Improving Soil Health and Crop Yields With Artificial Intelligence:
Innovations, Impacts, and Implementation

Ria Ghosh
https://orcid.org/0009-0004-7737-4171
University of Delhi, India

Rajeev Kumar
https://orcid.org/0009-0002-1655-6807
Independent Researcher, USA

ABSTRACT

Innovative and sustainable farming methods are desperately needed in light of the growing world's food demand, the depletion of soil resources, and the uncertain effects of climate change. In this chapter, we examine how AI is enabling data-driven, accurate, and adaptive farming practices, thereby transforming crop yield optimization and soil health management. Farmers and agronomists can now continuously monitor soil parameters, anticipate nutrient deficiencies, identify diseases early, and optimize irrigation and fertilization schedules with previously unheard-of precision by combining AI technologies like machine learning, deep learning, computer vision, and the Internet of Things (IoT). This chapter offers a thorough review of AI applications in agricultural production and soil analysis, along with case examples showing quantifiable increases in sustainability and productivity. It talks about how AI and precision agriculture work together, highlighting how intelligent systems may improve environmental results, increase food security, and reduce resource waste.

DOI: 10.4018/979-8-3373-4207-8.ch012

1. INTRODUCTION

An essential component of both agricultural production and environmental sustainability is soil health. It affects plant development, water retention, nutrient cycling, and resistance to disease and erosion. It includes the biological, chemical, and physical integrity of the soil. In addition to maintaining food yields, healthy soils can help mitigate climate change by storing carbon and promoting biodiversity. Growing populations, changing dietary needs, and dwindling arable land have all put increasing pressure on the world's agricultural systems to provide more food in recent decades. However, long-term soil degradation, an excessive reliance on chemical inputs, and unsustainable land use practices are frequently the costs of agricultural intensification (Nguyen & Tran, 2023; Mehta & Patel, 2023). Despite being based on centuries of experience, traditional agricultural techniques usually lack the accuracy required to handle the dynamic and complex character of soil systems. Traditional soil testing is usually limited in space, time, and frequency. Without a thorough grasp of particular field circumstances, farmers sometimes turn to uniform fertilizer and pesticide application, which over time results in resource inefficiencies, environmental contamination, and deteriorating soil fertility (Chowdhury & Islam, 2024). Furthermore, the vulnerabilities of conventional agricultural operations have been revealed by climate unpredictability and catastrophic weather occurrences, calling for more flexible, adaptable, and knowledgeable decision-making systems.

Artificial Intelligence (AI) has become a transformative instrument that has the potential to revolutionize current agriculture in this context. Through the utilization of big data, machine learning, computer vision, and real-time sensor technologies, artificial intelligence (AI) facilitates the transition from reactive to proactive and predictive agricultural methods (Malhotra et al., 2025). In order to provide accurate suggestions for crop rotation, fertilization, irrigation, and other practices, AI-powered systems, for example, are able to continuously evaluate soil moisture, texture, pH, nutrient content, and microbial activity. These realizations support the preservation of ecological balance, increased yields, and resource efficiency (Ghosh & Roy, 2024; O'Connor & Smith, 2023). In order to enable prompt interventions, AI can also analyze drone and satellite photography data to identify early indicators of nutrient depletion, compaction, and soil erosion over wide areas. AI has great potential for sustainable agriculture because of its scalability and adaptability in addition to its analytical capabilities. AI technologies are being adapted to a variety of agro-ecological environments, ranging from large-scale agribusinesses to smallholder farmers in poor countries. Cloud-based platforms, robotic soil samplers, autonomous tractors, and mobile-based decision support systems are all becoming more widely available and helping to close the digital divide in rural areas. Furthermore, to build smart farming ecosystems that guarantee traceability, transparency, and resilience in food production, AI applications are being combined with blockchain, geographic information systems (GIS), and Internet of Things (IoT) sensors.

AI integration in agriculture is not without its difficulties, though. Significant obstacles still exist in the areas of data quality, technology infrastructure, digital literacy, and ethical issues with data ownership and equity. Without proper training and support systems, AI solutions may not reach their full potential in areas with poor connection or technological capability. For AI-driven agricultural technologies to be successfully implemented and long-term adopted, it is imperative that researchers, technologists, politicians, and farmers collaborate across sectors. An extensive examination of how AI can be used to enhance crop yields and soil health is given in this chapter. It starts by breaking down the elements and markers of soil health and comparing conventional evaluation methods with AI-enhanced approaches. The subsequent sections explore particular AI technologies, including computer vision, deep learning models, and machine learning algorithms, and how they are used in precision agriculture and soil diagnostics (Sarkar & Banerjee, 2023; Thomas & Lee, 2023). Case studies from different parts of the world show real-life effects and practical implementations. The importance of AI in improving crop management techniques, such as water saving, nutrient balancing, and pest prediction, is covered in the sections that follow. The chapter also looks at how AI might enhance sustainable farming practices by boosting biodiversity and reducing the need for inputs. Implementation difficulties, such as socio-technical, financial, and legal obstacles, are examined closely. The chapter concludes with a discussion of new developments in digital farming, including the integration of AI with robotics and AI-assisted climate-smart agriculture.

The objectives of this chapter are fourfold:

1) To draw attention to the crucial connection, given the demands of contemporary agriculture, between crop productivity and soil health;
2) To evaluate the shortcomings of conventional crop and soil management techniques and show how artificial intelligence (AI) can provide more accurate, effective, and sustainable substitutes;
3) To showcase and assess state-of-the-art AI systems that are being utilized to improve crop yields and track soil conditions, with an emphasis on impact and scalability;
4) To shed light on potential future developments, obstacles, and policy ramifications related to the application of AI in agriculture, especially in resource-constrained and Global South contexts.

2. UNDERSTANDING SOIL HEALTH: COMPONENTS AND ASSESSMENT

The foundation of agricultural production and sustainability is soil health. The dynamic biological, chemical, and physical characteristics that promote plant growth, control water flow, cycle nutrients, and maintain biological activity are all included in soil health, as opposed to the more static idea of soil fertility. A thorough grasp of these elements is essential for both identifying soil deterioration and directing treatments that improve soil performance over the long run. Innovative assessment techniques, especially those driven by artificial intelligence (AI), are revolutionizing the monitoring, interpretation, and management of soil health in response to the growing needs for environmental preservation and food production.

2.1 Soil Physical, Chemical, and Biological Properties

The functionality of the soil is determined by a combination of physical, chemical, and biological markers that make up the complex concept of soil health. Texture, structure, porosity, bulk density, and water-holding capacity are all considered physical characteristics of the soil. Water infiltration, aeration, and root penetration are all impacted by these variables. In contrast to clayey soils, which retain water but can become compacted, hindering the growth of roots, sandy soils drain more quickly but contain fewer nutrients. Tilth and aggregation, which are markers of physical structure, are necessary to keep the soil environment friable and productive (Li & Zhang, 2023). Soil pH, electrical conductivity, cation exchange capacity (CEC), and nutrient availability are the main chemical characteristics. Plant health is greatly impacted by the concentration of micronutrients (zinc, copper, and boron) and macronutrients (nitrogen, phosphorus, and potassium). Crop yields can be directly impacted by imbalanced nutrient levels, which can result in toxicity or signs of deficiencies. In example, the pH of the soil affects nutrient solubility and microbial activity. The living elements of soil, such as fungi, earthworms, microbes, and plant roots, are referred to as biological properties. These elements aid in the breakdown of organic matter, the cycling of nutrients, and the prevention of disease. Important indicators of soil vitality are now understood to include microbial biomass, enzyme activity, and soil fauna variety (Singh et al., 2021). Microbial life abounds in healthy soils, facilitating the transformation of organic materials into easily assimilated plant nutrients. **Figure 1** infographic categorizes soil health into three core domains: physical properties (such as texture, structure, moisture, and density), chemical properties (including pH, nutrient content, cation exchange capacity, and salinity), and biological properties (featuring microorganisms, organic matter, earthworms, and respiration). Each component plays a critical role in maintaining fertile and productive soil for sustainable agriculture. Chemical testing was the main method used in laboratories to assess soil health in the past, frequently ignoring biological and certain physical factors. Understanding soil functionality under changing agro-ecological and climatic settings, however, requires a more comprehensive, integrated study.

Figure 1. Physical, chemical, and biological properties of soil

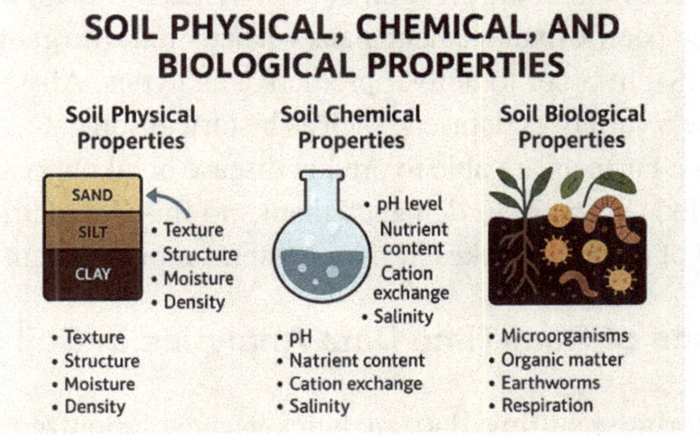

2.2 Conventional Soil Testing vs. AI-Powered Sensing Systems

Traditional soil testing techniques include expert interpretation, laboratory-based analysis, and manual sampling. This method is time-consuming, difficult, and frequently has limited spatial and temporal coverage, despite its reliability. Once or twice a year, farmers might examine a small number of samples from large fields, which could result in generalizations that don't account for within-field variability. Due to this lack of granularity, fertilizers and other amendments are typically applied excessively or insufficiently, which lowers efficiency and damages the ecosystem over time (Rana et al., 2022). AI-powered sensing systems, on the other hand, offer continuous, real-time soil data that is very accurate and resolved. Optical, electrochemical, dielectric, and mechanical sensor technologies are all integrated into these systems to track variables including temperature, salinity, pH, nutrient content, and soil moisture. After processing this data, AI systems produce insights that can be used in certain contexts and at particular moments (Nguyen & Tran, 2023). For instance, soil texture and organic matter content can be evaluated on-site using near-infrared (NIR) and mid-infrared (MIR) sensors, doing away with the requirement for laboratory examination. Such spectrum data can be quickly evaluated to forecast microbial activity and nutrient levels when combined with

AI (Kumar et al., 2023). Additionally, edge computing-enabled smart soil probes may evaluate data locally and send only the most important alerts or summaries, saving bandwidth and improving decision-making in real time. The use of satellites and drones for soil surface analysis is another innovation. Images are analyzed by AI-powered computer vision systems to find patterns of erosion, waterlogging, and soil compaction. Wide-area, non-invasive soil monitoring is made possible by these aerial platforms, which would be impossible with human methods (Martínez & Pérez, 2023). They aid in the creation of comprehensive maps of soil health that complement site-specific management plans when paired with geographic information systems (GIS). In order to deliver predictive analytics, AI-powered platforms can also combine a variety of datasets, such as historical soil data, crop models, and weather forecasts. Farmers are able to predict disease breakout risk zones, possible production results, and nutrient demands thanks to this. In addition to increasing output, this kind of foresight makes farming systems more resilient and sustainable.

2.3 Importance of Real-Time Data Analytics in Soil Monitoring

In soil monitoring, real-time data analytics has revolutionized the field by providing a flexible and dynamic method of agricultural decision-making. Continuous data collection, as opposed to standard periodic evaluations, makes it possible to identify minute changes in soil conditions that could otherwise go overlooked. In the management of irrigation and fertilizer inputs, where time is just as crucial as quantity, this temporal resolution is very crucial. In order to spot patterns, irregularities, and connections that human observers would overlook, artificial intelligence (AI) systems examine this stream of real-time data. For instance, automated irrigation systems can save water and avoid crop stress when in-ground sensors detect abrupt reductions in soil moisture. In a similar vein, farmers can plan targeted fertilization before crops show obvious symptoms by receiving early alerts about nutrient depletion (Islam & Rahman, 2023; Lee & Park, 2023). Furthermore, by integrating AI with IoT devices, feedback loops involving intelligent interactions between sensors, actuators, and decision-support systems can be created. By modifying irrigation schedules in response to anticipated rainfall and evapotranspiration rates, a smart agricultural system can maximize resource utilization and cut expenses. This accuracy reduces environmental externalities like runoff and leaching while simultaneously improving soil health. Adaptive management is also made possible via real-time monitoring. AI systems are able to adjust recommendations in response to changes in the environment brought about by farming operations or weather variations. This flexibility is essential for dealing with the growing unpredictability brought forth by climate change. Real-time analytics plays a significant role in the long-term monitoring of soil health. **Figure 2** bar graph shows the direct relationship between different soil health levels (poor, moderate, and healthy) and agricultural productivity. Maintaining ideal soil conditions is crucial for sustainable farming since crop yield potential increases significantly as soil health improves. These data can show patterns in microbial activity, organic matter accumulation, and nutrient cycling efficiency over time, providing important information on how well soil conservation techniques are working. This data can be used by researchers and decision-makers to assess the results of agricultural interventions and to direct extensive land management plans (Li et al., 2023). Despite these benefits, scaling up real-time soil monitoring still presents difficulties. Adoption may be constrained by high upfront costs, upkeep needs, and device interoperability problems. However, these systems are becoming more widely available, particularly for smallholder farmers, because to the quick development of inexpensive sensors, open-source software, and cloud-based platforms.

Figure 2. Correlation between soil health and agricultural productivity

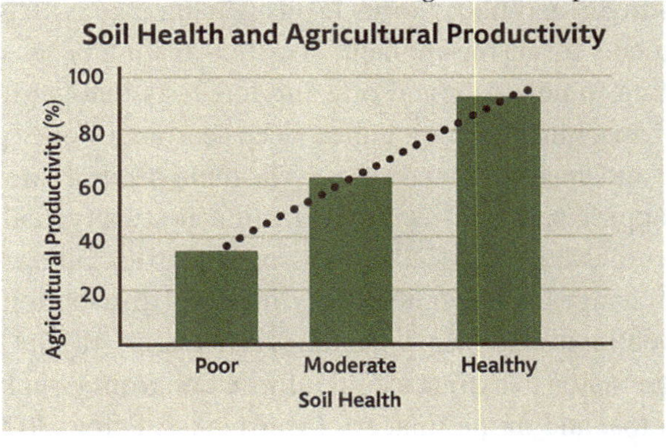

3. AI TECHNOLOGIES IN SOIL MONITORING AND ANALYSIS

Agriculture systems' methods for diagnosing and managing soil health are changing as a result of the incorporation of Artificial Intelligence (AI) into soil monitoring and analysis. Advanced computational models that can record, interpret, and react in real time to large and diverse datasets are replacing or supplementing traditional methods. Scalable, precise, and predictive assessments of soil conditions are made possible by the fundamental AI technologies of computer vision (CV), deep learning (DL), and machine learning (ML). AI has become a key component of contemporary soil diagnostics when paired with remote sensing, Internet of Things (IoT) technologies, and precision agricultural platforms.

3.1 Role of Machine Learning, Deep Learning, and Computer Vision

In machine learning, a branch of artificial intelligence, algorithms are trained to find patterns in data and generate classifications or predictions. Based on information gathered from sensors, satellites, and field observations, machine learning (ML) is used in soil science to forecast components such as microbial activity, pH, texture, nutrient content, and soil organic carbon levels. Support vector machines (SVMs), decision trees, random forests, and gradient boosting models are examples of frequently used techniques (Dutta & Banerjee, 2023; Singh & Kumar, 2023). For example, using data from soil sensors and climatic factors like temperature, precipitation, and vegetation indices, random forest models have shown useful in predicting soil nitrogen and phosphorus levels (Zhang et al., 2023). These models are useful instruments for in-the-moment decision-making because they get better with time, adjusting to new data and offering forecasts that get more precise. An advanced type of machine learning called deep learning works particularly well with complicated and unstructured data, such as multispectral satellite data or photographs. Soil mapping and land-use classification are two typical applications for Convolutional Neural Networks (CNNs), a form of deep learning architecture. They can detect minute changes in vegetation stress, moisture distribution, and soil surface textures that conventional statistical techniques can miss. Artificial Intelligence (AI) is used in computer vision to extract useful information from visual inputs including satellite data, photos, and drone imagery (Martínez & Pérez, 2023; Liu & Wang, 2023). When used for soil monitoring, CV algorithms can spot trends in salinization, compaction, or soil erosion over wide geographic regions. Techniques for object detection and image segmentation aid in differentiating between patches of healthy and damaged soil, enabling site-specific treatment. Furthermore, to automate field evaluations, these instruments are being utilized more and more in conjunction with robotics, such as autonomous rovers or drones that sample soil.

3.2 Remote Sensing, IoT, and Precision Agriculture Platforms

Large-scale soil study requires the use of remote sensing technologies, such as satellite and drone-based imagery. These systems gather hyperspectral and multispectral data that reveal information on soil fertility, organic matter, moisture content, and texture. This high-dimensional data is processed by AI algorithms to provide decision-support tools and thematic maps. Via vegetation vigor and coverage, spectral indexes like the Normalized Difference Vegetation Index (NDVI) and the Soil Adjusted Vegetation Index (SAVI) are used to indirectly estimate soil health. For example, soil salinity risk maps for arid and semi-arid regions in Africa and India have been produced using Landsat and Sentinel satellite imagery evaluated using machine learning approaches (Ahmed et al., 2022). With the use of these technologies, farmers and policymakers can take preventative measures like moving crops or implementing salinity-tolerant cultivars. IoT-based soil monitoring systems are made up of dispersed sensor networks that record data on soil characteristics in real time, including EC (electrical conductivity), pH, temperature, and moisture. Frequently, these sensors are linked to cloud-based systems where artificial intelligence analyzes the information and generates warnings or suggestions. Edge AI is becoming more prevalent in rural areas with poor internet connectivity, where analytics are done locally on the device rather than on a distant server. In order to optimize inputs and maximize outputs at precise spatial and temporal resolutions, precision agricultural platforms depend heavily on IoT devices. In order to generate personalized suggestions for tillage, fertilization, and irrigation, these platforms compile data from many sources, including satellite feeds, weather stations, and soil sensors, and input it into artificial intelligence algorithms (Watanabe & Tanaka, 2023). ML and remote sensing are used by industry-leading solutions like Microsoft Azure FarmBeats, IBM's Watson Decision Platform for Agriculture, and John Deere's Operations Center to enhance soil management. By using AI-generated insights, these systems assist farmers in determining when and where to apply particular inputs, greatly enhancing environmental sustainability and resource efficiency.

3.3 Case Studies: AI Applications in Real-World Soil Diagnostics

3.3.1 Case Study 1

The Netherlands' Accurate Soil Mapping Thousands of hectares of farmland in the Netherlands are being monitored for soil health by AI-powered devices. Hyper-localized fertility maps were produced by employing machine learning algorithms to examine the pH, moisture, and salinity data gathered by the installation of soil sensors. By using these maps to direct variable-rate fertilizer and lime treatments, farmers were able to increase wheat production by 10% while lowering input costs by 15% (Van Der Meer et al., 2023).

3.3.2 Case Study 2

Indian Soil Salinity Forecast Applying AI and Remote Sensing

Researchers in Gujarat, India, predicted regions with elevated soil salinity using SVM algorithms and Landsat-8 satellite data. Local authorities were able to identify vulnerable areas and carry out focused actions, such as gypsum treatment and crop rotation with salt-tolerant plants, thanks to the AI models' 91% accuracy (Choudhary et al., 2023). The initiative stopped long-term deterioration and enhanced crop performance.

3.3.3 Case Study 3

Using Deep Learning to Classify Australian Soil Texture

Using high-resolution drone photos, Australian researchers created a CNN model to categorize soil textures. With 95% accuracy, the model could distinguish between clay, loamy, and sandy soils, significantly lowering the need for laboratory testing and hand sampling. Later, the technique was incorporated into an autonomous drone platform that vineyard managers used to cut water use by 22% and optimize irrigation schedules (Nguyen et al., 2023).

3.3.4 Case Study 4

Sub-Saharan Africa's Use of IoT and AI for Soil Monitoring

A Kenyan entrepreneur created "SoilGuard," a low-cost AI-based soil health monitoring device. It integrates temperature, pH, and moisture sensors with machine learning algorithms that have been trained on local datasets. SoilGuard, which was implemented in areas where maize was grown, gave farmers SMS-based advice on the best times to plant and apply fertilizer. According to field tests, participating farmers' yields increased by 30% when compared to controls.

3.4 Use of Satellite Imagery and Drones for Soil Mapping

Soil mapping has been transformed by satellite imaging, which offers longitudinal, scaled, and uniform images of terrestrial surfaces. Soil properties can be linked to spectral signatures recorded by satellites like MODIS, Sentinel-2, and Landsat. This data is analyzed by AI systems to identify soil types, track erosion, and forecast fertility trends. Because it can capture hundreds of small spectral bands, hyperspectral imaging in particular provides precise information. These datasets are used to train machine learning algorithms that can accurately predict organic carbon, moisture, and mineral composition. In addition to individual farmers, policymakers organizing extensive land reclamation or conservation programs can benefit from these maps. Higher geographical resolution and more frequent updates are provided by drones, which enhance satellite photography. Drones using thermal or multispectral cameras can record soil characteristics at the field level. Drone data can be processed by AI algorithms to identify minute changes in soil moisture or compaction, allowing for incredibly accurate treatments (Araújo et al., 2023). For instance, CNN-processed drone-based data can detect compaction-stressed areas and suggest mechanical aeration. Additionally, some AI systems use data fusion, which combines data from drones and satellites to produce multi-scale maps of soil health. It is very successful for both macro- and micro-level soil diagnostics because of its integration, which allows for a balance between field-specific detail and large-area monitoring.

4. AI FOR OPTIMIZING CROP YIELDS

Particularly in the area of crop production optimization, artificial intelligence (AI) is drastically changing agricultural methods. Traditional farming has always depended on experience-based judgment, but artificial intelligence (AI) makes it possible for rapid, accurate, and data-driven interventions that have a direct effect on productivity. From real-time pest detection and decision support systems (DSS) for resource allocation to prediction models for crop health, artificial intelligence (AI) enables farmers to make better informed and sustainable decisions throughout the cultivation cycle. Through the integration of artificial intelligence (AI) with satellite imagery, agricultural databases, and Internet of Things (IoT) devices, farms can improve their ability to adapt to both microclimatic conditions and wider ecological dynamics.

4.1 Predictive Modelling for Crop Health and Yield

A key component of AI-driven agriculture is predictive modelling. These models anticipate crop growth stages, possible yields, and disease or environmental stress concerns using both historical and current data. Neural networks, ensemble methods, random forests, and regression analysis are some of the techniques. The forecasting ability of these systems is greatly increased by integrating weather data, crop phenological factors, remote sensing outputs, and soil health indicators (Zhao & Liu, 2023). For instance, using deep neural networks trained on sensor data and satellite photos, researchers have created models that estimate crop productivity. These models enable site-specific management by estimating yield variability across several zones within a field. **Figure 3** illustrates Performance comparison of various predictive modelling algorithms (Linear Regression, Random Forest, SVM, Neural

Network, XGBoost) in terms of Accuracy and R^2 Score for assessing crop health and yield. One prominent use is Google's TensorFlow-based artificial intelligence engine in sub-Saharan Africa, which forecasted maize yields up to 90 days before to harvest with an accuracy of over 85% by combining rainfall data and satellite NDVI indices (Singh et al., 2023). Furthermore, dynamic systems like crop development can be effectively modelled by time-series prediction algorithms like Long Short-Term Memory (LSTM) networks. In addition to eventual yields, LSTM models are able to forecast intermediate factors like as biomass buildup, canopy development, and flowering time. Input optimization is another benefit of predictive modelling. To find the ideal nitrogen level that increases yield while avoiding leaching and emissions, for example, models can simulate different fertilization scenarios. In addition to boosting output, this supports environmental sustainability objectives.

Figure 3. Performance comparison of various predictive modelling algorithms (Linear Regression, Random Forest, SVM, Neural Network, XGBoost) in terms of Accuracy and R^2 Score for assessing crop health and yield.

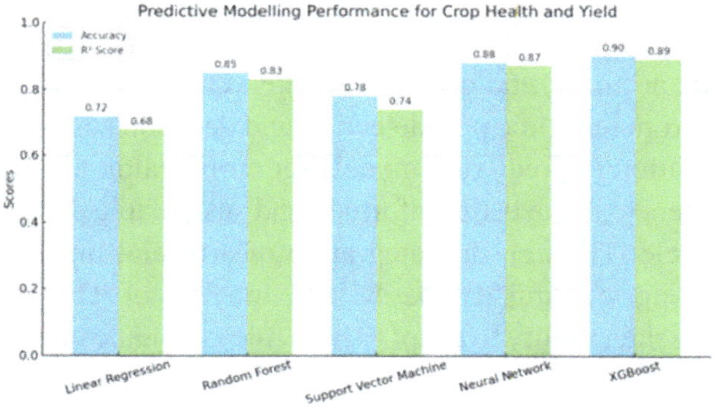

4.2 AI in Pest and Disease Detection

Globally, pest and disease outbreaks continue to be one of the biggest dangers to crop output. AI technologies have shown themselves to be quite useful in the field of early detection, which is crucial for efficient administration. Through the use of environmental modelling, anomaly detection, and image recognition, AI is able to spot infestation or infection symptoms before they are visibly apparent to the human eye (Gupta & Sharma, 2023; Jain & Kumar, 2023). Convolutional neural networks (CNNs)-powered computer vision (CV) systems are able to examine leaf photos and identify signs of common illnesses including blight, rust, and powdery mildew. Real-time field analysis is made possible by these technologies, which are frequently installed on self-governing drones and ground robots or integrated into smartphone apps. For example, even with variable light and angle settings, the PlantVillage Nuru app, which was created using deep learning models, can identify more than 30 distinct crop diseases with an accuracy rate of 90% (Kamilaris & Prenafeta-Boldú, 2023). To track the dynamics of pest populations, AI-driven pest monitoring systems combine thermal imaging, pheromone traps, and remote sensors. In order to minimize the usage of broad-spectrum insecticides, algorithms evaluate the collected data to predict outbreaks and provide treatment periods. AI systems like Trapview detect pests in traps and forecast infestation patterns in vineyards with high temporal resolution by using image processing and cloud-based analytics. Additionally, predictive AI models calculate the likelihood of disease onset based on soil and climate data. These models work especially well against bacterial and fungal illnesses that are sensitive to particular temperature and humidity ranges. For example, the spread of downy mildew in grapes or late blight in potatoes is predicted using logistic regression models and Bayesian networks. Unmanned Aerial Vehicles (UAVs) and AI combine to improve detection coverage. Drones equipped with thermal and multispectral cameras are able to scan vast areas of farmland and identify stress patterns that are not evident to the human eye. In order to enable focused intervention, these patterns are subsequently examined using machine learning algorithms to distinguish between biotic (disease/pest) and abiotic (nutrient, drought) stresses.

4.3 Decision Support Systems (DSS) for Irrigation, Fertilization, and Harvesting

Decision Support Systems (DSS) powered by AI have become indispensable instruments for maximizing agricultural output. These systems use AI algorithms to determine the best times and amounts for irrigation, fertilization, pesticide application, and harvesting by integrating a variety of data sources, including crop models, satellite imaging, weather forecasts, and soil sensors. Sustainable agriculture, particularly in areas with limited water resources, depends on efficient water usage. To decide when and how much to irrigate, AI-powered irrigation DSS use information from rainfall forecasts, evapotranspiration models, and soil moisture sensors. These technologies preserve the ideal soil moisture levels for crop growth while lowering runoff and overwatering. Israel's NetBeatTM system, which integrates edge computing and AI to provide real-time irrigation and fertigation decisions, is an example of a successful AI-based irrigation DSS deployment. Fruit orchard output increases of 15-20% and water reductions of up to 25% were recorded by farmers who used NetBeatTM (Levy et al., 2023). In order to suggest field-level irrigation schedules, KrishiBot, a low-cost AI-based irrigation advisor in India, combines soil data with vegetation indices and weather forecasts received from satellites (Ghosh & Roy, 2024; Kumar & Singh, 2023). By using KrishiBot, rice paddies have used more than 30% less water while maintaining or increasing yield levels. Based on crop uptake curves, environmental factors, and soil nutrient levels, AI-based DSS optimize fertilizer application and timing. In order to guarantee prompt replenishment and prevent both over- and under-application, machine learning algorithms evaluate real-time soil data and predict nutrient depletion rates. AI is used by systems like as CropX and Yara's Atfarm to produce spatial nutrient maps and suggest variable-rate treatment schedules. By using these technologies, the carbon footprint of fertilizer production and use is decreased and nutrient leaking into water bodies is prevented. Additionally, certain DSS technologies combine artificial intelligence (AI) with drone-based foliar nutrition monitoring, evaluating potassium, phosphorus, and nitrogen shortages through spectral analysis. In order to ensure that nutrient management is in line with real crop requirements, recommendations are produced using both soil factors and plant tissue analysis. Harvesting at the right time is essential for optimizing production and quality. To identify the optimal harvesting windows, AI-powered DSS examine market movements, crop maturity indexes, and weather forecasts. In order to forecast fruit ripeness or grain maturity, computer vision systems can evaluate color changes, moisture content, and structural integrity. For instance, deep learning algorithms are used in horticulture to accurately classify the ripeness of grapes, tomatoes, and apples using RGB and multispectral imagery. By feeding these insights onto robotic harvesting systems, post-harvest losses are decreased because only the most developed fruits are harvested. To identify peak harvest times for cereals and grains, machine learning algorithms examine regional climatic trends, remote sensing data, and growing degree days (GDD). AI-driven DSS reduced field time by 15% and improved nut quality uniformity in California's almond industry by enhancing harvest coordination and logistics (Rodriguez et al., 2023).

5. INTEGRATING AI WITH SUSTAINABLE FARMING PRACTICES

Sustainability has emerged as a critical necessity as environmental issues in global agriculture continue to grow. The goals of techniques like agroecology, organic farming, and conservation agriculture are to maintain biodiversity, lessen reliance on inputs, and enhance soil health. Adoption may be constrained, nevertheless, by the scalability and real-time decision-making requirements of sustainable farming. Artificial Intelligence (AI) is a key enabler in this situation. Farmers may maintain or increase yields while achieving greater efficiency, adaptive reactions, and ecological balance by incorporating AI into sustainable methods.

5.1 AI in Conservation Agriculture, Organic Farming, and Agroecology

The principles of conservation agriculture (CA) include diversified crop rotations, permanent soil cover, and minimal soil disturbance. With the use of machine learning models and real-time field data, AI makes it possible to precisely monitor and manage these principles. Based on slope, rainfall, and vegetation cover, AI-driven decision support systems (DSS) may simulate the effects of erosion, evaluate compaction concerns, and predict the ideal tillage intensity. For instance, computer vision models that use drone imagery can pinpoint areas that are prone to soil erosion and suggest mulching or strategically placed plant barriers. In order to maximize no-till operations and minimize soil erosion in maize fields by 40% over three seasons, a conservation agriculture project in Zambia combined AI algorithms with GPS and rainfall data (Mwansa et al., 2023). Organic systems mostly depend on ecological balances and human monitoring because they are not allowed to use synthetic inputs. Through pest identification, organic nutrient optimization, and precise scouting, AI systems can assist these processes. Image recognition programs, for example, can distinguish between nutritional deficits and pest-induced stress, allowing for the timely application of compost or biological pest control methods. Traceability and certification compliance are supported by smartphone apps that use natural language processing (NLP) to help smallholder organic farmers identify issues and provide remedies (Nguyen & Tran, 2023; Sarkar & Banerjee, 2023). Furthermore, using blockchain-backed traceability, AI can evaluate supply chain data to guarantee organic integrity. Agroecology is the application of ecological concepts to agricultural systems. AI helps with this by using bioinformatics and sensor data to identify pollinator patterns, plant-soil-microbe interactions, and soil microbial diversity. High-dimensional ecological data can be processed by deep learning algorithms to identify seasonal variations in functional resilience or biodiversity. An agroecological vineyard in France enhanced habitat conservation and reduced pesticide use by 60% by using AI to track beneficial insect populations through image capture and trained CNNs (Gauthier et al., 2023).

5.2 Enhancing Carbon Sequestration and Biodiversity through Intelligent Systems

The contribution of sustainable farming to climate mitigation through carbon sequestration and biodiversity conservation is one of its main tenets. These ecosystem services can be measured, validated, and improved at scale by AI technology. AI models estimate above-ground biomass and soil organic carbon (SOC) by analyzing multispectral and LiDAR data. This is essential for confirming results under regenerative agricultural projects and carbon credit schemes. By mapping SOC changes over time using Google Earth Engine and deep learning, areas with sequestration potential have been identified based on vegetation, soil type, and management history. Farmers can gain insight into the long-term effects on soil health by using AI-driven simulators like DayCent-AI, which can forecast carbon fluxes under different tillage, cover cropping, and manure application scenarios (Li & Zhang, 2023; Ghosh & Roy, 2024). Farmers in India's Deccan Plateau achieved a 17% rise in SOC over two years by monitoring carbon building in soil layers and adjusting organic matter inputs with the use of an AI-powered platform called SoilSense (Patil et al., 2023). By tracking migration patterns, predicting ecological relationships, and monitoring species variety, AI helps manage biodiversity. To evaluate biodiversity richness and ecological health, acoustic AI systems employ deep learning to analyze noises in several sectors, including insect activity, bird calls, and amphibian presence. Additionally, by suggesting hedgerows, biological corridors, and mixed-crop patterns, AI can maximize landscape variability, which is a major factor in biodiversity. Reinforcement learning systems provide management strategies that support conservation objectives by simulating the effects of land-use changes on biodiversity indexes. By balancing models of light, water, and nutrient competition, artificial intelligence (AI) aids in the selection of tree species and spatial layout in agroforestry systems to optimize productivity and biodiversity.

5.3 Smart Crop Rotation and Cover Cropping Using AI Tools

Crop rotation and cover crops are essential techniques for enhancing nutrient cycling, soil structure, and insect control. AI is able to optimize these tactics through the analysis of soil nutrient dynamics, insect cycles, climate data, and past crop performance. AI makes adaptive rotation planning based on probabilistic models possible, whereas traditional crop rotation follows set patterns. By using historical yield data and time-series satellite images, machine learning algorithms can determine the best rotation sequences to maximize nutrient availability and reduce insect accumulation. Utilizing reinforcement learning, programs like AgroAI Planner model various rotation scenarios across fields and years while taking soil pH, rainfall patterns, and residual nitrogen impacts into consideration. Both productivity and soil resilience objectives are integrated into the personalized advice these systems offer. For instance, compared to conventional corn-soybean monocultures, a three-year rotation model proposed by AI (corn-soybean-wheat with legume intercropping) in the Midwest US increased nitrogen efficiency by 35% and decreased aphid incidence by 25% (Anderson et al., 2024). Cover crops improve soil fertility, inhibit weed growth, and prevent erosion. With the use of remote sensing, soil sensor data, and weather forecasts, AI systems can assist in choosing the best cover crop species, sowing rates, and termination dates. In order to suggest cover crop mixtures that are appropriate for a given situation, multi-objective optimization algorithms assess trade-offs between biomass production, nitrogen fixation, and water consumption (Rahman & Chowdhury, 2023). To find the best termination dates that balance soil moisture retention and following crop planting, for instance, the Spanish-developed program CoverAI employs soil moisture sensors and machine learning-based prediction. Dynamic management is made possible by the real-time assessment of ground covering and cover crop biomass made possible by convolutional networks and drone-mounted multispectral imagery. This guarantees that cover crops are not only sown but also successfully support ecological processes. AI also facilitates the incorporation of cover crops with other environmentally friendly methods. For example, it can determine how livestock grazing and cover crops work together in mixed systems to maximize carbon return and land production.

6. BENEFITS AND CHALLENGES OF AI ADOPTION IN AGRICULTURE

Artificial Intelligence (AI) is revolutionizing agriculture in ways that go beyond technology advancement; it is a paradigm shift in the way that food is produced, resources are managed, and sustainability objectives are sought. Even while AI has many advantages for agriculture, there are also significant obstacles that must be overcome, such as concerns about infrastructure, ethics, regulation, and equity. The economic, environmental, and productivity benefits of integrating AI are examined in this section, which is followed by a thorough examination of the obstacles to its widespread use.

6.1 Economic, Environmental, and Productivity Benefits

Through improved market access, loss reduction, and input usage optimization, AI technology can dramatically increase farm profitability. Farmers may save operating costs and increase crop efficiency by using predictive analytics to make data-driven decisions about sowing, irrigation, fertilization, and harvesting. Using AI-enabled irrigation systems, a 2023 pilot research in Brazil found that net farm income increased by 28% while water and fertilizer use decreased by 35% and 22%, respectively (de Oliveira et al., 2023). In a similar vein, AI-based agricultural price prediction systems assist farmers in selecting the best times to sell, reducing the need for middlemen and increasing market returns. Robotic weed eaters and autonomous tractors are examples of mechanized AI solutions that lessen the need for human labour, especially in high-income countries where there is a labour shortage. By making precision agriculture possible, AI improves environmental stewardship by coordinating resource use with plant requirements. This lessens nutrient leaching, chemical runoff, and soil degradation-all of which are significant causes of pollution in the environment (Patel & Desai, 2023; Lee & Park, 2023). Additionally, AI helps reduce greenhouse gas (GHG) emissions by predicting carbon sequestration, minimizing tillage, and applying fertilizer. According to Müller et al. (2023), nitrous oxide emissions in wheat production systems in Germany were lowered by 20% with the use of smart nitrogen sensors in conjunction with AI prediction tools. Additionally, artificial intelligence is essential for monitoring habitat changes, conserving biodiversity, and directing sustainable land-use planning. AI algorithms are able to suggest land distributions that maximize production zones while preserving natural corridors. AI helps to increase yield in both quantitative and qualitative ways. Through early-warning systems and real-time monitoring, it reduces losses brought on by diseases, pests, and drought. Convolutional neural networks (CNNs), for instance, are able to detect insect infestations days before visual signs manifest, allowing for preventative treatment (Gupta & Sharma, 2023). By evaluating genome-wide data and environmental interactions, AI speeds up genetic selection in precision breeding, enabling the creation of crops that are adaptable to climate change. AI technologies also assist in scheduling harvesting at the right moment, which lowers post-harvest losses. According to the World Bank (2024), AI-supported agronomic advising services through mobile apps raised maize yields in sub-Saharan Africa by 21% over two seasons.

6.2 Data Availability, Digital Divide, Farmer Training, and Ethical Concerns

AI systems require timely, geo-specific, high-quality data to function well. However, in agricultural industries around the world, data fragmentation, lack of standardization, and shortage are prevalent. In order to produce rich statistics, many farms lack digital infrastructure like drones, soil sensors, and Internet of Things devices. Furthermore, biased datasets (such as underrepresentation of particular crops, climates, or areas) may produce AI suggestions that are inaccurate or ineffectual. Additionally, there is an increasing demand for open-access agricultural databases that allow algorithmic innovation while protecting data privacy (Nguyen & Tran, 2023; Chowdhury & Islam, 2024). A significant obstacle is the disparity in access to AI technologies between large-scale and smallholder farms, as well as between developed and developing countries. There are various ways that the digital divide appears: Infrastructure deficiency (poor electricity, internet access), Gap in devices (no smartphones or contemporary farm equipment), Affordability gap (exorbitant prices for AI subscriptions or services). Less than 30% of rural areas in Southeast

Asia and sub-Saharan Africa have dependable broadband, which restricts the use of real-time AI applications (UN FAO, 2023). AI runs the risk of increasing inequality in the absence of focused public-private partnerships and funding for digital infrastructure in rural areas. For AI to be used effectively, farmers must be able to understand and follow digital advice. A large portion of the rural population lacks the digital literacy required to use AI-based systems or decipher model results. Additional obstacles to adoption include non-intuitive user interfaces and language limitations. Extension programs must transition from conventional forms to advisory models powered by AI, combining localized human interaction with virtual support. It is imperative that government and nonprofit organizations make investments in farmers' and extension officers' digital skill development. Weather, pest, and crop management information is provided by voice-based AI systems in local languages through India's "AI Krishi Sakhi" initiative, which increases user engagement among female farmers (Narayan et al., 2024). The application of AI in agriculture presents ethical questions, particularly with relation to algorithmic openness, data ownership, and privacy. The farmers who produce the data frequently have no say in how it is used, preserved, or made money off of. Asymmetries in the extraction of data value may result from companies gathering data maintaining exclusive rights (Jain & Kumar, 2023; Thomas & Lee, 2023). Furthermore, opaque AI models, sometimes known as "black-box" systems, might produce suggestions that are hard to interpret or validate, which damages user confidence. Explainable AI (XAI), which enables people to understand the decision-making process, is being promoted as a solution to this problem. Creating models that are equitable, non-discriminatory, and considerate of the needs of underserved areas is another aspect of ethical AI. Policies should encourage farmers to co-create AI technologies that are appropriate for their economic, cultural, and environmental conditions.

6.3 Policy Frameworks and Funding Challenges

Enabling policies that establish data governance guidelines, offer incentives, and promote public-private collaborations are essential to the development of AI in agriculture. Clear regulations pertaining to data ownership, cross-border data flows, and agricultural technology interoperability standards must be established by governments. Nations such as the Netherlands and Canada have developed National AI AgTech policies that combine digital innovation with sustainability objectives, financing R&D, startup ecosystems, and AI-based extension systems. Additionally, international collaboration is essential, particularly for shared ecosystems and global crops. Effective AI solutions are supported by initiatives like GODAN (Global Open Data for Agriculture and Nutrition), which encourage data exchange for nutrition and food security. Large investments in R&D, infrastructure, and human resources are necessary for the large-scale deployment of AI tools. Even if agri-tech businesses are increasingly being funded by venture capital, public investment is essential for the fair distribution of technology. There are funding needs for interdisciplinary research that connects agronomy and AI, rural internet and sensor infrastructure, localized AI model training, and open-source AI tools for smallholders. Financial institutions must also encourage investment in smart technologies by creating credit and insurance models that take AI's application in farm risk assessment into account. Small farmers in Kenya were able to obtain loans and crop insurance through microcredit programs connected to AI-based yield estimates, which resulted in a 19% increase in agricultural investment (Odhiambo et al., 2023).

7. FUTURE DIRECTIONS AND INNOVATIONS

Modern agriculture is changing as a result of the convergence of developing technology and artificial intelligence (AI). The future of AI in agriculture depends on multifaceted innovation as issues related to climate change, food security, and environmental degradation become more complicated. The most exciting developments in AI-driven farming are examined in this part, including the use of AI to robots, blockchain, and customized agronomy for climate-resilient farming.

7.1 AI-Robotics Convergence in Agriculture

With intelligent, self-governing gear, the fusion of robotics and artificial intelligence is transforming agricultural operations. AI-enabled robotic tractors, seeders, and harvesters can operate in challenging environments, make choices in real time, and carry out agricultural duties with little assistance from humans. These devices identify anomalies, adjust to crop conditions, and maximize resource utilization through the use of computer vision and sensor fusion. For example, precise herbicide targeting is made possible by AI-powered robotic weeders, such those created by Blue River Technology, which can cut chemical usage by up to 90% (Singh et al., 2023). Convolutional neural networks (CNNs) are used by these devices to differentiate between weeds and crops in milliseconds, a feat that is practically difficult to accomplish at scale with manual techniques. Robotic swarms, which are networks of tiny autonomous units cooperating on activities like planting, weeding, and pest detection, will be used in future deployments. These robots, which are driven by Edge AI, process data locally, guaranteeing low latency and independence from constant internet connection. This is particularly beneficial in isolated or impoverished agricultural areas. Because choices can be made directly on the machine using real-time sensor data, edge AI in robotics reduces energy usage and increases reliability (Araújo et al., 2023; Singh & Kumar, 2023). In Japan and the Netherlands, agricultural swarm robots is being tested for rice and strawberry cultivation, and it has early promise for increasing labour efficiency (Watanabe et al., 2023). Human-robot cooperation will be essential as farming shifts from manual to intelligent mechanization. Co-bot models, in which people handle complicated decision-making while robots help human workers by taking over dangerous or monotonous jobs, will be supported by future technologies. Figure 4 depicts AI-Robotics Convergence in Agriculture. AI translators will make speech, gesture, and augmented reality (AR) control interfaces more user-friendly.

Figure 4. AI-Robotics convergence in agriculture

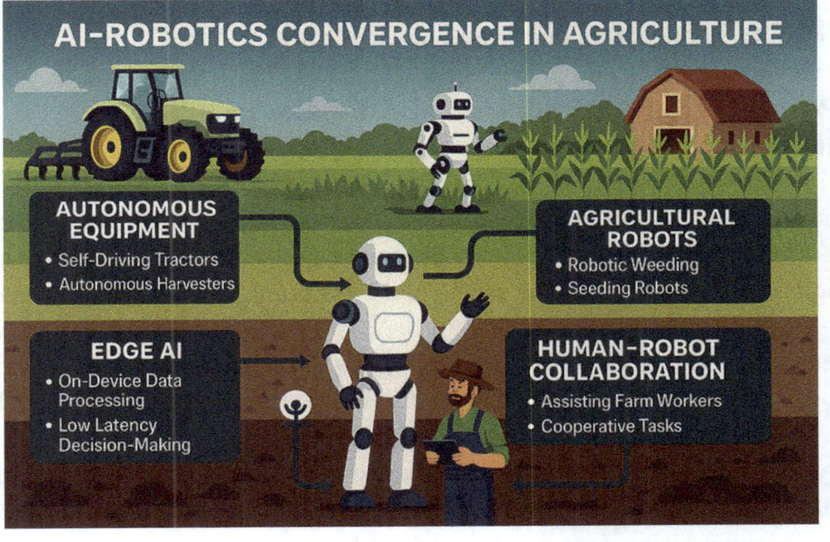

7.2 Blockchain-AI Synergy for Supply Chain Traceability

Blockchain technology and artificial intelligence together provide previously unheard-of levels of accountability and transparency throughout agricultural value chains. Blockchain guarantees unchangeable records, traceability, and confidence in the production, distribution, and consumption of food, while artificial intelligence (AI) offers data analytics and prediction. Blockchain and AI integration, for instance, makes it possible to trace product from farm to table, guaranteeing freshness, organic certification, and origin verification. IBM's Food Trust network reduces fraud and boosts consumer confidence by using blockchain technology and AI-driven supply chain monitoring to track food origins in a matter of seconds

(Chowdhury et al., 2023). Blockchain records can be examined by AI algorithms for variations in temperature, transit time, or documents, which could indicate fraud or non-compliance. In real time, machine learning, for example, can identify trends of inconsistent pesticide use or mislabelled products. AI helps smart contracts reduce bureaucracy and improve supply chain efficiency by automating transaction execution and compliance verification. For ongoing environmental monitoring of perishables, future implementations will combine blockchain-AI systems with IoT sensors in packaging. New models for green rewards are made possible by blockchain's capacity to tokenize actions. Blockchain-based carbon credits or sustainability tokens can be awarded to farmers that implement AI-driven sustainable practices, such as reducing chemical inputs or enriching soil carbon (Fosso Wamba & Queiroz, 2022; Jain & Kumar, 2023). By trading or using these credits for subsidies, a positive feedback loop of ecologically conscious farming is established.

7.3 Personalized Agriculture and Climate-Resilient Farming via AI

The transition from generic farming advice to customized agronomy will be fuelled by AI developments. Using genotype-environment interactions, soil microbiomes, microclimate patterns, and historical yield data, machine learning models can offer hyper-localized farming recommendations. AI systems such as CropX and PEAT use soil sensors and satellite photos to deliver real-time recommendations for particular fields or zones, directing pest control, nutrient management, and watering. In the following stage, genetic data will be integrated to match crop types with site-specific biophysical circumstances, guaranteeing resilience and maximum yield. AI plays a key role in creating climate-smart farming practices as climate variability becomes the new normal. Using generative models, AI can model how crops will react to anticipated weather extremes and suggest adaptive strategies like choosing early-maturing or drought-tolerant cultivars, recommending adjustments to planting dates or density, and adjusting irrigation schedules in response to soil moisture predictions. Rice farmers in Bangladesh were able to alter transplanting dates with the use of an AI-powered flash flood early warning system, which reduced crop losses by more than 35% in 2023 (Rahman et al., 2024). AI systems will help with dynamic crop planning more and more, including not only environmental factors but also labour availability, market demand projections, and policy requirements. By bringing production and consumption into line, waste is decreased and farmer incomes rise. For example, farmers might choose crops that offer higher returns by using AI models to forecast future market values based on patterns. These features are being integrated by platforms such as Agroscout to assist smallholders in diversifying their revenue streams and adjusting to shifting economic and climatic situations. Natural language processing (NLP) and conversational AI are revolutionizing extension services by providing accessible, context-specific guidance in regional languages (Chowdhury et al., 2023; Zhang & Chen, 2023). These systems are available around-the-clock to respond to inquiries about financial planning, pest outbreaks, weather information, and crop care. In the future, wearable technology like augmented reality glasses that provide in-field advice through AI overlays may be integrated, decreasing the need for cell-phones and improving usability for farmers with poor literacy levels (O'Connor & Smith, 2023; Liu & Wang, 2023).

8. CONCLUSION

The 21st-century approach to crop yield and soil health has undergone a paradigm shift with the introduction of artificial intelligence (AI) into agriculture. The multifaceted role AI plays in improving agricultural sustainability, efficiency, and resilience has been thoroughly examined in this chapter. AI has shown its revolutionary potential in rethinking conventional farming paradigms, from optimizing inputs through predictive analytics to diagnosing soil conditions using sensors driven by AI. The chapter emphasized how AI provides strong mechanisms to analyze complicated soil parameters and deliver real-time, data-driven insights for soil monitoring and management through machine learning, deep learning, remote sensing, and precision agricultural tools. It became evident from a comparison of traditional soil testing with AI-powered sensing devices that AI enables more accurate, scalable, and continuous evaluation of the physical, chemical, and biological characteristics of soil. Additionally, combining AI with drones, satellite imagery, and Internet of Things (IoT) devices improves diagnostic capabilities, facilitates timely intervention decision-making, and develops predictive modelling for yield optimization, disease and pest management, irrigation scheduling, and nutrient application. Case studies showed how AI-driven technologies are already making a big difference in actual agricultural settings, increasing sustainability and efficiency. The conversation also explored how AI and sustainable agricultural methods, such crop rotation, conservation agriculture, and organic farming, may work together to maintain biodiversity, improve carbon sequestration, and use less chemicals. The chapter does, however, also recognize important obstacles to the adoption of AI, such as unequal access to data, gaps in digital literacy, ethical issues, and limitations in infrastructure and legislation. Investing in farmer-centric AI platforms, fostering digital literacy through focused training, promoting public-private partnerships for inclusive technology deployment, and creating clear policy frameworks for data governance, sustainability incentives, and funding support are some of the strategic recommendations to get past these obstacles and optimize the advantages of AI. Future prospects are bright as AI, robotics, blockchain, and genomics come together to provide autonomous operations, safe and traceable food systems, and customized, climate-resilient farming practices. AI stands out as a revolutionary force that empowers stakeholders across the agricultural value chain, from farmers and agronomists to policymakers and consumers, as agriculture faces growing complexity and urgency. In addition to technological advancement, ethical stewardship, equitable access, and ecologically intelligent sustainable practices will be essential to agriculture's future. We can usher in a new era of intelligent, just, and resilient agriculture that can feed a growing population while preserving the health of the earth by carefully utilizing AI.

REFERENCES

Araújo, R., Silva, F., & Costa, L. (2023). Mobile robotics in smart farming: Current trends and applications. *Frontiers in Artificial Intelligence*, *6*, 1213330. DOI: 10.3389/frai.2023.1213330 PMID: 37719082

Chowdhury, H., Paul Argha, D. B., & Ahmed, M. A. (2023). Artificial intelligence in sustainable vertical farming. *arXiv*. https://arxiv.org/abs/2312.00030arXiv

Chowdhury, S., & Islam, M. (2024). Towards sustainable agriculture: Harnessing AI for global food security. *Artificial Intelligence in Agriculture*, *4*(3), 100123. DOI: 10.1016/j.aiia.2024.04.003

Dutta, S., & Banerjee, A. (2023). Enhancing precision agriculture: A comprehensive review of machine learning and AI applications. *Agricultural Technology*, *5*(2), 100483. DOI: 10.1016/j.atech.2024.100483

Fosso Wamba, S., & Queiroz, M. M. (2022). Artificial intelligence and blockchain implementation in supply chains: A case study of the tuna fish supply chain in Thailand. *Annals of Operations Research*, *308*(1), 1–25. DOI: 10.1007/s10479-021-04450-0 PMID: 34931103

Ghosh, A., & Roy, S. (2024). AI-driven future farming: Achieving climate-smart and sustainable agriculture. *Agriengineering*, *7*(3), 89. DOI: 10.3390/agriengineering7030089

Gupta, R., & Sharma, P. (2023). The innovative role of blockchain in agri-food systems: A literature review. *Food Control*, *145*, 110603. DOI: 10.1016/j.foodcont.2024.110603

Islam, M. S., & Rahman, M. M. (2023). AI- and data-driven pre-crop values and crop rotation matrices. *European Journal of Agronomy*, *135*, 126949. DOI: 10.1016/j.eja.2023.126949

Jain, V., & Kumar, S. (2023). The role of blockchain technology in the sustainability of supply chains. *Cleaner Logistics and Supply Chain*, *5*, 100113. DOI: 10.1016/j.clscn.2023.100113

Kumar, R., & Singh, A. (2023). Artificial intelligence tools for the agriculture value chain. *Electronics (Basel)*, *13*(22), 4362. DOI: 10.3390/electronics13224362

Lee, H., & Park, J. (2023). Computer vision in smart agriculture and precision farming. *Artificial Intelligence in Agriculture*, *5*(1), 100234. DOI: 10.1016/j.aiia.2024.06.004

Li, X., & Zhang, Y. (2023). Artificial intelligence and its role in soil microbiology and agricultural sustainability. *Microbiology, 12*, 100006. DOI: 10.1016/bs.mim.2024.05.006

Liu, Y., & Wang, H. (2023). Enhancing global agricultural monitoring systems for climate-smart agriculture. *Climate Services for Agriculture*, *2*(1), 100037. DOI: 10.1016/j.csag.2024.100037

Malhotra, M., Ghosh, R., Kumar, N., & C, K. K. R. (2025). AI-Powered Harmony. In *Advances in social networking and online communities book series* (pp. 95-126). https://doi.org/DOI: 10.4018/979-8-3373-0543-1.ch004

Martínez, L., & Pérez, R. (2023). Transforming agriculture with machine learning, deep learning, and IoT. *Smart Agriculture*, *4*(2), 66–78. DOI: 10.1007/s44279-024-00066-7

Mehta, S., & Patel, D. (2023). *Sustainable artificial intelligence solutions for agricultural efficiency*. IntechOpen., DOI: 10.5772/intechopen.112996doi.org

Nguyen, T., & Tran, L. (2023). Artificial intelligence for sustainable agriculture: A comprehensive review. *Sustainability (Basel)*, *17*(5), 2281. DOI: 10.3390/su17052281

O'Connor, D., & Smith, J. (2023). AIIRA: AI Institute for Resilient Agriculture. *AI Magazine*, *44*(1), 51. DOI: 10.1002/aaai.12151

Patel, K., & Desai, M. (2023). Artificial intelligence in sustainable vertical farming. *arXiv*. https://arxiv.org/abs/2312.00030arXiv

Rahman, M., & Chowdhury, H. (2023). Blockchain and artificial intelligence for ensuring the authenticity of organic legumes. *Legume Research*, *46*(5), 786. DOI: 10.18805/LRF-786

Sarkar, D., & Banerjee, S. (2023). Sustainable development of AI applications in agriculture: A review. *Procedia Computer Science*, *219*, 350–359. DOI: 10.1016/j.procs.2023.10.350

Singh, A., & Kumar, P. (2023). Leveraging imitation learning in agricultural robotics. *Frontiers in Robotics and AI*, *10*, 1441312. DOI: 10.3389/frobt.2024.1441312 PMID: 39483488

Thomas, G., & Lee, S. (2023). Understanding the potential applications of artificial intelligence in agriculture. *Advances in Agricultural Computing*, *2*(3), 100001. DOI: 10.1016/j.aac.2022.10.001

Watanabe, K., & Tanaka, Y. (2023). Towards autonomous mapping in agriculture: A review of supportive technologies. *Robotics and Autonomous Systems*, *12*(1), 104514. DOI: 10.1016/j.robot.2023.104514

Zhang, L., & Chen, Y. (2023). Trends in the sustainable use of artificial intelligence: A bibliometric analysis. *Sustainable Computing : Informatics and Systems*, *5*(2), 1222–1235. DOI: 10.1007/s43621-025-01222-9

Zhao, X., & Liu, H. (2023). Artificial intelligence in agricultural technology: Enhancing crop harvest prediction. *International Journal of Agricultural Sciences*, *10*(2), 22240. DOI: 10.21275/SR24825222240

Chapter 13
Innovations and Challenges in Smart Agriculture:
A Case Study on IoT and Machine Learning Applications

Jagadeshwari Puttanapura
https://orcid.org/0009-0008-9607-7438
University of South Alabama, USA

T. Monika Singh
https://orcid.org/0009-0006-8534-9130
Annamalai University, India

Shugufta Fatima
https://orcid.org/0009-0005-4396-4947
Annamalai University, India

R. Madana Mohana
https://orcid.org/0000-0002-7652-6266
Stanley College of Engineering and Technology for Women, India

Anindya Nag
https://orcid.org/0000-0001-6518-8233
Northern University of Business and Technology, Khulna, Bangladesh

ABSTRACT

This chapter explores the evolving future of data-driven decision-making through the convergence of the Internet of Things (IoT) and Machine Learning (ML), fo-

DOI: 10.4018/979-8-3373-4207-8.ch013

cusing on applications in smart agriculture. IoT enables the real-time collection of vast data from connected devices, while ML provides the intelligence to interpret and act on this information. Their integration enhances automation, efficiency, and predictive insights in agricultural systems. Core concepts of IoT and ML are presented together, highlighting their role in optimizing crop monitoring, precision irrigation, disease detection, and yield forecasting. A case study illustrates how ML applied to IoT data improves operational performance and decision-making. The chapter also addresses benefits, challenges such as data integration, privacy, and connectivity, and explores emerging trends like edge computing and federated learning. Overall, it offers a holistic view of how IoT and ML are shaping the future of smart agriculture and related domains.

1. INTRODUCTION

The Internet of Things (IoT) and Machine Learning (ML) are construed as transformative technologies for industries and ways of life. IoT is the world of devices that are interconnected, collecting data, exchanging data, and processing data for automated routine decision-making purposes. Smart home appliances, industrial sensor devices, and many others generate real-time data in huge quantities. But employing this data calls for effective computational techniques.

Machine Learning, which is a sub-field of Artificial Intelligence (AI), teaches a system the kind of prediction or decision without explicitly programming it to learn. Machine Learning algorithms analyze complex data sets, identify trends, and convert them into actionable insights. The combined usage of IoT with ML leads to intelligent automation, ensuring optimal resource use and better efficiency of the system.This gain from incorporating machine learning into any IoT or Internet of Things system is to add enhanced capabilities for predictive analytics, anomaly detection, and decision-making applications. This has been around and incorporated in various sectors of health care, smart cities, and theirs has been industrial automated. And creation of smarter, data driven solutions. The smart combination of real-time IoT-generated data and machine learning-driven analytics set up a strong ecosystem that may prove itself in revolutionising various domains.

1.1 Importance of Convergence

The fusion of IoT and ML becomes very relevant because the traditional IoT systems hardly process the voluminous streaming data. The basic IoT application uses mainly rule-defined automation, which cannot much adapt to the dynamic environment. ML thus adds intelligence to IoT by analyzing patterns, making forecasts, and optimizing processes in real time.Among the most significant benefits of this merging is real-time decision-making. Traditional IoT systems collect and send data for humans to interpret, whereas ML-driven IoT solutions can analyze and act immediately. As an example in predictive maintenance, ML models can analyze sensor data from industrial machines to anticipate failures prior to their occurrence, thus reducing downtime and repair costs.Alongside this, it has also improved efficiency, security, and personalization(Sharma & Nandal, 2019). For instance, ML models detect anomalies in IoT networks to avert attacks, optimize energy consumption in smart grids, and personalize the experiences of users in smart homes. The merging of IoT and ML promises organizations and individuals' greater automation, savings, and improvement in operational efficiency.

Table 1 shows the major contrasts and collaborations between IoT and ML are elaborated in this table. Hybridization of data collection and aggregation with connectivity and monitoring for real-time analytics. Machine Learning aims at data analysis: prediction and automation based on previous learning-while making possible learning intelligent IoT systems capable of learning through data interpretation and optimized process and better decision making.

Table 1. Comparison of IoT and machine learning

Feature	Internet of Things (IoT)	Machine Learning (ML)	IoT-ML Convergence
Primary Focus	Data collection & transmission	Data analysis & prediction	Intelligent decision-making
Key Components	Sensors, actuators, networks, cloud	Algorithms, models, training data	AI-powered IoT devices
Data Type	Time-series, real-time	Structured, unstructured	Multimodal (sensor + AI insights)
Processing Location	Edge, cloud	Cloud, edge	Hybrid (Edge AI)
Applications	Smart homes, healthcare, industry	NLP, computer vision, analytics	Smart cities, predictive maintenance

1.3 Real-World Applications

The possible visible impact of convergence IoT with ML extends over a variety of industries. In the area of health care, for example, the wearable devices are embedded with sensors where the part of IoT is used to monitor the vital biometrics of patients, while the usage of ML is embedded to analyze trends that help in pronouncing early detection of diseases or lifestyle-related changes. This would have added better possibilities for patient monitoring through the reduction of hospital visits and improvement in health outcomes.

Traffic systems such as intelligent IoT-enabled traffic management systems use traffic optimization techniques based on ML to smoothen traffic flow that ensure signals reduced congestion while improving efficiency in public transport(Lin et al., 2022). Such systems include surveillance ones, where ML-enabled facial recognition and anomaly detection boosts the safety of such environments Irrespective, all innovations lend themselves to making cities safer and more efficient.

In industrial automation, Internet of Things sensors monitor the manufacturing machines, and ML models predict their failure before the actual collapse. This predictive maintenance approach lowers operational costs and increases productivity. Likewise, the intelligent farming methods make use of IoT sensors -considering the soil condition- while the ML model recommends the irrigation and fertilization schedule in order to increase crop yield and resource efficiency(Tzafestas, 2018).

2. INTERNET OF THINGS (IOT) FUNDAMENTALS

The Internet of Things (IoT) refers to the entire system consisting of interconnected devices which collect, transport, and analyze data for the automating, making decisions, and performance improvement across various areas. These devices can be simple smart sensors or very complex measurements taken at the industrial level or used in autonomous vehicles. IoT is a whole into which every other connected communication technology feeding into embedded technologies and wireless protocols connects. It is vital for health care, agriculture, smart cities, and manufacturing where real-time monitoring in predictive analytics becomes possible. Knowing the important aspects of IoT, such as its architecture, communication protocols, sensors, and data transmission mechanisms, would facilitate designing efficient and scalable IoT solutions (Singh, T. M., et al., 2024).

2.1 IoT Architecture and Components

The architecture of the Internet of Things includes many layers that facilitate the connection between real things and virtual platforms. Usually, IoT architecture is built on a four-layer design: the perception layer (with sensors and their associated actuators); the network layer (that concerns the different known communication protocols and the transmission of data); the edge computing layer (which concerns the data found right at the edge); and the application layer (user interface and analytics). Each of these layers contributes its share to the realization of efficient data flow and real-time decision-making(Kumar & Mallick, 2018).

The perception layer contains numerous sensors and actuators to sense real-world data, including temperature, humidity, motion, and pressure. The physical sense parameters are converted into digital signals destined for the next layer. The network layer is responsible for transmitting these data through wired or wireless technologies, such as Wi-Fi, Bluetooth, Zigbee, or a cellular network.

Figure 1. IoT architecture: Five-Layer model

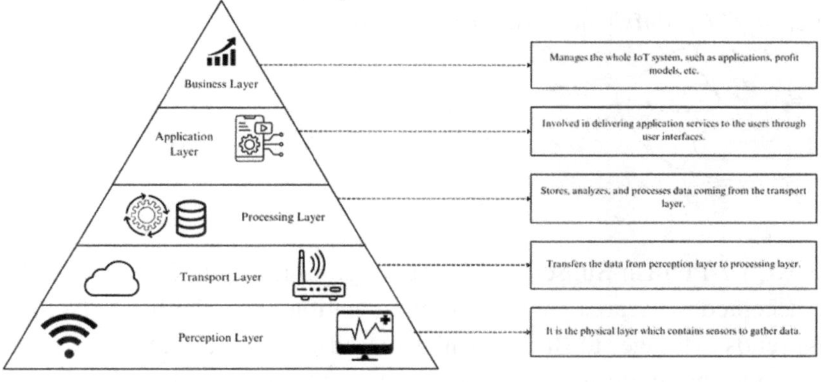

The five-layer architecture of IoT is illustrated in the figure 1, facilitating seamless data collection, processing, and decision-making. The Perception Layer contains sensors that collect real-world data, which the Transport Layer transmits using communication technologies, such as5Gand Wi-Fi. The Processing layer analyzes and stores this data in the cloud or on edge computing. On the other end, the Application Layer provides services for users through apps and dashboards, while the Business Layer manages systems, decision-making, and profit-making activities. This well-structured methodology guarantees efficient functioning of the IoT in various domains.

At the edge computing layer, raw data would be processed locally before further transmission into the cloud or data centers, thus reducing latency and bandwidth usage. This application layer includes such item such as user interfaces, dashboards for the clients, and insights driven by AI, thereby making all that data actionable. It is essential to understand how these parts interact as they are all functional in creating scalable security systems such as IoT systems.

2.2 IoT Communication Protocols (MQTT, CoAP, HTTP, etc.)

IoT devices are based on specific communication protocols to effectively transmit data. Unlike traditional ones, these are not energy-wasteful and consume less bandwidth so that their devices could use them for real-time communication. Some of the well-known protocols under this group include MQTT(Message Queuing Telemetry Transport), CoAP (Constrained Application Protocol), and HTTP(HyperText Transfer Protocol).

Figure 2. MQTT publish-subscribe architecture in IoT

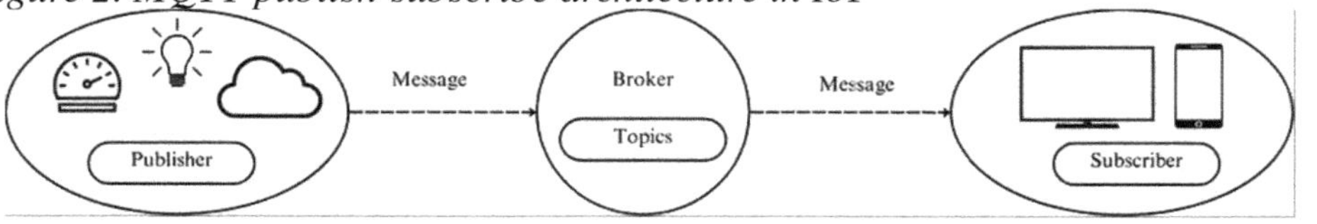

The MQTT Publish-Subscribe Model Diagram in figure 2, is one of the most widely accepted standards for IoT communication. The Publisher (sensors, IoT devices) sends messages to Broker, which classifies messages into Topics, and forwards them to the interested Subscribers (smartphones, dashboards). Supplying a value-added, lightweight, and scaled solution for communication, this characteristic model becomes more prominent in real-time data transfer scenarios dealing with harsh network constraints.

MQTT is a lightweight publish-subscription messaging protocol above and beyond TCP/IP. It has established its popularity among IoT applications because of low bandwidth shortages and the fact that it can still work in areas of poor connectivity(Dinculeană & Cheng, 2019). In addition, MQTT guarantees messages being delivered reliably with three different Quality of Service (QoS) levels so developers can optimize performance for any use cases based on network conditions.

Below table 2 describes the special communication protocol, with criterions like speed, security, and power efficiency, marks the difference among IoT devices. A few comparative points regarding the usages between some common IoT protocols such as MQTT, CoAP, and HTTP and also including LoRaWAN and Zigbee, have been furnished in the table.

Table 2. IoT protocols and their use cases

Protocol	Type	Speed	Security	Power Efficiency	Best Use Cases
MQTT	Publish-subscribe	High	Moderate	High	Smart homes, industrial automation
CoAP	RESTful API	Moderate	High	High	Low-power IoT devices, wearables
HTTP	Request-response	Slow	High	Low	Web-based IoT applications
LoRaWAN	LPWAN	Low	Moderate	Very High	Smart agriculture, remote monitoring
Zigbee	Mesh networking	Moderate	High	Moderate	Smart lighting, home automation

CoAP, however, is a web protocol that is specifically for resource-constrained IoT devices. It differs from MQTT which, unlike CoAP, is broker-dependent and is very much like HTTP but carries relatively lighter loads in terms of communication to perform a request-response cycle(Rahman & Shah, 2016). Further, while HTTP has remained dominant as a protocol in web-based IoT applications, the power consumption and expensive bandwidth costs make it ineffective for use in battery-driven devices. Choice of communication protocol will therefore depend on energy efficiency and real-time processing applications as well as network reliability(Daud et al., 2017).

2.3 IoT Sensors and Devices

IoT systems anchor sensors as the basis of which the technology can be used to further explore and deploy data in real life. Most sensor types are also found in IoT devices: temperature sensors, humidity sensors, motion detectors, pressure sensors, and biometric sensors. All these IoT sensors are available for environmental parameter monitoring and data continuously transmitted to analysis and supported decision-making.For example, temperature and humidity sensors find wide applications in smart homes, weather stations, and industrial automation. Motion sensors include the infrared and the ultrasonic types, which find applications in security and smart lighting. Pressure sensors are widely used in automotive and health industries, for instance, in measuring tire pressure and blood pressure in medical devices.IoT devices include actuators whose function is converting electronic signals to physical actions. Examples include smart locks, automatic irrigation systems, robotic arms into manufacturing. Seamless integration of sensors and actuators, therefore, allows IoT systems to automate tasks, increase efficiency, and improve decision-making in a wide range of applications(Sehrawat & Gill, 2019).

2.4 Data Generation and Transmission in IoT

IoT devices generate copious data; efficient transmission can, therefore, become one integral source of IoT systems. Initially, data are collected at the perception layer where sensors touch and sense environmental values. Data will either be locally processed or transmitted to a cloud server for further analysis. The aspect of such highly volume, velocity, diversity in data that they generate in IoT would provide tremendous challenges and opportunities in managing and processing this information(Izaddoost & Siewierski, 2020).

Bandwidth efficiency and latency are among the prime aspects in data transmission. An IoT network employs both wired (Ethernet, fiber optics) and wireless (Wi-Fi, LoRaWAN, Zigbee) communications for information relay. While wired connectivity enhances reliability and security, mobility and flexibility favor the wireless transmission preference.

With edge computing, customers will be able to use devices close to the location where the IoT devices are installed. This will reduce response time between cloud servers because it processes data closer to the source. Though, cloud-based architecture is where most historical stored data, running machine learning models, and analytics dashboards are still in place. The appropriate mixture of edge and cloud processing must match the real-time needs of the application and the infrastructure capabilities.

3. MACHINE LEARNING (ML) FUNDAMENTALS

Machine learning (ML) is a form of artificial intelligence (AI) that lets systems learn from data and subsequently improve without requiring explicit programming. Unlike rule-based approaches, ML algorithms use analytical techniques to examine vast amounts of information, so they can identify patterns and then make predictions. The integration of ML with the Internet of Things (IoT) has produced intelligent decision-making systems that can analyze sensor data, understand anomalies, and optimize operations in real time. The section covers fundamental ML concepts, such as the different categories of ML, feature engineering, and the IoT data-related challenges.

3.1 Basics of Machine Learning and AI

Machine learning is a subset of AI that concentrates on the creation of algorithms that learn from the data and make decisions. AI pervades in much broaden concept like expert systems, processing natural languages (NLP), robotics, while ML can only be on predictive modeling and pattern recognition. ML integrated IoT system supplements the automation provided by IoTs to adaptively change environments.

Data, algorithms, and models make up the three main components of ML(Sarkar et al., 2018). At the bottom of the structure is data, which then feeds the algorithms-folks like neural networks, decision trees and support vector machines. These trained models are then attached to using the algorithms in undertaking tasks such as classification, regression and clustering. With IoT, it is much challenging since there will be a continuous data stream in real time, which necessitates more efficient preprocessing techniques.

Various Machine learning algorithms fulfill various IoT applications. The following table 3 lists some of the preferred ML techniques, along with their strengths and weaknesses while applied to IoT, such as Decision Trees for anomaly detection, CNNs for images-based IoT, and LSTMs for time-series analyses.

Table 3. ML algorithms for IoT applications

Algorithm	Type	Suitable for IoT Applications	Advantages	Limitations
Decision Trees	Supervised	Predictive maintenance, anomaly detection	Easy to interpret, fast	Overfitting with deep trees
Random Forest	Supervised	Sensor-based health monitoring	Handles missing data well	Computationally expensive
CNN	Deep Learning	Smart surveillance (image-based IoT)	High accuracy in image tasks	Needs high processing power
LSTM	Deep Learning	IoT time-series analysis (e.g., weather forecasting)	Handles sequential data well	Requires large datasets
K-Means	Unsupervised	IoT traffic pattern analysis	Works well for unlabeled data	Sensitive to initial cluster selection

Machine learning is mainly used in the fields of anomaly detection,predictive-maintenanceand smart automation in an IoT applications. For example, analyzing energy consumption in smart grids, fault detection in industrial equipment, and traffic control in smart cities may be achieved using ML models. Capable of scaling up and efficient in processing vast amounts of sensor data, ML becomes increasingly important in the application of IoT.

3.2 Supervised, Unsupervised, and Reinforcement Learning

3.2.1 Supervised Learning

The supervised learning in machine learning basically lets the models learn from labeled datasets, which means that in every example found in training data, there is a corresponding correct output. In supervised learning algorithms, the relation between input variables and output labels is established by minimizing the prediction error(Hastie et al., 2009). Some popular examples of supervised learning algorithms are decision trees,linear regression, random forests, or deep neural networks. In most common IoT applications, supervised learning finds use for predictive maintenance where sensor data from industrial equipment is labeled by history of failures. Hence, the model is capable of predicting failures. This, as an example, is found in health care applications of IoT where wearables use supervised learning to classify an abnormal heart rate and early detection of diseases.

Supervised learning generalized itself through training for large and diverse datasets; the main downside is cost and time inefficacy in acquiring a labeled dataset especially in environments that employ real-time sensor data streaming like IoT. In addition, models may not perform well on changes as they were trained under old conditions, so they require constant retraining to ensure their results remain accurate. Nevertheless, supervised learning is, on the whole, the foundation of IoT-automation-from predicting traffic flows in smart cities to real-time fraud detection in an interconnected financial system.

3.2.2 Unsupervised Learning

Unsupervised learning, in contrast to supervised learning, is applied to unlabeled data sets, wherein models need to learn the patterns and structure by themselves. It usually does not map out any input to a specific output; rather, it tries to group similar data points or reduce the complexity of the data space. With these clustering techniques, most of them such as K-means, DBSCAN, or hierarchical clustering, are commonly used to segment IoT-generated data. One of these examples is the grouping of customers based on their energy consumption for smart homes. Similarly, reduction techniques such as PCA (Principal Component Analysis) enable optimization in terms of storage and processing of high-dimensional IoT sensor data.

The major advantage of unsupervised learning is that it can reveal some hidden patterns and discover anomalies without any prior assumption(Dayan et al., 1999). Hence, it is of great use for anomaly detection, such as discovering unauthorized access attempts in smart security systems or faulty sensors in industrial IoT networks. The interpretability of unsupervised learning models, however, tends to be very ambiguous, as these models do not have explicit rules set for them and, hence, require domain expertise on the part of the user to validate the findings. Furthermore, they usually fail with very high-dimensional and noisy data in the IoT sense, which emphasizes the pre-processing of the input data for improving outcomes.

3.2.3 Reinforcement Learning

Reinforcement Learning (RL) is a kind of interactive learning in which an agent learns by trial and error from interacting with the environment and earning rewards or penalties for its actions. Instead of just pattern recognition, RL emphasizes making decisions and optimizing towards a goal for the longer term, contrary to supervised or unsupervised learning(Harmon et al., 1996). The application areas of RL include robotics and autonomous vehicles as well as smart grid optimization: For example, in industrial IoT, RL-based robotic arms will learn efficient and precise motions for manufacturing; RL-enabled smart meters will optimize electricity consumption in energy management toward cost and environmental efficiency.

Figure 3. Reinforcement learning-based data processing in IoT

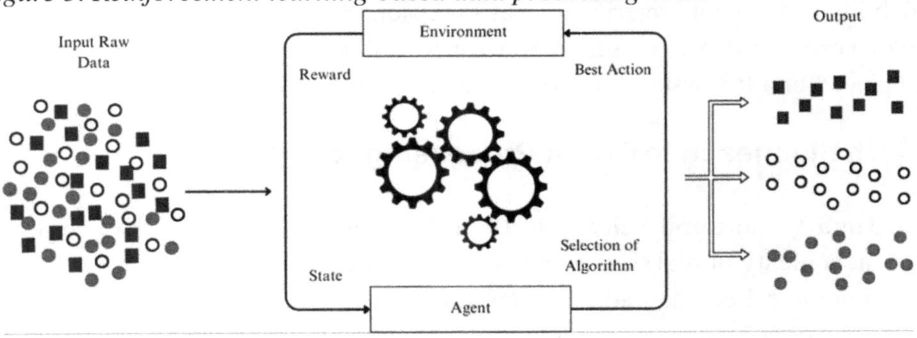

This diagram exemplifies how Reinforcement Learning (RL) fits into the processing of IoT data. Raw input data goes into the agent that selects an algorithm according to what the environment states. Then, the environment gives feedback in terms of rewards to lead the agent toward the best action for optimal classification or clustering of data. Data output after structure and categorization stands to enhance decision-making in IoT systems. This increases adaptability, autonomy, and efficiency in machine learning applications based on the IoT.

The advantage of RL especially lies in its power to adapt to different and unpredictable environments prevailing in IoT, where the existing rules would not apply. However, even after extensive training, specialized RL models tend to perform better when simulated environments are present and optimized. This would mean using great computational resources (Rithika, A et al., 2025). Defining an appropriate reward function is tricky, and if the reward provided is not good enough, unintended behaviors may result from the agent. Yet it holds a very high promise in autonomous IoT systems, from self-learning HVAC controls in smart buildings to intelligent traffic management in connected cities.

3.3 Feature Engineering and Model Training

Feature engineering is the translating of raw data into understandable and meaningful features that help models perform better. Extracting useful features from sensor data is vital in IoT domains because model accuracy is ruined by noisy and irrelevant features. Some methods used in relevant feature extraction include Fourier transforms, wavelet analysis, and statistical summarization. After feature engineering, there is ML model training in the context of historical data with an algorithm for data feeds to adjust the model parameters pertaining to the error minimization. Training of models in IoT may take place in the clouds due to high computational power or it may be done at the edge for real-time decision making(Edge AI).Depending on the need of the design regarding latency and computational resources. Evaluation is the assessment of an ML model to ensure it meets its accuracy and reliability criteria with metrics such as Precision, Recall, F1-score, and root mean squared error (RMSE) to assess performance. Following this, models are validated and placed into IoT systems, continuously adapting and learning as they develop.

3.4 Challenges in IoT Data Processing for ML

High Volume and Velocity of Data: Internet of Things (IoT) devices generate a plenty of real time data which are difficult to be stored and processed, moreover, because traditional machine learning frameworks could not cope with such high-speed streaming data. Distributed computing (like Apache Spark and Hadoop) and online learning algorithms are some of the solutions to such scalability problems costs above.

Data Quality and Noise: Internet of Things (IoT) data is very frequently lacking, unreliable, or noisy in regime of measuring or electrical faults, or inconsistencies in communication. Preprocessing steps such as normalization, outlier detection, and interpolation are important before the introduction of data into the ML models.

Security and Privacy Concerns: ML models must be designed with robust privacy-preserving techniques like federated learning, homomorphic encryption, and differential privacy to ensure their safe and ethical integration into systems. IoT systems are therefore vulnerable to heavy data threats such as data breaches and adversarial attacks, along with violations of privacy(Rahman et al., 2023).

4. CONVERGENCE OF IOT AND MACHINE LEARNING

The integration of the Internet of Things and Machine Learning is bringing significant transformation in automation which creates the possibility of intelligent actions and real-time decision-making, enabling predictive analytics to transform industries. Creating a humungous data pool from sensors, devices, and connected systems without being able to derive meaningful information would always remain the prelude. In other words, ML can supplement IoT in a big way by devices learning and detecting the anomaly in all relative cases and automating the response without human intervention. It is, therefore, bringing a major change in healthcare: on smart cities, it has brought an altogether new economy of industrial automations, and much more. IoT will create dynamic, adaptive, efficient, and intelligent systems.

4.1 Why ML is Essential for IoT

Data with high volumes is characterized by machine learning because handles the high-volume data streams efficiently. Traditional IoT systems are typically rule-based systems. However, this becomes impractical when it comes to processing high-volume complex data. While machine learning algorithms may analyze pattern, trend, and anomaly-in-time data so that such information can feed into a smarter decision-making scheme-Machine Learning can really scale-up IoT. An example is predictive maintenance, which employs machine learning for industrial IoT applications by determining the failure of machines even before their occurrence, therefore minimizing downtime and operating costs.

Figure 4. IoT-Driven machine learning workflow

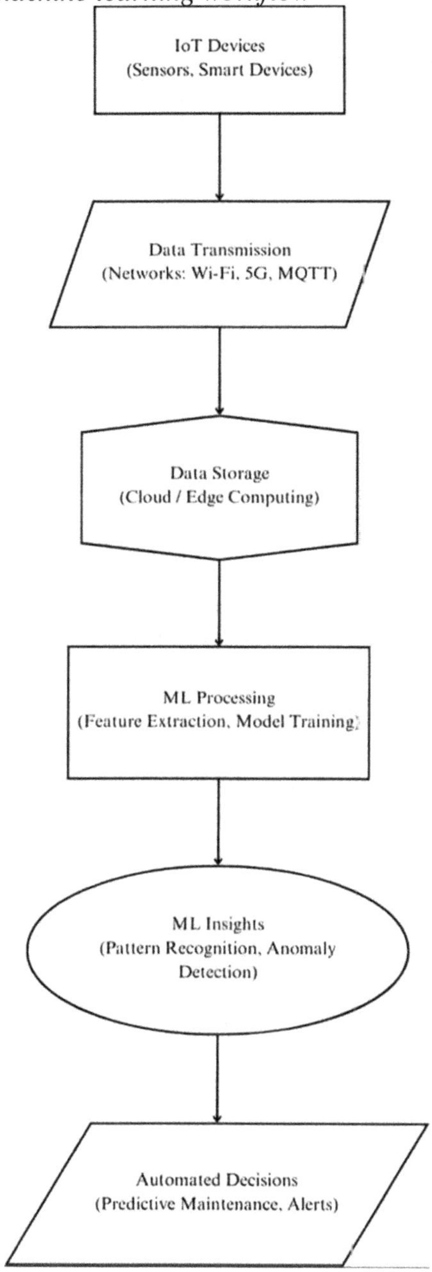

Figure 4 depicts a flow chart showing the entering process in a chain from IoT data handling to its end decision-making employing machine learning (ML). IoT devices like sensors and smart devices collect data and share it through networks like Wi-Fi, 5G, or MQTT. The data is stored in either cloud or edge computing environments waiting for processing. Machine learning techniques are used to extract features with model learning, which results in pattern recognition and anomaly detection. These insights get used in automated decisions such as predictive maintenance and alerts for increasing efficiency and real-time responsiveness of an IoT system.

Machine Learning is another important part of IoT due to its ability to manage automatic responses from real-time data analysis. Unlike traditional rules pre-defined, they enable evolving responses to changing conditions. For example, if a person lives in a smart home where an IoT system is run on machine learning, it will learn its user's preferences and function optimally, thereby enhancing energy usage, security, and comfort. It helps the user manually but will give them a better experience at the end.

Apart from that, ML plays significant roles in securing IoT networks against anomaly detections and overlaps of potential cyber threats(Bout et al., 2021). It indeed connects billions of devices, thus making the IoT ecosystem vulnerable to cyberattacks. Traditional security means fail to provide effective remedies against sophisticated threats; however, security systems based on ML can continuously monitor the network traffic for suspicious behavior and thus prevent possible breaches. So, that makes it an inseparable tool to secure infrastructures for IoT.

4.2 Edge vs. Cloud Computing in IoT-ML Integration

Analogous to the Internet of Things, a vast amount of data is being generated, and handling all the generated data is a major concern.Traditionally, this data was sent to centralized cloud servers for analysis. Cloud-based ML offers the advantages of massive scalability, high computational resources, and access to state-of-the-art AI models(Bajpai et al., 2023). However, this solution also involves latency, considerable bandwidth costs, and exposure to privacy violations that make it unsuitable for time-critical IoT applications, such as autonomous vehicles and industrial automation.

Edge computing, in contrast, allows for processing of data nearer to its source, right at the edge of the network. This reduces the need to transport mountains of data to the cloud, helping improve response times and also saving bandwidth(Awad et al., 2023). With this deployment, edge ML models can make real-time, faster, and more reliable decisions. For example, in healthcare, edge artificial intelligence in wearables could analyze patient data at the source and communicate to doctors immediately if there are signs of abnormality, rather than depending on cloud servers.

Table 4 compares edge computing, cloud computing, and hybrid computing to help determine the best deployment strategy for IoT-ML systems.IoT-ML solutions can be deployed separately at the edge (device-level), in the cloud, or in hybrid environments. This is a summary to contrast the trade-offs-such as latency, computation power, or security-in order to determine which is the best architecture for IoT-ML systems.

Table 4. Edge vs. cloud computing in IoT-ML integration

Feature	Edge Computing	Cloud Computing	Hybrid Computing
Latency	Very low (real-time)	High	Balanced
Computational Power	Limited	High	Moderate
Data Privacy	High (localized)	Low (data sent to cloud)	Moderate
Power Consumption	Low	High	Moderate
Best Suited For	Industrial automation, healthcare devices	Big data analysis, AI model training	Smart cities, autonomous systems

And yet, as advantageous as it may be, edge computing does possess its own limitations in terms of not matching the high computational power and storage space available through cloud platforms. A combination of both edge and cloud computing is often the solution that is most indicated under the circumstances. It is then that edge devices perform real-time processing of data, while critical processing is done by cloud systems. This makes for the optimal balance between efficiency, security, and cost-effectiveness while making IoT-ML integration much more robust and scalable.

4.3 Benefits and Challenges

The unison of IoT and ML can avail numerous benefits among them improved decision making, automation and operational efficiency. The biggest benefit would be predictive analytics whereby organizations can really predict equipment failure, demand and supply fluctuations as well as security threats in advance(Koul, 2023). An example of this is in agriculture where the IoT systems powered by ML would predict weather patterns, soil conditions and crop health, and derive an optimal farming strategy that results in higher yields and resource savings.

Personalization and adaptability are a really another great advantage. The machine learning trains the IoT systems to learn from user interactions and tailor the services based on it. The use of machine learning algorithms in smart cities will optimize traffic signals in line with real-time data regarding congestion reducing travel time and amount of fuel consumed. In retail, for example, customer behavior using IoT sensors in combination with ML may be analyzed and then give personalized shopping experiences and, therefore, improve customer satisfaction and sales as well.

Figure 5. Integration of IoT and machine learning

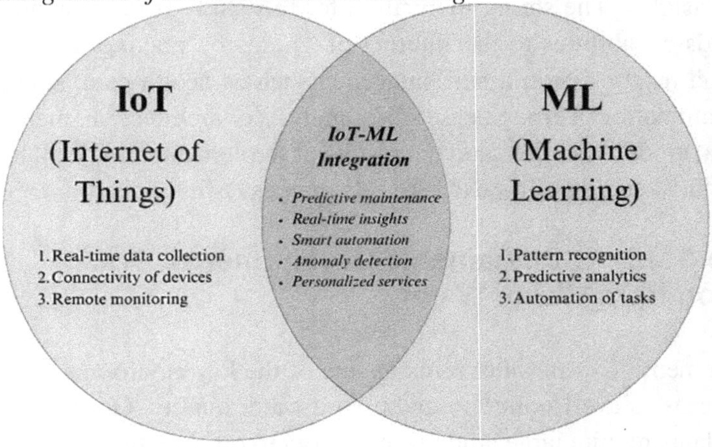

The Venn diagram in figure 5 shows the collusion between IoT with Machine Learning. Data collection in real-time, connectivity of devices, and remote monitoring are the responsibilities of IoT, while ML is competent in pattern recognition, predictive analysis, and task automation. Together they enable predictive maintenance, real-time insight, smart automation, anomaly detection, and personalization. When IoT is combined with ML, the result will allow industries to become more efficient, automated, and intelligent decision-makers.

There might be several challenges in IoT-ML integration despite various advantages. The biggest issues related to it are privacy and security, as IoT devices primarily use sensitive information and may have a chance of prosecution. Moreover, it uses more computational resource approaches, which may cost higher to an organization if applied to the real world. Integration as a challenge has been raised by the common lacking standards across various IoT platforms, making that very interoperable between devices. These issues would be met through continuous research, good security standards, and development of lightweight ML models optimized for different edges.

5. CASE STUDY

The convergence of the Internet of Things and machines with learning is heralding in a new wave of industries-smart decision making, greater automation, and predictive analysis. However, there is a need to address some challenges associated with real-time processing and system scalability in order to implement the said IoT system with machine learning capabilities. Real-world applications of gathering data from IoT devices are analyzed with the help of machine learning models to derive actionable insights. The strong motivation for this study is to show how machine learning adds capabilities to the Internet of Things by making it more efficient, accurate, and responsive in a particular area such as health care, smart cities, or industrial automation. Through case study analysis of architecture, methods of processing data, model selection, and the results of implementation, we'll have a more thorough understanding of how the IoT-ML process is integrated.

5.1. Case Study 1: Sustainable Green Energy Generation from Waste Water

The treatment of wastewater remains one of the key environmental health and safety concerns to avert pollution and assure water quality. Given, however, that treatment plants require large amounts of energy to operate, their operational costs and carbon emissions can be quite substantial. As a remedy, scientists have set their sights on renewable energy sources, including anaerobic digestion, Microbial Fuel Cells (MFCs), and geothermal desalination. Anaerobic digestion simply defines the metabolism of microorganisms to break down organic material from wastewaters with biogas being the useful by-productin the absence of oxygen. Biogas is then often used for electricity and heat generation. Hence, anaerobic digestion serves to manage organic waste while at the same time harnessing energy obtainable from it, thereby lessening reliance on non-renewable resources. When applied to the treatment of wastewater, in MFCs the electricity is generated by means of the metabolic activity of microorganisms such that sustainable energy production is coupled with wastewater treatment. Geothermal desalination emphasizes a range of possibilities whereby geothermal energy could potentially be implemented to desalinate seawater or brackish water into freshwater so that we may cut down on fossil fuel use and simultaneously address the issue of water scarcity(Kumar et al., 2023).

Advent of IoT technology has made greater contributions towards the optimization of energy use in bioenergy sources through real-time monitoring, data capturing, and optimizing processes in anaerobic digestions, MFCs, or geothermal desalination systems. It allows the operators to remotely follow how the system performs, early identification of problems, and optimization of energy production, which in turn translates into savings in costs and possible environmental benefits. Most of these technologies are reducing operational expenditure as well as carbon emissions but all have the potential to "transform" the wastewater treatment plants from energy consumers to energy producers. Their success sometimes outstands technological advancement, financial efficiency, flexibility, and supportive frameworks for their implementation. Awareness from the public, supporting policies, and stakeholders stand necessary in this approach to acceptance. Integrated through optimizations using IoT-driven renewables with those in the future going forward will realize a low carbon economy and increased energy security as the evolution of wastewater treatment advances into a hub for generating sustainable energy.

5.2. Case Study 2: Prediction of Diabetes Using Internet of Things (IoT) and Decision Trees

Diabetes is one of the most serious diseases afflicting humanity today, caused by an inadequate response to insulin—a hormone that aids in the conversion of sugar into energy. This disease can then lead to complications such as kidney failure, heart failure, retinopathy, neuropathy, and peripheral vascular disorders. Therefore, we propose a decision-tree-based system, called SLDPS (Supervised Learning Diabetes Prediction System), to predict diabetes based on parameters collected from IoT sensors to enhance prediction accuracy. After rebalancing the dataset, our model displayed an improved accuracy of 94.63%, which is a perspective compared to other classification models. IoT systems provide an ever-so-important means of monitoring diabetes by enabling a non-stop collection of physiological data through wearable devices. Timely analysis of this real-time incoming data becomes the essence of risk assessment and disease management, making healthcare even more proactive(Allugunti et al., 2021).

Decision trees have been widely used in diabetes prediction because they are simple, interpretable, and handle categorical and numerical data. Various studies have combined them with other algorithms to enhance the accuracy and reliability of predictions. Internet of Things-based solutions help provide healthcare access to the economically weaker segment of society without becoming unmanageable or incurring additional costs. Data privacy and security, along with developing a strong infrastructure, need to be considered, which will be essential for broader implementation. These systems will also require validation in the real-world scenario through extensive clinical trials to demonstrate practical efficacy. Overall, despite these challenges, integration of the IoT with decision tree models holds great promise for improving health outcomes through diabetes prediction and management.

5.3. Case Study 3: Building a Smart Healthcare System Using Internet of Things and Machine Learning

Continuous patient monitoring is enabled by the great features of the Internet of Things (IoT) and Machine Learning (ML). IoT systems employ wearable sensors to detect abnormalities and provide signals in time for medical intervention. In this way, these systems augment patient care by offering real-time health monitoring, remote diagnostics, and automated emergency response mechanisms. With ML, significant contributions have been made to medical diagnosis by commercializing new models that are sensitive to subtle changes in a patient's condition. With the ability to handle enormous amounts of medical data, ML algorithms can detect patterns, predict diseases, and suggest personalized treatment plans. Both sets of technologies, IoT and ML, support the healthcare processes with instant access to the latest medical data to assist the health professional in making an informed decision; patients can also benefit positively from it. Another advantage of IoT in reducing healthcare expenditures is achieved by automating routine tasks, streamlining processes, and improving patient satisfaction through enhanced medical services efficiency.

Considering the speedy progress in 5G technology and cloud computing, the integration of IoT and ML has now manifested themselves in smart healthcare systems(Kute et al., 2022). ML analytics usually process a high volume of medical data from IoT devices to guarantee improved patient outcomes. Enabling faster and more prompt diagnoses is one of the ways in which technologies indirectly affect remote patients: telemedicine and predictive analytics for disease prevention-increasingly-would be inconvenient without tele-monitoring. Such enormous potentials in utilizing this technology come along with numerous hurdles including data security, privacy concerns, and infrastructure requirements that need to be managed before it can become broadly accepted. It would provide applications on IoT and ML in health, with the use of the underpinnings given by 5G network and cloud computing, which further enhances the efficiency and adaptabilityof smart healthcare adaptations. These technologies combined reflect a transformation to more effective, less expensive, and more patient-centric healthcare solutions (Puttanapura, J et al., 2024).

5.4. Case Study 4: IIoT Framework Based ML Model to Improve Automobile Industry Product

The validation of the proposed predictive maintenance system shall be conducted qualitatively as well as quantitatively, with an encompassing comparison against other machine learning approaches, including unsupervised algorithms (K-Means & Agglomerative clustering) and supervised algorithms (Random Forest & SVM). The experimental results in this study are divided into two main parts. Part A analyzes the production success of the manufacturing machine module during operation and assures high-quality products when they exit the production line, thus engaging qualitative means of analyzing the performance of the said predictive maintenance system. Part B works through the evaluative process of the real-time product data against historical record data to prognosticate early product failures quantitively. The feature extraction module tracks the product attributes such as shape and weight, and stores such information locally and in the cloud for cases of future product failure claims by customers(Gopalakrishnan & Kumaran, 2022). The prototype model is designed in MATLAB using the SimEvent toolbox, allowing for qualitative and quantitative performance analysis. Many open small-scale industries lack integrated electronic diagnostics setups at every stage of the production line. They rely on human supervision, which leads to inconsistencies and failure most of the time. Automated systems can execute dynamic and static tests on the product unlike the manual checks done on a stationary product. By simulating various scenarios to test the product, the collected data are stored in the cloud for industries to scrutinize failure trends, optimize their manufacturing processes, and enhance product reliability. Using the parallel computing toolbox, model training and optimization on datasets are processed. These models statistically evaluate and help to identify trends and inefficiencies.

With Industry 4.0, innovative predictive maintenance systems are transforming factory operation from continuous machine performance monitoring and quality assurance(Chen & Wan, 2019). Smart sensors and the Industrial Internet of Things (IIoT) come into play enhancing real-time data collection, thus requiring a coordinated and interconnected infrastructure supported by wireless sensor networks. This configuration helps enhance connectivity between machines, supervisors, and external networks to enable seamless data communication. Real-time monitoring of machine conditions takes place while the processing of internal machine data happens through a programmable logic controller (PLC) for initiating fault detection and predictive analysis. Since there is a vast collection of data generated from the sensors, a storage and processing mechanism has to be highly efficient in working. A double-layer processing framework (Sensor-to-Fog and Fog-to-Fog) is proposed in the system further reducing the cloud storage demand by filtering critical data prior to cloud transmission. The IIoT network architecture improves energy efficiency and prediction accuracy, especially for environments needing constant monitoring. The architectural structure comprises three layers: device connectivity, data analytics, and business connectivity. The sensor gateway collects machine data, compares it with historical records in local servers, and transfers only the essential data for further analysis to the Azure cloud. A machine learning algorithm is then applied to train and optimize predictive models that ensure real-time monitoring of machine quality and product quality. Finally, notifications concerning anomaly detection are sent to the maintenance system to enact the required corrective action to restore performance calibration.

6. COMPARATIVE ANALYSIS AND INSIGHTS

Many traditional IoT solutions have become reinvigorated by the element of added machine learning (ML)-intelligent decisions, programming automation, and prediction capabilities. Most traditional IoT systems, on the other hand, use preset rules and threshold limits. In contrast, truly intelligent ML-based IoT solutions automatically track changes in the environment by relying on real-time data. Here, traditional and modern machine learning-based IoT differences are discussed, performance evaluation metrics are enumerated and compared, and insights are derived from comparisons.

6.1 Traditional vs. ML-Driven IoT Solutions

The rule-based logic with which a traditional IoT system operates still collects data from sensors and processes them according to pre-set conditions. Incorporates simple threshold alerts and structured data processing. Though well suited to basic monitoring, this approach proves inadequate in managing the scattered complex data patterns. Besides, traditional IoT would not be self-learning and would thus require constant manual tuning to maintain efficiency(Bhanu Chander et al., 2020).

However, the IoT in Machine Learning is an application of machine learning algorithms to assess data, recognize patterns, and make predictions in real time. Unlike rules based on fixed values, the ML models learn and teach themselves through past and real-time data to enhance their resourceful decisions. An anomaly can be detected; predictive maintenance may occur, or a machine might operate autonomously. ML-enabled IoT provides another promising step towards edge intelligence in enabling real-time computation near the data source, which in turn will reduce latency and improve response times(Kazi & Mahant, 2025).

The traditional methods are surpassed by the intelligent solutions in scalability, adaptability, and accuracy of decision-making. Traditional IoT systems are easy to deploy and consume limited computational power but fail to perform complex analytics and automation. Conversely, ML-based IoT systems demand more computation and provide better insights, efficiency, and predictions. A clear shift is happening from traditional IoT to ML-based applications for smart cities, health care, and industrial automation, where decisions have to be made in real time.

6.2 Performance Metrics and Evaluation

The most important benefit of IoT associated with Machine Learning is its ability to forecast and recognize patterns accurately. In contrast to conventional IoT systems that only react to a few predefined conditions, ML is capable of foreseeing the future trends based on the historical data. Evaluation metrics like precision, recall, and F1-score put measurement to determine the accuracy of the ML models process on IoT data.

Typical traditional IoT solutions indeed have lower computation and consume rule-based straightforward mechanisms for operation. However, they do not have a very effective response time in handling large amounts of raw data. The response times of ML-driven IoT solutions, particularly one leveraging edge computing, are closer to real-time decisions made locally than via processing in the cloud. Here, the evaluation of response times, processing speeds, and computational efficiencies provides fair trade-offs between traditional IoT systems and ML-powered IoT from a practical perspective(Ismail et al., 2018).

Traditional IoT architectures have been quite ineffective at scaling as they require too much human input to update the rules or change the thresholds, while ML-assisted IoT systems can handle large-scale deployment with minimum manual tuning of the devices and systems. Metrics such as system uptime, adaptability to new data, and learning rate efficiency are important for evaluating scalability. It allows the models to be retrained and improved over time, rendering them more effective for use in dynamic environments where frequent changes occur with respect to patterns and behaviors.

6.3 Key Learnings

When using an ML-influenced IoT, it does present advanced capabilities; however, the combination of rule-based logic with ML models offers a more optimal and balanced system. This would be required when explainability and deterministic decision-making become critical - especially in medical IoT scenarios - for such a combination to ensure it is both reliable and intelligent(Li et al., 2021).

Another important factor is that machine learning becomes effective in IoT when the data collected is good and plenty. Bad tagging, noise, and incompleteness will lead to wrong predictions and unreliable insights. Therefore, robust data preprocessing techniques, such as feature engineering and anomaly detection, must not be ignored if one wants to maximize the potential of machine-learning-based IoT solutions.

Yet another evolutionary step is introduced by the junction point of IoT and ML, which would introduce additional eminent developments in the next generation of smart systems such as federated learning, tiny ML, and self-learning edge devices. These developments are the results of both increased computational power and efficient ML algorithms that will highly promise advanced applications that will mark revolution of autonomous IoT in terms of automation, efficiency, and decision-making accuracy.

7. CHALLENGES AND FUTURE DIRECTIONS

While the integration of IoT and Machine Learning yields several dividends, it is also full of challenges that need to be addressed before their widespread acceptance. Such challenges include security, privacy, scalability, and deployment problems. With technological evolution also come new trends and advancements towards the future of IoT combined with ML. This section thereby adds the most important challenges and the direction the field would probably take in the future.

Table 5 explains how the inclusion of ML into IoT systems has its challenges including privacy of data, energy consumption and scalability. Some potential remedies are mentioned in the following table such as federated learning, edge AI, and model compression which can make upcoming systems much more efficient.

Table 5. Key challenges in IoT-ML integration and possible solutions

Challenge	Description	Possible Solution
Data Privacy	IoT devices collect sensitive user data	Federated learning, secure AI models
Model Deployment	ML models require frequent updates	Edge AI, continuous learning techniques
Scalability	Handling millions of IoT devices	Distributed computing, 5G networks
Energy Consumption	ML processing on IoT devices drains power	Low-power AI chips, model compression

7.1 Security and Privacy Issues

Security as well as privacy form some of the top-most issues concerning IoT and ML integration. IoT devices are capable of collecting and transmitting sensitive information at rates that make them appealing targets for many forms of cyberattacks. Most traditional security measures fail to safeguard resource-constrained IoT devices, leading to such vulnerabilities as data breaches, unauthorized access, and identity theft. Mitigation strategies include implementing strong encryption and authentication mechanisms and decentralized security models like blockchain(Bharati et al., 2022).

Figure 6. IoT security and privacy threats

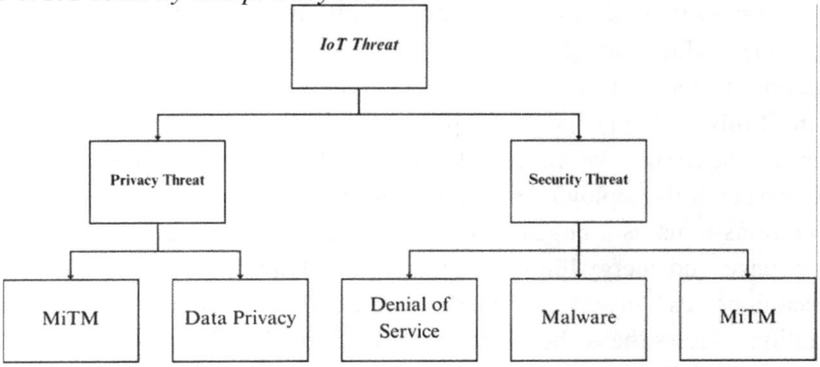

Figure 6 showcases the two major types of threats: privacy threats and security threats in the Internet of Things (IoT) ecosystem. Privacy threats include Man-in-the-Middle (MiTM) attacks, and data privacy breaches such familiarize an external person with a user. Security threats like denial-of-service (DoS) attacks, malware infections, and MiTM attacks disrupt access to the system and lead to unauthorized access. Understanding these threats is integral to implementing a robust security mechanism to secure data and its integrity and to ensure privacy and efficient system reliability within any IoT network.

In addition to these challenges, there exist privacy regulations, such as GDPR and CCPA, at the national and state levels. All these are strict concerning data collection, storage, and processing. Machine Learning models consume lots of data for efficient operation; however, compliance with such legislation while maintaining their usefulness becomes very difficult. Federated learning and differential privacy are some of the techniques that provide interesting solutions models learning from decentralized data sources without the exposition of raw data.

Thus, while IoT messages are increasing in important areas of application like healthcare, finance, and smart cities, more end-to-end security is required. Regular security updates, continuous monitoring, and using AI for anomaly detection have to be integrated into the IoT-ML ecosystems to prevent potential cyber threats. Future research must concern the investigation of AI attacks and the improvement of adaptable security solutions in relation to threats.

7.2 Scalability and Deployment Challenges

Other challenges present in machine learning models in IoT locations are the scalability requirements. This is partly because such networks consist of a few million connected devices generating a huge data stream. Processing these data must be done quickly while simultaneously making real-time decisions, and thus, scalable infrastructure is required. However, a lot of traditional cloud-based solutions create latency issues, which would contradict their purpose with respect to time-critical applications such as self-driving vehicles and industrial automation.

To meet this challenge, edge computing has been seen as a promising solution for enabling the running of machine learning (ML) models directly on IoT devices or edge servers. Still, deploying and managing ML models on a variety of hardware configurations remains a huge challenge. IoT devices differ in processing power, memory space, and energy limits, and they demand optimized, nimble ML models. Such techniques as TinyML and model compression techniques like pruning and quantization address these shortcomings(Schizas et al.,2022).

Dynamism in IoT environments is another major challenge. Unlike static computing systems that remain constant, devices come and bolt the network continuously in an open environment. This complexity can be understood regarding deploying ML models that must continuous adaptation based on this real-time change. Continuous model update, automated retraining, and self-learning systems are essential for improving model accuracy and performance in any of the sizable IoT deployments.

7.3 Future Trends in IoT-ML Integration

The imminent trend comprises the merger of IoT and ML to follow host of new trends in its trudging of the future. One eminent trend to highlight self-learning IoT devices; systems which turn out to be more efficient with the passing time under no human care. Reinforcement learning and online learning techniques help IoT devices make wiser choices by real-time data processing.

AI-enabled energy conservation is becoming a pattern in the further developments in IoT applications. Energy consumption has become a major constraint in IoT networks, especially in battery-operated devices (Das et al., 2023). This is why most researchers are working on making energy-aware machine learning algorithms more efficient in terms of power consumption without compromising accuracy. Neuromorphic computing and event-based AI computation are expected to bring improvements in this regard in future IoT deployments.

Integration of quantum computing into IoT and ML is, finally, gaining steam. Quantum algorithms hold the potential to process vast amounts of IoT data at speeds far beyond that of classical computation, solving complex optimization problems unimaginable by conventional computation(Gowda et al., 2025). While still incipient, quantum-enhanced applications of IoT-ML are on the verge of disrupting areas like cryptography, predictive maintenance, and large-scale data analysis.

8. CONCLUSION

Machine learning and the internet of things have completely transformed different industrial sectors by enabling real-time data processing, intelligent decisions, and automation. These are the devices that continuously generate data, and machine learning models bring an interpretation of this data to enable predictive maintenance, anomaly detection, and automation. The research case study proved that machine learning based on IoT technology improves efficiency and correctness, presented how this took place with a prominent role for edge computing that makes latency reduced. However, there are obstacles, such as data privacy, increases, and computation limits, that have to be crossed for general adaption to it. The integrated modules will keep changing the dynamics in healthcare, Industry 4.0, smart cities, and environment monitoring due to increasing IoT networks and evolving machine learning algorithms.

The road ahead is brimming with open ends, as edge AI, as well as federate learning, is going to be ushered with the influx of new-age connectivity, such as 5G, opening up avenues of possibilities down the line for the next surge of development in IoT-ML. Of course, several challenges persist - security-risk business models and infrastructure requirements, to name two very important considerations - but the most vital aspect will indeed be research, taking into account that adjacent technologists will come together with politicians and industry leaders. Finally, the future will promise even more automation, smarter intelligence, and much greater scalability in the way of convergence of IoT and ML-these are shapers of tomorrow's digital transformation of industries and also equalizers in quality of life.

REFERENCES

Allugunti, V. R., Reddy, C. K. K., Elango, N. M., & Anisha, P. R. (2021). Prediction of diabetes using Internet of Things (IoT) and decision trees: SLDPS. In Intelligent Data Engineering and Analytics: Frontiers in Intelligent Computing: Theory and Applications (FICTA 2020), Volume 2 (pp. 453-461). Springer Singapore.

Awad, A. I., Fouda, M. M., Khashaba, M. M., Mohamed, E. R., & Hosny, K. M. (2023). Utilization of mobile edge computing on the Internet of Medical Things: A survey. *ICT Express*, *9*(3), 473–485. DOI: 10.1016/j.icte.2022.05.006

Bajpai, A., Tolani, M., & Balodi, A. (2023). Smart Healthcare System Using IoT, Cloud and AI/ML. *Journal of Engineering Science & Technology Review*, *16*(6), 117–123. DOI: 10.25103/jestr.166.14

Bhanu Chander, & Gopalakrishnan, K. (2020). Security vulnerabilities and issues of traditional wireless sensors networks in IoT. In S. L. Peng, S. Pal, & L. Huang (Eds.), Principles of Internet of Things (IoT) ecosystem: Insight paradigm (Vol. 174). Springer, Cham. https://doi.org/DOI: 10.1007/978-3-030-33596-0_21

Bharati, S., & Podder, P. (2022). Machine and deep learning for IoT security and privacy: Applications, challenges, and future directions. Security and Communication Networks, 2022(1). https://doi.org/DOI: 10.1155/2022/8951961

Bout, E., Loscri, V., & Gallais, A. (2021). How machine learning changes the nature of cyberattacks on IoT networks: A survey. *IEEE Communications Surveys and Tutorials*, *24*(1), 248–279. DOI: 10.1109/COMST.2021.3127267

Chen, B., & Wan, J. (2019). Emerging trends of ML-based intelligent services for industrial Internet of Things (IIoT). 2019 Computing, Communications and IoT Applications (ComComAp), 135-139.

Das, S. K., Benkhelifa, F., Sun, Y., Abumarshoud, H., Abbasi, Q. H., & Mohjazi, L. (2023). Comprehensive review on ML-based RIS-enhanced IoT systems: Basics, research progress and future challenges. *Computer Networks*, *224*, 109581. DOI: 10.1016/j.comnet.2023.109581

Daud, M. A., & Suhaili, W. S. H. (2017). Internet of things (IoT) with CoAP and HTTP protocol: A study on which protocol suits IoT in terms of performance. In Computational Intelligence in Information Systems: Proceedings of the Computational Intelligence in Information Systems Conference (CIIS 2016) (pp. 165-174). Springer International Publishing.

Dayan, P., Sahani, M., & Deback, G. (1999). Unsupervised learning. The MIT encyclopedia of the cognitive sciences, 857-859.

Dinculeană, D., & Cheng, X. (2019). Vulnerabilities and limitations of MQTT protocol used between IoT devices. *Applied Sciences (Basel, Switzerland)*, *9*(5), 848. DOI: 10.3390/app9050848

Gopalakrishnan, S., & Kumaran, M. S. (2022). IIoT framework-based ML model to improve automobile industry product. *Intelligent Automation & Soft Computing*, *31*(3), 1435–1449. DOI: 10.32604/iasc.2022.020660

Gowda, V. D., Rajalakshmi, J., Guruprakash, B., Hariram, V., & Prasad, K. D. V. (2025). Integrating IoT, blockchain, and quantum machine learning. In Lilhore, U. K., Kumar, A., Vyas, N., Simaiya, S., & Dutt, V. (Eds.), *Multimodal data fusion for bioinformatics artificial intelligence*. Wiley., DOI: 10.1002/9781394269969.ch6

Harmon, M. E., & Harmon, S. S. (1996). Reinforcement learning: A tutorial. WL/AAFC. *WPAFB Ohio*, *45433*, 237–285.

Hastie, T., Tibshirani, R., & Friedman, J. (2009). Overview of supervised learning. In *The elements of statistical learning: Data mining, inference, and prediction* (pp. 9–41). Springer. DOI: 10.1007/978-0-387-84858-7_2

Ismail, A. A., Hamza, H. S., & Kotb, A. M. (2018). Performance evaluation of open source IoT platforms. In *2018 IEEE Global Conference on Internet of Things (GCIoT)* (pp. 1-5). IEEE. https://doi.org/DOI: 10.1109/GCIoT.2018.8620130

Izaddoost, A., & Siewierski, M. (2020). Energy efficient data transmission in iot platforms. *Procedia Computer Science*, *175*, 387–394. DOI: 10.1016/j.procs.2020.07.055

Kazi, K. S. L., & Mahant, M. A. (2025). Machine Learning-Driven Internet of Things (MLIoT)-Based Healthcare Monitoring System. In Digitalization and the Transformation of the Healthcare Sector (pp. 205-236). IGI Global Scientific Publishing.

Koul, R. M. (2023). Convergence of machine learning and IoT: Towards intelligent sensing and decision-making. IJEETE Journal of Research, 1(1).

Kumar, B. M., Kumar, K. K., Sasikala, P., Sampath, B., Gopi, B., & Sundaram, S. (2023). Sustainable green energy generation from wastewater: IoT and ML integration. In Mishra, B. (Ed.), *Sustainable science and intelligent technologies for societal development* (pp. 440–463). IGI Global Scientific Publishing., DOI: 10.4018/979-8-3693-1186-8.ch024

Kumar, N. M., & Mallick, P. K. (2018). The Internet of Things: Insights into the building blocks, component interactions, and architecture layers. *Procedia Computer Science*, *132*, 109–117. DOI: 10.1016/j.procs.2018.05.170

Kute, S., Tyagi, A. K., Sahoo, R., & Malik, S. (2022). Building a smart healthcare system using Internet of Things and machine learning. In *Big data management in sensing* (pp. 159–177). River Publishers. DOI: 10.1201/9781003337355-11

Li, W., Chai, Y., Khan, F., Jan, S. R. U., Verma, S., Menon, V. G., Kavita, , & Li, X. (2021). A Comprehensive Survey on Machine Learning-Based Big Data Analytics for IoT-Enabled Smart Healthcare System. *Mobile Networks and Applications*, *26*(1), 234–252. DOI: 10.1007/s11036-020-01700-6 PMID: 40477400

Lin, K., Xu, X., & Xiao, F. (2022). MFFusion: A multi-level features fusion model for malicious traffic detection based on deep learning. *Computer Networks*, *202*, 108658. DOI: 10.1016/j.comnet.2021.108658

Puttanapura, J., Singh, T. M., Reddy, C. K. K., & Doss, S. (2024). Improvised prediction of rainfall using random forest classifier over Taiwan. In *Proceedings of the 2024 4th International Conference on Soft Computing for Security Applications (ICSCSA)* (pp. 152–156). IEEE. DOI: 10.1109/ICSCSA64454.2024.00031

Rahman, R. A., & Shah, B. (2016, March). Security analysis of IoT protocols: A focus in CoAP. In 2016 3rd MEC international conference on big data and smart city (ICBDSC) (pp. 1-7). IEEE. DOI: 10.1109/ICBDSC.2016.7460363

Rahman, S. M., Ibtisum, S., Podder, P., & Hossain, S. M. (2023). Progression and challenges of IoT in healthcare: A short review. arXiv preprint arXiv:2311.12869.

Rithika, A., Singh, T. M., Reddy, C. K. K., & Doss, S. (2025). Leveraging generative AI for designing next-generation haptic systems: Advancements, challenges, and applications. In Mamodiya, U., Kalam, A., Bossoufi, B., Awad, M., & Somwanshi, D. (Eds.), *Integrating AI with haptic systems for smarter healthcare solutions* (pp. 327–350). IGI Global., DOI: 10.4018/979-8-3373-2307-7.ch015

Sarkar, D., Bali, R., Sharma, T., Sarkar, D., Bali, R., & Sharma, T. (2018). Machine learning basics. Practical Machine Learning with Python: A Problem-Solver's Guide to Building Real-World Intelligent Systems, 3-65.

Schizas, N., Karras, A., Karras, C., & Sioutas, S. (2022). TinyML for ultra-low power AI and large-scale IoT deployments: A systematic review. *Future Internet*, *14*(12), 363. DOI: 10.3390/fi14120363

Sehrawat, D., & Gill, N. S. (2019, April). Smart sensors: Analysis of different types of IoT sensors. In 2019 3rd International Conference on Trends in Electronics and Informatics (ICOEI) (pp. 523-528). IEEE.

Sharma, K., & Nandal, R. (2019). A literature study on machine learning fusion with IoT. 2019 3rd International Conference on Trends in Electronics and Informatics (ICOEI), 1440–1445. https://doi.org/DOI: 10.1109/ICOEI.2019.8862656

Singh, T. M., Reddy, C. K. K., & Lippert, K. (2024). The revolution and future of blockchain technology in cybersecurity: A comprehensive analysis. In *Artificial intelligence for blockchain and cybersecurity powered IoT applications* (1st ed., p. 28). CRC Press., DOI: 10.1201/9781003497585-5

Tzafestas, S. G. (2018). Synergy of IoT and AI in modern society: The robotics and automation case. *Robot. Autom. Eng. J*, *31*(5), 1–15. DOI: 10.19080/RAEJ.2018.03.555621

Compilation of References

Abayaratne, S., Su, D., & Qiao, Y. (2024). "CropFinder: AI-based Detection and Tracking of Crops for Precision Agriculture," 2024 IEEE 33rd International Symposium on Industrial Electronics (ISIE), Ulsan, Korea, Republic of, pp. 1-6, DOI: 10.1109/ISIE54533.2024.10595716

Abbes, A. B., Jarray, N., & Farah, I. R. (2024). Advances in remote sensing based soil moisture retrieval: Applications, techniques, scales and challenges for combining machine learning and physical models. *Artificial Intelligence Review*, *57*(9), 224. DOI: 10.1007/s10462-024-10734-1

Abdoh, S. A. (2024). Art and sustainability: Can digital technologies achieve sustainability? *Journal of Cultural Heritage Management and Sustainable Development*, *14*(4), 555–562. DOI: 10.1108/JCHMSD-03-2022-0038

Abdulai, A. R., Tetteh Quarshie, P., Duncan, E., & Fraser, E. (2023). Is agricultural digitization a reality among smallholder farmers in Africa? Unpacking farmers' lived realities of engagement with digital tools and services in rural Northern Ghana. *Agriculture & Food Security*, *12*(1), 11. DOI: 10.1186/s40066-023-00416-6

Abdul-Qawy, A. S., Pramod, P. J., Magesh, E., & Srinivasulu, T. (2015). The internet of things (iot): An overview. *International Journal of Engineering Research and Applications*, *5*(12), 71–82.

Adamu, A. Y., & Ahmad, A. (2024). Machine learning: Algorithm, real-world influence, and path to innovation. ISAR Journal of Multidisciplinary Research and Studies, ISAR International Journal of Mathematics and Computing Techniques, 70-77.

Aggarwal, A. (2025). Evolution of recommendation systems in the age of Generative AI. *International Journal of Science and Research Archive*, *14*(1), 485–492. DOI: 10.30574/ijsra.2025.14.1.0061

Aijaz, N., Lan, H., Raza, T., Yaqub, M., Iqbal, R., & Pathan, M. S. (2025). Artificial intelligence in agriculture: Advancing crop productivity and sustainability. *Journal of Agriculture and Food Research, 10*, 101762. DOI: 10.1016/j.jafr.2025.101762

Ajith, S., Vijayakumar, S., & Elakkiya, N. (2025). Yield prediction, pest and disease diagnosis, soil fertility mapping, precision irrigation scheduling, and food quality assessment using machine learning and deep learning algorithms. *Discover Food, 5*(1), 1–23. DOI: 10.1007/s44187-025-00338-1

Akkem, Y., Biswas, S. K., & Varanasi, A. (2023). Smart farming using artificial intelligence: A review. *Engineering Applications of Artificial Intelligence, 120*, 105899. DOI: 10.1016/j.engappai.2023.105899

Akkem, Y., Biswas, S. K., & Varanasi, A. (2024). A comprehensive review of synthetic data generation in smart farming by using variational autoencoder and generative adversarial network. *Engineering Applications of Artificial Intelligence, 131*, 107881. DOI: 10.1016/j.engappai.2024.107881

Alakkari, K., & Ali, B. (2025). Artificial Intelligence of Things: A Review. *Babylonian Journal of Internet of Things, 2025*, 113–120. DOI: 10.58496/BJIoT/2025/006

Albanese, A., Nardello, M., & Brunelli, D. (2021). Automated Pest Detection with DNN on the Edge for Precision Agriculture. *arXiv preprint arXiv:2108.00421*. https://arxiv.org/abs/2108.00421

Albreem, M. A., El-Saleh, A. A., Isa, M., Salah, W., Jusoh, M., Azizan, M. M., & Ali, A. (2017, November). Green internet of things (IoT): An overview. In 2017 IEEE 4th International Conference on Smart Instrumentation, Measurement and Application (ICSIMA) (pp. 1-6).IEEE.

Alexander, P., Arneth, A., Henry, R., Maire, J., Rabin, S., & Rounsevell, M. D. (2023). High energy and fertilizer prices are more damaging than food export curtailment from Ukraine and Russia for food prices, health and the environment. *Nature Food, 4*(1), 84–95. DOI: 10.1038/s43016-022-00659-9 PMID: 37118577

Alharithi, F. S., & Alzahrani, A. A. (2024). Enhancing environmental sustainability with federated LSTM models for AI-driven optimization. *Alexandria Engineering Journal, 108*, 640–653. DOI: 10.1016/j.aej.2024.09.058

Ali, A., Tajamul, H., & Azlan, Z. (2025). Smart Irrigation Technologies and Prospects for Enhancing Water Use Efficiency for Sustainable Agriculture. *AgriEngineering, 7*(4), 106. DOI: 10.3390/agriengineering7040106

Ali, Z., Muhammad, A., Lee, N., Waqar, M., & Lee, S. W. (2025). Artificial intelligence for sustainable agriculture: A comprehensive review of AI-driven technologies in crop production. *Sustainability (Basel), 17*(5), 2281. DOI: 10.3390/su17052281

Allen, R. G., Pereira, L. S., Raes, D., & Smith, M. (1998). Crop evapotranspiration-Guidelines for computing crop water requirements-FAO Irrigation and drainage paper 56. *Fao, Rome, 300*(9), D05109.

Allugunti, V. R., Reddy, C. K. K., Elango, N. M., & Anisha, P. R. (2021). Prediction of diabetes using Internet of Things (IoT) and decision trees: SLDPS. In Intelligent Data Engineering and Analytics: Frontiers in Intelligent Computing: Theory and Applications (FICTA 2020), Volume 2 (pp. 453-461). Springer Singapore.

Almukhalfi, H., Noor, A., & Noor, T. H. (2024). Traffic management approaches using machine learning and deep learning techniques: A survey. *Engineering Applications of Artificial Intelligence, 133*, 108147. DOI: 10.1016/j.engappai.2024.108147

Al-Zu'bi, M., Mabhaudhi, T., Daher, B., & Brouziyne, Y. (2025). Inclusive policy development from the ground up: Insights from the household water-energy-food nexus. *Environmental Science & Policy, 169*, 104084. DOI: 10.1016/j.envsci.2025.104084

Amiri-Zarandi, M., Dara, R. A., Duncan, E., & Fraser, E. D. G. (2022). Big Data Privacy in Smart Farming: A Review. *Sustainability (Basel), 14*(15), 9120. DOI: 10.3390/su14159120

Antora, N. A., Rahman, M. A., Mosharraf, A. A., Ehsan, M. I., Alve, M., & Elahi, M. M. (2022). Design and implementation of a smart bin using IoT for an efficient waste management system. In *2022 25th International Conference on Computer and Information Technology (ICCIT)* (pp. 774-779). IEEE. DOI: 10.1109/ICCIT57492.2022.10055998

Araújo, R., Silva, F., & Costa, L. (2023). Mobile robotics in smart farming: Current trends and applications. *Frontiers in Artificial Intelligence, 6*, 1213330. DOI: 10.3389/frai.2023.1213330 PMID: 37719082

Ariesen-Verschuur, N., Verdouw, C., & Tekinerdogan, B. (2022). Digital Twins in greenhouse horticulture: A review. *Computers and Electronics in Agriculture, 199*, 107183. DOI: 10.1016/j.compag.2022.107183

Arif, M., Khan, M. A., Rehman, A. U., Jan, M. A., Almogren, A., & Kumar, R. (2024). Resource-efficient ubiquitous sensor networks for smart agriculture: A survey. *IEEE Access : Practical Innovations, Open Solutions, 12*, 193332–193364. DOI: 10.1109/ACCESS.2024.3516814

Arularasan, A. N., Kalaiyarasi, D., Ashokkumar, P., Suresh, K., Rajasekar, M., & Prasad, A. R. (2025). Utilization of Advanced IoT and High-Performance Computing for Enhanced Agricultural Systems. In *Integrating Artificial Intelligence Into the Energy Sector* (pp. 467–492). IGI Global Scientific Publishing. DOI: 10.4018/979-8-3693-7112-1.ch022

Asimiyu, Z. (2024). *Data-driven decision-making in automotive manufacturing*. Retrieved from https://www.researchgate.net/profile/Zainab-Asimiyu/publication/387668758_Data-Driven_Decision-Making_in_Automotive_Manufacturing/links/677723ea00aa3770e0d307e1/Data-Driven-Decision-Making-in-Automotive-Manufacturing.pdf

Assimakopoulos, F., Vassilakis, C., Margaris, D., Kotis, K., & Spiliotopoulos, D. (2025). AI and related technologies in the fields of smart agriculture: A review. *Information (Basel)*, *16*(2), 100. DOI: 10.3390/info16020100

Atila, Ü., Uçar, M., Akyol, K., & Uçar, E. (2021). Plant leaf disease classification using EfficientNet deep learning model. *Ecological Informatics*, *61*, 101182. DOI: 10.1016/j.ecoinf.2020.101182

Awad, A. I., Fouda, M. M., Khashaba, M. M., Mohamed, E. R., & Hosny, K. M. (2023). Utilization of mobile edge computing on the Internet of Medical Things: A survey. *ICT Express*, *9*(3), 473–485. DOI: 10.1016/j.icte.2022.05.006

Awais, M., Wang, X., Hussain, S., Aziz, F., & Mahmood, M. Q. (2025). Advancing precision agriculture through digital twins and smart farming technologies: A review. *AgriEngineering*, *7*(5), 137. DOI: 10.3390/agriengineering7050137

Ayim, C., Kassahun, A., Tekinerdogan, B., & Addison, C. (2020). Adoption of ICT innovations in the agriculture sector in Africa: A Systematic Literature Review. *arXiv preprint* arXiv:2006.13831.

Bac, C. W., van Henten, E. J., Hemming, J., & Edan, Y. (2014). Harvesting Robots for High-value Crops: State-of-the-art Review and Challenges Ahead. *Journal of Field Robotics*, *31*(6), 888–911. DOI: 10.1002/rob.21525

Badiger, M., & Mathew, J. A. (2023). Tomato plant leaf disease segmentation and multiclass disease detection using hybrid optimization enabled deep learning. *Journal of Biotechnology*, *374*, 101–113. DOI: 10.1016/j.jbiotec.2023.07.011 PMID: 37543108

Baeten, J., Donné, K., Boedrij, S., Beckers, W., & Claesen, E. (2008). Autonomous Fruit Picking Machine: A Robotic Apple Harvester. In Laugier, C., & Siegwart, R. (Eds.), *Field and Service Robotics. Springer Tracts in Advanced Robotics* (Vol. 42). Springer., DOI: 10.1007/978-3-540-75404-6_51

Bahari, M., Arpaci, I., Der, O., Akkoyun, F., & Ercetin, A. (2024). Driving Agricultural Transformation: Unraveling Key Factors Shaping IoT Adoption in Smart Farming with Empirical Insights. *Sustainability (Basel)*, *16*(5), 2129. DOI: 10.3390/su16052129

Bais, D., Tiwari, V., Kolhe, S., Jain, S. (2023). IoT and AI Enabled Framework to Monitor Soil and Crop Health for Sustainable Development in Precision Farming-A Study.

Bajpai, A., Tolani, M., & Balodi, A. (2023). Smart Healthcare System Using IoT, Cloud and AI/ML. *Journal of Engineering Science & Technology Review*, *16*(6), 117–123. DOI: 10.25103/jestr.166.14

Balyan, S., Jangir, H., Tripathi, S. N., Tripathi, A., Jhang, T., & Pandey, P. (2024). Seeding a Sustainable Future: Navigating the Digital Horizon of Smart Agriculture. *Sustainability (Basel)*, *16*(2), 475. DOI: 10.3390/su16020475

Bambam, A. K., & Gajrani, K. K. (2023). Challenges in achieving sustainability during manufacturing. In *Sustainable Materials and Manufacturing Technologies* (pp. 108–124). CRC Press. DOI: 10.1201/9781003291961-9

Bang, J., & Han, J. W. (2025). Factors Influencing Farmers' Motivation to Adopt Smart Farm Technology in South Korea. *arXiv preprint* arXiv:2504.01795.

Batan, A. (2025). Assessing the Role of Public-Private Partnerships in Advancing Sustainable Development Goals Through Agri-Tech and Renewable Energy Solutions. *Transactions on Social Innovation, Digital Inclusion, and Ethical AI*, *15*(2), 1–11.

Batchu, C., & Satya, V. V. (2024). *Generative AI: Evolution and its Future*. International Journal For Multidisciplinary Research., DOI: 10.36948/ijfmr.2024.v06i01.12046

Batra, I., Sharma, C., Malik, A., Sharma, S., Kaswan, M. S., & Garza-Reyes, J. A. (2024). Industrial revolution and smart farming: A critical analysis of research components in Industry 4.0. *The TQM Journal*. Advance online publication. DOI: 10.1108/TQM-10-2023-0317

Bauer, B., Schulz, M., Hufnagel, T., Schramm, D., & Wittenberg, C. (2025). Reinforcement Learning for Robust Control of Individual Wheel Drive Mobile Robots with Passive Articulated Steering for Reverse Maneuvering. In Degen, H., & Ntoa, S. (Eds.), *Artificial Intelligence in HCI* (pp. 123–137). Springer Nature Switzerland., DOI: 10.1007/978-3-031-93429-2_8

Baxter, P., Cielniak, G., Hanheide, M., & From, P. (2018).. . *Safe Human-Robot Interaction in Agriculture.*, *59-60*, 59–60. Advance online publication. DOI: 10.1145/3173386.3177072

Bazargani, K., & Deemyad, T. (2024). Automation's Impact on Agriculture: Opportunities, Challenges, and Economic Effects. *Robotics (Basel, Switzerland)*, *13*(2), 33. DOI: 10.3390/robotics13020033

Bechar, A., & Vigneault, C. (2016). Agricultural robots for field operations: Concepts and components. *Biosystems Engineering*, *149*, 94–111. DOI: 10.1016/j.biosystemseng.2016.06.014

Belli, L., Cilfone, A., Davoli, L., Ferrari, G., Adorni, P., Di Nocera, F., Dall'Olio, A., Pellegrini, C., Mordacci, M., & Bertolotti, E. (2020). IoT-enabled smart sustainable cities: Challenges and approaches. *Smart Cities*, *3*(3), 1039–1071. DOI: 10.3390/smartcities3030052

Bellini, P., Nesi, P., & Pantaleo, G. (2022). IoT-enabled smart cities: A review of concepts, frameworks and key technologies. *Applied Sciences (Basel, Switzerland)*, *12*(3), 1607. DOI: 10.3390/app12031607

Benke, K., & Tomkins, B. (2017). Future food-production systems: vertical farming and controlled-environment agriculture. *Sustainability: Science. Practice and Policy*, *13*(1), 13–26.

Best Practice AI. (n.d.). *Johns Hopkins Hospital improves patient monitoring and resource allocation in ER and critical care units in real-time using machine learning*. Retrieved from https://www.bestpractice.ai/ai-case-study-best-practice/johns_hopkins_hospital_improves_patient_monitoring_and_resource_allocation_in_er_and_critical_care_units_in_real-time_using_machine_learning

Bhangar, N. A., & Shahriyar, A. K. (2024). *IoT and AI for Next-Generation Farming: Opportunities, Challenges, and Outlook.* Creative Commons International License.

Bhanu Chander, & Gopalakrishnan, K. (2020). Security vulnerabilities and issues of traditional wireless sensors networks in IoT. In S. L. Peng, S. Pal, & L. Huang (Eds.), Principles of Internet of Things (IoT) ecosystem: Insight paradigm (Vol. 174). Springer, Cham. https://doi.org/DOI: 10.1007/978-3-030-33596-0_21

Bharati, S., & Podder, P. (2022). Machine and deep learning for IoT security and privacy: Applications, challenges, and future directions. Security and Communication Networks, 2022(1). https://doi.org/DOI: 10.1155/2022/8951961

Bhatti, M. A., Zeeshan, Z., Bhatti, U. A., Khan, A., Ghadi, Y. Y., Alsenan, S., & Afzal, T. (2024). Advanced plant disease segmentation in precision agriculture using optimal dimensionality reduction with fuzzy c-means clustering and deep learning. *IEEE Journal of Selected Topics in Applied Earth Observations and Remote Sensing*. Advance online publication. DOI: 10.1109/JSTARS.2024.3437469

Binswanger, H. (1986). Agricultural mechanization: A comparative historical perspective. *The World Bank Research Observer*, *1*(1), 27–56. DOI: 10.1093/wbro/1.1.27

Bisen, W. H., & Agrawal, A. J. (2022). review on natural language generation. *International Journal of Health Sciences*, 10365–10376. Advance online publication. DOI: 10.53730/ijhs.v6nS1.7489

Bishop, C. M. (2006). *Pattern recognition and machine learning*. Springer.

Biswas, A., Sarkar, S., Das, S., Dutta, S., Choudhury, M. R., Giri, A., & Paul, D. (2025). Water scarcity: A global hindrance to sustainable development and agricultural production–A critical review of the impacts and adaptation strategies. *Cambridge Prisms: Water*, *3*, e4. DOI: 10.1017/wat.2024.16

Blei, D. M., Kucukelbir, A., & McAuliffe, J. D. (2017). Variational inference: A review for statisticians. *Journal of the American Statistical Association*, *112*(518), 859–877. DOI: 10.1080/01621459.2017.1285773

Bommasani, R., Hudson, D. A., Adeli, E., Altman, R., Arora, S., von Arx, S., . . . Liang, P. (2021). On the opportunities and risks of foundation models. arXiv preprint arXiv:2108.07258. /arXiv.2108.07258DOI: 10.48550

Botero-Valencia, J., García-Pineda, V., Valencia-Arias, A., Valencia, J., Reyes-Vera, E., Mejia-Herrera, M., & Hernández-García, R. (2025). Machine Learning in Sustainable Agriculture: Systematic Review and Research Perspectives. *Agriculture*, *15*(4), 4. Advance online publication. DOI: 10.3390/agriculture15040377

Bout, E., Loscri, V., & Gallais, A. (2021). How machine learning changes the nature of cyberattacks on IoT networks: A survey. *IEEE Communications Surveys and Tutorials*, *24*(1), 248–279. DOI: 10.1109/COMST.2021.3127267

Change, I. C. (2013). The physical science basis.

Chataut, R., Phoummalayvane, A., & Akl, R. (2023). Unleashing the Power of IoT: A Comprehensive Review of IoT Applications and Future Prospects in Healthcare, Agriculture, Smart Homes, Smart Cities, and Industry 4.0. *Sensors (Basel)*, *23*(16), 7194. DOI: 10.3390/s23167194 PMID: 37631731

Chataut, S., Bhatta, S., Dahal, B., Ojha, G., Subedi, B., & Bastakoti, B. (2024). Advancements and Applications of Generative AI in Healthcare. *European Journal of Theoretical and Applied Sciences*, *2*(6), 873–895. DOI: 10.59324/ejtas.2024.2(6).77

Chavula, P., Kayusi, F., & Juma, L. (2024). Leveraging Artificial Intelligence for Enhancing Wheat Yield Resilience Amidst Climate Change in Sub-Saharan Africa. *LatIA*, *2*, 88. DOI: 10.62486/latia202588

Chen, B., & Wan, J. (2019). Emerging trends of ML-based intelligent services for industrial Internet of Things (IIoT). 2019 Computing, Communications and IoT Applications (ComComAp), 135-139.

Chen, Z., Li, X., Zhao, T., Wei, X., Song, Z., Wang, J., & Jian, J. (2025). PhenoBee: Drone-based robot for advanced field in vivo contact-based phenotyping in agriculture. *Computers and Electronics in Agriculture*, *236*, 110483. DOI: 10.1016/j.compag.2025.110483

Chevalier, A., Copot, C., De Keyser, R., Hernandez, A., & Ionescu, C. (2015). A Multi-Agent System for Precision Agriculture. In Busoniu, L., & Tamás, L. (Eds.), *Handling Uncertainty and Networked Structure in Robot Control* (pp. 361–386). Springer International Publishing., DOI: 10.1007/978-3-319-26327-4_15

Chlingaryan, A., Sukkarieh, S., & Whelan, B. (2018). Machine learning approaches for crop yield prediction and nitrogen status estimation in precision agriculture: A review. *Computers and Electronics in Agriculture*, *151*, 61–69. DOI: 10.1016/j.compag.2018.05.012

Choi, K., Yi, J., Park, C., & Yoon, S. (2021). Deep learning for anomaly detection in time-series data: Review, analysis, and guidelines. *IEEE Access : Practical Innovations, Open Solutions*, *9*, 120043–120065. DOI: 10.1109/ACCESS.2021.3107975

Cho, S. B., Soleh, H. M., Choi, J. W., Hwang, W. H., Lee, H., Cho, Y. S., Cho, B.-K., Kim, M. S., Baek, I., & Kim, G. (2024). Recent Methods for Evaluating Crop Water Stress Using AI Techniques: A Review. *Sensors (Basel)*, *24*(19), 6313. DOI: 10.3390/s24196313 PMID: 39409355

Chowdhury, H., Paul Argha, D. B., & Ahmed, M. A. (2023). Artificial intelligence in sustainable vertical farming. *arXiv*. https://arxiv.org/abs/2312.00030arXiv

Chowdhury, S., & Islam, M. (2024). Towards sustainable agriculture: Harnessing AI for global food security. *Artificial Intelligence in Agriculture*, *4*(3), 100123. DOI: 10.1016/j.aiia.2024.04.003

Cob-Parro, A. C., Lalangui, Y., & Lazcano, R. (2024). Fostering Agricultural Transformation through AI: An Open-Source AI Architecture Exploiting the MLOps Paradigm. *Agronomy (Basel)*, *14*(2), 2. Advance online publication. DOI: 10.3390/agronomy14020259

Cohen, A., & Dalyot, S. (2020). Machine-learning prediction models for pedestrian traffic flow levels: Towards optimizing walking routes for blind pedestrians. *Transactions in GIS*, *24*(5), 1264–1279. DOI: 10.1111/tgis.12674

Compare, M., Baraldi, P., & Zio, E. (2019). Challenges to IoT-enabled predictive maintenance for industry 4.0. *IEEE Internet of Things Journal*, *7*(5), 4585–4597. DOI: 10.1109/JIOT.2019.2957029

Cuena, J., Hernández, J., & Molina, M. (1996). An intelligent model for road traffic management in the motorway network around Barcelona. In *Proceedings of the 1996 International Conference on Road Traffic Management* (pp. 173-180). Springer. https://doi.org/DOI: 10.1007/978-0-387-34979-4_20

Dai, C.-P., & Ke, F. (2022). Educational applications of artificial intelligence in simulation-based learning: A systematic mapping review. *Computers and Education: Artificial Intelligence*, *3*, 100087. DOI: 10.1016/j.caeai.2022.100087

Daraghmah, R. H., & Younis, N. R. (2025). A Comprehensive Review of Machine Learning Algorithms in Autonomous Robotics: Challenges and Future Prospects. *Estonian Journal of Engineering*, *1*(1), 1. Advance online publication. DOI: 10.63496/eje.Vol1.Iss1.35

Dara, R., Hazrati Fard, S. M., & Kaur, J. (2022). Recommendations for ethical and responsible use of artificial intelligence in digital agriculture. *Frontiers in Artificial Intelligence*, *5*, 884192. DOI: 10.3389/frai.2022.884192 PMID: 35968036

Das, S. K., & Nayak, P. (2024). Integration of IoT-AI powered local weather forecasting: A Game-Changer for Agriculture. *arXiv preprint arXiv:2501.14754*. https://arxiv.org/abs/2501.14754

Das, S. K., Benkhelifa, F., Sun, Y., Abumarshoud, H., Abbasi, Q. H., & Mohjazi, L. (2023). Comprehensive review on ML-based RIS-enhanced IoT systems: Basics, research progress and future challenges. *Computer Networks*, *224*, 109581. DOI: 10.1016/j.comnet.2023.109581

Daud, M. A., & Suhaili, W. S. H. (2017). Internet of things (IoT) with CoAP and HTTP protocol: A study on which protocol suits IoT in terms of performance. In Computational Intelligence in Information Systems: Proceedings of the Computational Intelligence in Information Systems Conference (CIIS 2016) (pp. 165-174). Springer International Publishing.

Dayan, P., Sahani, M., & Deback, G. (1999). Unsupervised learning. The MIT encyclopedia of the cognitive sciences, 857-859.

Demestichas, K., Peppes, N., Alexakis, T., & Adamopoulou, E. (2020). Blockchain in agriculture traceability systems: A review. *Applied Sciences (Basel, Switzerland)*, *10*(12), 4113. DOI: 10.3390/app10124113

Demircioglu, P., Bogrekci, I., Durakbasa, M. N., & Bauer, J. (2024). Autonomation, Automation, AI, and Industry-Agriculture 5.0 in Sustainable Agro-Ecological Food Production. In Durakbasa, N. M., & Gençyılmaz, M. G. (Eds.), *Industrial Engineering in the Industry 4.0 Era* (pp. 545–556). Springer Nature Switzerland., DOI: 10.1007/978-3-031-53991-6_42

Devarajan, Y. (2025). Investigation of Emerging Technologies in Agriculture: An In-depth Look at Smart Farming, Nano-agriculture, AI, and Big Data. *Journal of Biosystems Engineering*, •••, 1–23.

Dhanaraju, M., Chenniappan, P., Ramalingam, K., Pazhanivelan, S., & Kaliaperumal, R. (2022). Smart Farming: Internet of Things (IoT)-Based Sustainable Agriculture. *Agriculture*, *12*(10), 1745. DOI: 10.3390/agriculture12101745

Dhavale, C., Pawar, T., Singh, A., Pole, S., & Sabat, K. (2024). "Revolutionizing Farming: GAN-Enhanced Imaging, CNN Disease Detection, and LLM Farmer Assistant"," 2024 2nd International Conference on Computer, Communication and Control (IC4), Indore, India,, pp. 1-6, DOI: 10.1109/IC457434.2024.10486501

Dinculeană, D., & Cheng, X. (2019). Vulnerabilities and limitations of MQTT protocol used between IoT devices. *Applied Sciences (Basel, Switzerland)*, *9*(5), 848. DOI: 10.3390/app9050848

Dinis, I., Santos, D., & Mendes-Moreira, P. (2025). Revitalising Traditional Cereals in Portugal: Challenges, Opportunities, and Strategies for Value Chain Development. Sustainability (2071-1050), 17(6).

Doersch, C. (2016). Tutorial on variational autoencoders. arXiv preprint arXiv:1606.05908

Du, B. et al., (2024). "YOLO-Based Semantic Communication With Generative AI-Aided Resource Allocation for Digital Twins Construction," in IEEE Internet of Things Journal, vol. 11, no. 5, pp. 7664-7678, 1, .DOI: 10.1109/JIOT.2023.3317629

Duckett, T., Pearson, S., Blackmore, S., Grieve, B., Chen, W.-H., Cielniak, G., & Yang, G.-Z. (2018). Agricultural Robotics: The Future of Robotic Agriculture. *arXiv [Cs.RO]*. Retrieved from http://arxiv.org/abs/1806.06762 DOI: 10.31256/WP2018.2

Dudhe, P. V., Kadam, N. V., Hushangabade, R. M., & Deshmukh, M. S. (2017, August). Internet of Things (IOT): An overview and its applications. In 2017 International conference on energy, communication, data analytics and soft computing (ICECDS) (pp. 2650-2653).IEEE.

Dusadeerungsikul, P. O., & Nof, S. Y. (2024). Precision agriculture with AI-based responsive monitoring algorithm. *International Journal of Production Economics*, *271*, 109204. DOI: 10.1016/j.ijpe.2024.109204

Dutta, S., & Banerjee, A. (2023). Enhancing precision agriculture: A comprehensive review of machine learning and AI applications. *Agricultural Technology*, *5*(2), 100483. DOI: 10.1016/j.atech.2024.100483

Elavarasan, D., & Vincent, P. M. D. (2020). Crop Yield Prediction Using Deep Reinforcement Learning Model for Sustainable Agrarian Applications. *IEEE Access : Practical Innovations, Open Solutions*, *8*, 86886–86901. DOI: 10.1109/ACCESS.2020.2992480

Elmessery, W. M., Maklakov, D. V., El-Messery, T. M., Baranenko, D. A., Gutiérrez, J., Shams, M. Y., El-Hafeez, T. A., Elsayed, S., Alhag, S. K., Moghanm, F. S., Mulyukin, M. A., Petrova, Y. Y., & Elwakeel, A. E. (2024). Semantic segmentation of microbial alterations based on SegFormer. *Frontiers in Plant Science*, *15*, 1352935. DOI: 10.3389/fpls.2024.1352935 PMID: 38938642

Etezadi, H., & Eshkabilov, S. (2024). A comprehensive overview of control algorithms, sensors, actuators, and communication tools of autonomous all-terrain vehicles in agriculture. *Agriculture*, *14*(2), 163. DOI: 10.3390/agriculture14020163

Etheredge, L. M. (2025). *AI in agriculture*. ResearchGate. Li, H., Guo, Y., Zhao, H., Wang, Y., & Chow, D. (2021). Towards automated greenhouse: A state of the art review on greenhouse monitoring methods and technologies based on internet of things. *Computers and Electronics in Agriculture*, *191*, 106558.

Evenson, R. E., & Gollin, D. (2003). Assessing the impact of the Green Revolution, 1960 to 2000. *Science*, *300*(5620), 758–762. DOI: 10.1126/science.1078710 PMID: 12730592

Farhangmehr, V., Imanian, H., Mohammadian, A., Cobo, J. H., Shirkhani, H., & Payeur, P. (2025). A spatiotemporal CNN-LSTM deep learning model for predicting soil temperature in diverse large-scale regional climates. *The Science of the Total Environment, 968*, 178901. DOI: 10.1016/j.scitotenv.2025.178901 PMID: 39987832

Fasal. (2023). Smart agriculture platform: Revolutionizing farming with AI and IoT. https://www.fasal.co

Fasciolo, B., Panza, L., & Lombardi, F. (2024). Exploring the Integration of Industry 4.0 Technologies in Agriculture: A Comprehensive Bibliometric Review. *Sustainability (Basel), 16*(20), 8948. DOI: 10.3390/su16208948

Ferroni, M., & Castle, P. (2011). Public-private partnerships and sustainable agricultural development. *Sustainability (Basel), 3*(7), 1064–1073. DOI: 10.3390/su3071064

Floridi, L., & Cowls, J. (2019). A unified framework of five principles for AI in society. *Harvard Data Science Review, 1*(1). Advance online publication. DOI: 10.1162/99608f92.8cd550d1

Fosso Wamba, S., & Queiroz, M. M. (2022). Artificial intelligence and blockchain implementation in supply chains: A case study of the tuna fish supply chain in Thailand. *Annals of Operations Research, 308*(1), 1–25. DOI: 10.1007/s10479-021-04450-0 PMID: 34931103

Fountas, S., Carli, G., Sørensen, C. G., Tsiropoulos, Z., Cavalaris, C., Vatsanidou, A., Liakos, B., Canavari, M., Wiebensohn, J., & Tisserye, B. A. (2015). Farm management information systems: Current situation and future perspectives. *Computers and Electronics in Agriculture, 115*, 40–50. DOI: 10.1016/j.compag.2015.05.011

Fountas, S., Espejo-García, B., Kasimati, A., Gemtou, M., Panoutsopoulos, H., & Anastasiou, E. (2024). Agriculture 5.0: Cutting-Edge Technologies, Trends, and Challenges. *IT Professional, 26*(1), 40–47. DOI: 10.1109/MITP.2024.3358972

Fuentes, F., Gutter, K., Vega, R., & Carrasco Silva, G. (2024). Transformative Technologies in Digital Agriculture: Leveraging Internet of Things, Remote Sensing, and Artificial Intelligence for Smart Crop Management. *J. Sens. Actuator Netw., 13*(4), 39. DOI: 10.3390/jsan13040039

Fuhrmann-Aoyagi, M. B., Miura, K., & Watanabe, K. (2024). Sustainability in Japan's Agriculture: An Analysis of Current Approaches. *Sustainability (Basel), 16*(2), 596. DOI: 10.3390/su16020596

Furendal, M., & Jebari, K. (2023). The future of work: Augmentation or stunting? *Philosophy & Technology, 36*(2), 36. DOI: 10.1007/s13347-023-00631-w

Gadekallu, T. R., Kumar Reddy Maddikunta, P., Boopathy, P., Deepa, N., Chengoden, R., Victor, N., Wang, W., Wang, W., Zhu, Y., & Dev, K. (2025). XAI for Industry 5.0—Concepts, Opportunities, Challenges, and Future Directions. *IEEE Open Journal of the Communications Society*, *6*, 2706–2729. DOI: 10.1109/OJCOMS.2024.3473891

Ganaou, N. I., & Salaou, A. I. (2025). Communication Technologies and Protocols in IoT Systems: Principles, Standards, and Security. In Enhancing Data-Driven Electronics Through IoT (pp. 323-390). IGI Global Scientific Publishing.

Gangula, R., Thirupathi, L., Parupati, R., Sreeveda, K., & Gattoju, S. (2021). Ensemble machine learning based prediction of dengue disease with performance and accuracy elevation patterns, Materials Today: Proceedings, , ISSN 2214-7853, DOI: 10.1016/j.matpr.2021.07.270

Garcia, E., Calvet, L., Carracedo, P., Serrat, C., Miró, P., & Peyman, M. (2024). Predictive analyses of traffic level in the city of Barcelona: From ARIMA to eXtreme Gradient Boosting. *Applied Sciences (Basel, Switzerland)*, *14*(11), 4432. DOI: 10.3390/app14114432

Garzón, J., Montes, L., Garzón, J., & Lampropoulos, G. (2023). Systematic review of technology in aeroponics: Introducing the technology adoption and integration in sustainable agriculture model. *Agronomy (Basel)*, *13*(10), 2517. DOI: 10.3390/agronomy13102517

Getahun, S., Kefale, H., & Gelaye, Y. (2024). Application of Precision Agriculture Technologies for Sustainable Crop Production and Environmental Sustainability: A Systematic Review. *TheScientificWorldJournal*, *2024*(1), 2126734. DOI: 10.1155/2024/2126734 PMID: 39421732

Ghetiya, R., Kumari, P., Vekariya, P., & Kumar, M. S. (2025). Artificial Intelligence in Achieving Sustainable Development Goals: Models and Approaches. In AI in Agriculture for Sustainable and Economic Management (pp. 12-24). CRC Press.

Ghosh, A., & Roy, S. (2024). AI-driven future farming: Achieving climate-smart and sustainable agriculture. *Agriengineering*, *7*(3), 89. DOI: 10.3390/agriengineering7030089

Ghumman, S., Di Troia, F., Andreopoulos, W., Stamp, M., & Rai, S. (2025). AGRO: An autonomous AI rover for precision agriculture. *arXiv preprint*, arXiv:2505.01200.

Giuffrida, I. (2019). Liability for AI decision-making: Some legal and ethical considerations. *Fordham Law Review*, *88*, 439.

Gkoulis, D. (2025). Creating interpretable synthetic time series for enhancing the design and implementation of Internet of Things (IoT) solutions. *Internet of Things : Engineering Cyber Physical Human Systems, 30*, 101500. DOI: 10.1016/j.iot.2025.101500

Goncalves, J. P., Pinto, F. A., Queiroz, D. M., Villar, F. M., Barbedo, J. G., & Del Ponte, E. M. (2021). Deep learning architectures for semantic segmentation and automatic estimation of severity of foliar symptoms caused by diseases or pests. *Biosystems Engineering, 210*, 129–142. DOI: 10.1016/j.biosystemseng.2021.08.011

Goodfellow, I. J., Pouget-Abadie, J., Mirza, M., Xu, B., Warde-Farley, D., Ozair, S., Bengio, Y. (2014). Generative Adversarial Networks. *arXiv [Stat.ML].*

Goodfellow, I., Pouget-Abadie, J., Mirza, M., Xu, B., Warde-Farley, D., Ozair, S., . . . Bengio, Y. (2014). Generative adversarial nets. In *Advances in Neural Information Processing Systems* (pp. 2672–2680). https://proceedings.neurips.cc/paper_files/paper/2014/file/5ca3e9b122f61f8f06494c97b1afccf3-Paper.pdf

Goodfellow, I. J., Pouget-Abadie, J., Mirza, M., Xu, B., Warde-Farley, D., Ozair, S., & Bengio, Y. (2014). Generative adversarial nets. *Advances in Neural Information Processing Systems*, 27.

Goodfellow, I., Bengio, Y., & Courville, A. (2016). *Deep learning*. MIT Press.

Gopalakrishnan, S., & Kumaran, M. S. (2022). IIoT framework-based ML model to improve automobile industry product. *Intelligent Automation & Soft Computing, 31*(3), 1435–1449. DOI: 10.32604/iasc.2022.020660

Gopi, A., L.R, S., & S., I. T. J. (. (2025). Enhancing soil data classification for cassava crop productivity using swarm robotics optimization (SRO) and convolutional neural networks. *International Journal of System Assurance Engineering and Management, 16*(4), 1409–1423. DOI: 10.1007/s13198-025-02727-2

Gowda, V. D., Rajalakshmi, J., Guruprakash, B., Hariram, V., & Prasad, K. D. V. (2025). Integrating IoT, blockchain, and quantum machine learning. In Lilhore, U. K., Kumar, A., Vyas, N., Simaiya, S., & Dutt, V. (Eds.), *Multimodal data fusion for bioinformatics artificial intelligence*. Wiley., DOI: 10.1002/9781394269969.ch6

Gracia-Rodriguez, C., Martínez-Medina, A. E., Torres-Cosio, L., Lopez-Ortiz, C., Nimmakayala, P., Luévanos-Escareño, M. P., Hernández-Almanza, A. Y., Castro-Alonso, M. J., Sosa-Martínez, J. D., Reddy, U. K., & Balagurusamy, N. (2025). Can the molecular and transgenic breeding of crops be an alternative and sustainable technology to meet food demand? *Functional & Integrative Genomics, 25*(1), 1–32. DOI: 10.1007/s10142-025-01594-1 PMID: 40205022

Gregor, K., Danihelka, I., Graves, A., Rezende, D. J., & Wierstra, D. (2015). DRAW: A recurrent neural network for image generation. In *Proceedings of the 32nd International Conference on Machine Learning* (pp. 1462–1471). https://proceedings.mlr.press/v37/gregor15.html

Gupta, S. K., & Singh, N. (2023).AI and IoT-Enabled Technologies and Applications for Smart City.In Artificial Intelligence in Cyber-Physical Systems (pp. 175-193). CRC Press. DOI: 10.1201/9781003248750-10

Gupta, B. B., & Quamara, M. (2020). An overview of Internet of Things (IoT): Architectural aspects, challenges, and protocols. *Concurrency and Computation*, *32*(21), e4946. DOI: 10.1002/cpe.4946

Gupta, G., & Kumar Pal, S. (2025). Applications of AI in precision agriculture. *Discover Agriculture*, *3*(1), 61. DOI: 10.1007/s44279-025-00220-9

Gupta, R., & Sharma, P. (2023). The innovative role of blockchain in agri-food systems: A literature review. *Food Control*, *145*, 110603. DOI: 10.1016/j.foodcont.2024.110603

H2O.ai. (2020). *Driving away fraudsters at PayPal: Case study*. https://h2o.ai/content/dam/h2o/en/marketing/documents/2020/01/PayPal-Customer-Case-Study-rnd2-1.pdf

Habil, S., El-Deeb, S., & El-Bassiouny, N. (2023). AI-based recommendation systems: The ultimate solution for market prediction and targeting. In Wang, C. L. (Ed.), *The Palgrave handbook of interactive marketing* (pp. 1–15). Palgrave Macmillan., DOI: 10.1007/978-3-031-14961-0_30

Han, H., & Hawken, S. (2018). Introduction: Innovation and identity in next-generation smart cities. *City, Culture and Society*, *12*, 1–4. DOI: 10.1016/j.ccs.2017.12.003

Harmon, M. E., & Harmon, S. S. (1996). Reinforcement learning: A tutorial. WL/AAFC. *WPAFB Ohio*, *45433*, 237–285.

Hasan, M. M., Islam, M. U., & Sadeq, M. J. (2022). Towards technological adaptation of advanced farming through AI, IoT, and Robotics: A Comprehensive overview. *arXiv preprint* arXiv:2202.10459.

Hashem, T. N., Joudeh, J. M., & Zamil, A. M. (2024). Smart farming (AI-generated) as an approach to better control pest and disease detection in agriculture: POV agricultural institutions. *Migration Letters : An International Journal of Migration Studies*, *21*(S1), 529–547. DOI: 10.59670/ml.v21iS1.6178

Hastie, T., Tibshirani, R., & Friedman, J. (2009). Overview of supervised learning. In *The elements of statistical learning: Data mining, inference, and prediction* (pp. 9–41). Springer. DOI: 10.1007/978-0-387-84858-7_2

Hebsale Mallappa, V. K., & Pathak, T. B. (2023). Climate smart agriculture technologies adoption among small-scale farmers: A case study from Gujarat, India. *Frontiers in Sustainable Food Systems*, *7*, 1202485. DOI: 10.3389/fsufs.2023.1202485

Heikonen, S., Heino, M., Jalava, M., Siebert, S., Viviroli, D., & Kummu, M. (2025). Climate change threatens crop diversity at low latitudes. *. *Nature Food*, *6*(4), 1–12. DOI: 10.1038/s43016-025-01135-w PMID: 40038529

Hemming, S., de Zwart, F., Elings, A., Righini, I., & Petropoulou, A. (2019). Remote control of greenhouse vegetable production with artificial intelligence—Greenhouse climate, irrigation, and crop production. *Sensors (Basel)*, *19*(8), 1807. DOI: 10.3390/s19081807 PMID: 31014024

Heschl, A., Murillo, M., Najafian, K., & Maleki, F. (2024). SynthSet: Generative diffusion model for semantic segmentation in precision agriculture. *arXiv preprint arXiv:2411.03505*. https://arxiv.org/abs/2411.03505

Hiran, K. K., Khazanchi, D., Vyas, A. K., & Padmanaban, S. (Eds.). (2021). *Machine learning for sustainable development* (Vol. 9). Walter de Gruyter GmbH & Co KG. DOI: 10.1515/9783110702514

Holm, S., Pedersen, S. M., & Tamirat, T. W. (2024). Robots in agriculture–A case-based discussion of ethical concerns on job loss, responsibility, and data control. *Smart Agricultural Technology*, *9*, 100633. DOI: 10.1016/j.atech.2024.100633

Holzinger, A., Saranti, A., Angerschmid, A., Retzlaff, C. O., Gronauer, A., Pejakovic, V., Medel-Jimenez, F., Krexner, T., Gollob, C., & Stampfer, K. (2022). Digital transformation in smart farm and forest operations needs human-centered AI: Challenges and future directions. *Sensors (Basel)*, *22*(8), 3043. DOI: 10.3390/s22083043 PMID: 35459028

Hong, W. (2001). *Modeling, Estimation, and Control of Robot-Soil Interactions.* https://dspace.mit.edu/handle/1721.1/7075

Hoque, A., & Padhiary, M. (2024). Automation and AI in Precision Agriculture: Innovations for Enhanced Crop Management and Sustainability. *Asian J. Res. Comput. Sci*, *17*(10), 95–109. DOI: 10.9734/ajrcos/2024/v17i10512

Horabik, J., Parafiniuk, P., & Molenda, M. (2016). Experiments and discrete element method simulations of distribution of static load of grain bedding at bottom of shallow model silo. *Biosystems Engineering*, *149*, 60–71. DOI: 10.1016/j.biosystemseng.2016.06.012

Hou, G., Niu, R., Chen, H., Wang, Y., Zhu, Z., Zhang, J., Ma, Y., & Li, T. (2025, May 15). A Robotic AI Algorithm for Fusing Generative Large Models in Agriculture Internet of Things. *IEEE Internet of Things Journal*, *12*(10), 13511–13524. Advance online publication. DOI: 10.1109/JIOT.2024.3516729

Huang, J., Xu, Y., Wang, Q., Wang, Q., Liang, X., Wang, F., Zhang, Z., Wei, W., Zhang, B., Huang, L., Chang, J., Ma, L., Ma, T., Liang, Y., Zhang, J., Guo, J., Jiang, X., Fan, X., An, Z., & Fei, A. (2025). Foundation models and intelligent decision-making: Progress, challenges, and perspectives. *Innovation (Cambridge (Mass.))*, *6*(6), 100948. Advance online publication. DOI: 10.1016/j.xinn.2025.100948 PMID: 40528892

Hussein, A. H. A., Jabbar, K. A., Mohammed, A., & Al-Jawahry, H. M. (2024). AI and IoT in Farming: A Sustainable Approach. E3S Web of Conferences, 491, 01020. DOI: 10.1051/e3sconf/202449101020

Hussein, A., Jabbar, K., Mohammed, A., & Alzubaidi, L. (2024). Harvesting the Future: AI and IoT in Agriculture. E3S Web of Conferences, 477. DOI: 10.1051/e3sconf/202447700090

Isinkaye, F. O., Olusanya, M. O., & Akinyelu, A. A. (2025). A multi-class hybrid variational autoencoder and vision transformer model for enhanced plant disease identification. *Intelligent Systems with Applications*, *26*, 200490. DOI: 10.1016/j.iswa.2025.200490

Islam, M. H., Anam, M. Z., Hoque, M. R., Nishat, M., & Bari, A. M. (2024). Agriculture 4.0 adoption challenges in the emerging economies: Implications for smart farming and sustainability. *Journal of Economy and Technology*, *2*, 278–295. DOI: 10.1016/j.ject.2024.09.002

Islam, M. S., & Rahman, M. M. (2023). AI- and data-driven pre-crop values and crop rotation matrices. *European Journal of Agronomy*, *135*, 126949. DOI: 10.1016/j.eja.2023.126949

Ismail, A. A., Hamza, H. S., & Kotb, A. M. (2018). Performance evaluation of open source IoT platforms. In *2018 IEEE Global Conference on Internet of Things (GCIoT)* (pp. 1-5). IEEE. https://doi.org/DOI: 10.1109/GCIoT.2018.8620130

Issa, A., Majed, S., Ameer, S., & Al-Jawahry, H. (2024). Farming in the Digital Age: Smart Agriculture with AI and IoT. E3S Web of Conferences, 477, 00081. DOI: 10.1051/e3sconf/202447700081

Izaddoost, A., & Siewierski, M. (2020). Energy efficient data transmission in iot platforms. *Procedia Computer Science*, *175*, 387–394. DOI: 10.1016/j.procs.2020.07.055

Izatt, R. M., Izatt, S. R., Bruening, R. L., Izatt, N. E., & Moyer, B. A. (2014). Challenges to achievement of metal sustainability in our high-tech society. *Chemical Society Reviews*, *43*(8), 2451–2475. DOI: 10.1039/C3CS60440C PMID: 24464346

Jaber, M. (2023). IoT and machine learning for enabling sustainable development goals. *Frontiers in Communications and Networks*, *4*, 1219047. DOI: 10.3389/frcmn.2023.1219047

Jabrane, K., & Bousmah, M. (2021). A new approach for training cobots from small amount of data in Industry 5.0. *International Journal of Advanced Computer Science and Applications*, *12*(10). Advance online publication. DOI: 10.14569/IJACSA.2021.0121070

Jägermeyr, J., Gerten, D., Heinke, J., Schaphoff, S., Kummu, M., & Lucht, W. (2015). Water savings potentials of irrigation systems: Global simulation of processes and linkages. *Hydrology and Earth System Sciences*, *19*(7), 3073–3091. DOI: 10.5194/hess-19-3073-2015

Jain, V., & Kumar, S. (2023). The role of blockchain technology in the sustainability of supply chains. *Cleaner Logistics and Supply Chain*, *5*, 100113. DOI: 10.1016/j.clscn.2023.100113

Jana, P. (2024). AI-powered IoT solutions for sustainable water management in cities. *Uncertainty Discourse and Applications*, *1*(2), 2. Advance online publication. DOI: 10.48313/uda.v1i2.37

Javaid, M., Haleem, A., Khan, I. H., & Suman, R. (2023). Understanding the potential applications of Artificial Intelligence in Agriculture Sector. *Advanced Agrochem*, *2*(1), 15–30. DOI: 10.1016/j.aac.2022.10.001

Javidan, S. M., Banakar, A., Vakilian, K. A., & Ampatzidis, Y. (2023). Diagnosis of grape leaf diseases using automatic *K*-means clustering and machine learning. *Smart Agricultural Technology*, *3*, 100081. DOI: 10.1016/j.atech.2022.100081

Jayachandran, M., Gatla, R. K., Rao, K. P., Rao, G. S., Mohammed, S., Milyani, A. H., Azhari, A. A., Kalaiarasy, C., & Geetha, S. (2022). Challenges in achieving sustainable development goal 7: Affordable and clean energy in light of nascent technologies. *Sustainable Energy Technologies and Assessments*, *53*, 102692. DOI: 10.1016/j.seta.2022.102692

Jin, X., Liu, S., Baret, F., Hemerlé, M., & Comar, A. (2017). Estimates of plant density of wheat crops at emergence from very low altitude UAV imagery. *Remote Sensing of Environment*, *198*, 105–114. DOI: 10.1016/j.rse.2017.06.007

Jockusch, O., Hossain, M. Z., Imteaj, A., & Shahid, A. R. (2024). "Generative AI-based Land Cover Classification via Federated Learning CNNs: Sustainable Insights from UAV Imagery," *2024 IEEE Conference on Technologies for Sustainability (SusTech)*, Portland, OR, USA, , pp. 356-361, DOI: 10.1109/SusTech60925.2024.10553449

Judith, J., Tamilselvi, R., Beham, M. P., Lakshmi, S. S. P., Panthakkan, A., Al Mansoori, S., & Al Ahmad, H. (2025). Remote Sensing Based Crop Health Classification Using NDVI and Fully Connected Neural Networks. *arXiv preprint arXiv:2504.10522*. https://arxiv.org/abs/2504.10522

Jumper, J., Evans, R., Pritzel, A., Green, T., Figurnov, M., Ronneberger, O., Tunyasuvunakool, K., Bates, R., Žídek, A., Potapenko, A., Bridgland, A., Meyer, C., Kohl, S. A. A., Ballard, A. J., Cowie, A., Romera-Paredes, B., Nikolov, S., Jain, R., Adler, J., & Hassabis, D. (2021). Highly accurate protein structure prediction with AlphaFold. *Nature*, *596*(7873), 583–589. DOI: 10.1038/s41586-021-03819-2 PMID: 34265844

Jung, A. (2022). *Machine learning: the basics*. Springer Nature. DOI: 10.1007/978-981-16-8193-6

Kabir, S., Gope, P., & Mohanty, S. P. (2023). A security-enabled safety assurance framework for IoT-based smart homes. *IEEE Transactions on Industry Applications*, *59*(1), 6–14. DOI: 10.1109/TIA.2022.3176257

Kale, R. B., Dangi, P. A., Nikam, V., Khandagale, K., Gadge, S. S., & Mahajan, V. (2025). How are new collective models transforming support for smallholder farmers? Empirical evidence from thematic and sentiment content analysis. *Frontiers in Sustainable Food Systems*, *8*, 1542676. DOI: 10.3389/fsufs.2024.1542676

Kalusivalingam, A. K., Sharma, A., Patel, N., & Singh, V. (2020). Enhancing Predictive Maintenance in Manufacturing Using Machine Learning Algorithms and IoT-Driven Data Analytics. *International Journal of Artificial Intelligence and Machine Learning*, *1*(3).

Kamilaris, A., Kartakoullis, A., & Prenafeta-Boldú, F. X. (2018). A review on the practice of big data analysis in agriculture. *Computers and Electronics in Agriculture*, *143*, 23–37. DOI: 10.1016/j.compag.2017.09.037

Kamilaris, A., & Prenafeta-Boldú, F. X. (2018). Deep learning in agriculture: A survey. *Computers and Electronics in Agriculture*, *147*, 70–90. DOI: 10.1016/j.compag.2018.02.016

Kant, J. K., Sripaad, M., Bharadwaj, A., Rajashekhar, V. S., & Sundaram, S. (2023). An Autonomous Hybrid Drone-Rover Vehicle for Weed Removal and Spraying Applications in Agriculture. *arXiv preprint arXiv:2308.04794*. https://arxiv.org/abs/2308.04794 DOI: 10.1109/AGRETA57740.2023.10262416

Karunathilake, E. M. B. M., Le, A. T., Heo, S., Chung, Y. S., & Mansoor, S. (2023). The Path to Smart Farming: Innovations and Opportunities in Precision Agriculture. *Agriculture*, *13*(8), 1593. DOI: 10.3390/agriculture13081593

Kataray, T., Nitesh, B., Yarram, B., Sinha, S., Cuce, E., Shaik, S., Vigneshwaran, P., & Roy, A. (2023). Integration of smart grid with renewable energy sources: Opportunities and challenges–A comprehensive review. *Sustainable Energy Technologies and Assessments*, *58*, 103363. DOI: 10.1016/j.seta.2023.103363

Kaur, A., Bhatt, D. P., & Raja, L. (2024). Developing a hybrid irrigation system for smart agriculture using IoT sensors and machine learning in Sri Ganganagar, Rajasthan. *Journal of Sensors*, *2024*(1), 6676907. DOI: 10.1155/2024/6676907

Kaur, P., Harnal, S., Gautam, V., Singh, M. P., & Singh, S. P. (2024). Performance analysis of segmentation models to detect leaf diseases in tomato plants. *Multimedia Tools and Applications*, *83*(6), 16019–16043. DOI: 10.1007/s11042-023-16238-4

Kavipriya, J., & Vadivu, G. (2024). "Exploring Crop Yield Prediction with Remote Sensing Imagery and AI," *2024 International Conference on Advances in Computing, Communication and Applied Informatics (ACCAI)*, Chennai, India , pp. 1-5, DOI: 10.1109/ACCAI61061.2024.10601854

Kazi, K. S. L., & Mahant, M. A. (2025). Machine Learning-Driven Internet of Things (MLIoT)-Based Healthcare Monitoring System. In Digitalization and the Transformation of the Healthcare Sector (pp. 205-236). IGI Global Scientific Publishing.

Kendall, H., Clark, B., Li, W., Jin, S., Jones, G. D., Chen, J., Taylor, J., Li, Z., & Frewer, L. J. (2022). Precision agriculture technology adoption: A qualitative study of small-scale commercial "family farms" located in the North China Plain. *. *Precision Agriculture*, *23*(1), 1–33. DOI: 10.1007/s11119-021-09839-2

Kerdegari, H., Razaak, M., Argyriou, V., & Remagnino, P. (2019). Smart monitoring of crops using generative adversarial networks. In Vento, M., & Percannella, G. (Eds.), Lecture notes in computer science: Vol. 11678. *Computer analysis of images and patterns. CAIP 2019* (pp. 517–528). Springer., DOI: 10.1007/978-3-030-29888-3_45

Khan, A. O. R., Islam, S. M., Sarkar, A., Islam, T., Paul, R., & Bari, M. S. (2024). Real-time predictive health monitoring using AI-driven wearable sensors: Enhancing early detection and personalized interventions in chronic disease management. *International Journal for Multidisciplinary Research.*

Khan, F. U., Nouman, M., Negrut, L., Abban, J., Cismas, L. M., & Siddiqi, M. F. (2024). Constraints to agricultural finance in underdeveloped and developing countries: A systematic literature review. *International Journal of Agricultural Sustainability*, *22*(1), 2329388. DOI: 10.1080/14735903.2024.2329388

Khan, K., Khan, R. U., Albattah, W., & Qamar, A. M. (2022). End-to-End Semantic Leaf Segmentation Framework for Plants Disease Classification. *Complexity*, *2022*(1), 1168700. DOI: 10.1155/2022/1168700

Kim, J., Shah, P., Gaskell, J. C., & Prasann, A. (2020). *Scaling Up Disruptive Agricultural Technologies in Africa*. World Bank Publications. DOI: 10.1596/978-1-4648-1522-5

Kingma, D. P., & Welling, M. (2022). Auto-Encoding Variational Bayes. *arXiv [Stat.ML]*. Retrieved from http://arxiv.org/abs/1312.6114

Kishor Kumar Reddy, C., Bhavya, T., Anisha, P. R., & Hanafiah, M. M. (2025). Text-Based Prediction and Classification Model of Stress, Anxiety and Depression Among Indians. In Mishra, B. K., Mallik, S., & Le, D.-N. (Eds.), *Smart Sensors for Industry 4.0.*, DOI: 10.1002/9781394214723.ch9

Klair, Y. S., Agrawal, K., & Kumar, A. (2024). "Impact of Generative AI in Diagnosing Diseases in Agriculture," 2024 2nd International Conference on Disruptive Technologies (ICDT), Greater Noida, India, pp. 870-875, DOI: 10.1109/ICDT61202.2024.10489759

Klerkx, L., Jakku, E., & Labarthe, P. (2019). A review of social science on digital agriculture, smart farming and agriculture 4.0: New contributions and a future research agenda. *NJAS-Wageningen Journal of Life Sciences, 90*, 100315.

Kolhalkar, N. R., Pandit, A. A., Kedar, S. A., & Yedukondalu, G. (2025). "Artificial Intelligence Algorithms for Robotic Harvesting of Agricultural Produce," 2025 1st International Conference on AIML-Applications for Engineering & Technology (ICAET), Pune, India, pp. 1-6, DOI: 10.1109/ICAET63349.2025.10932313

Korthals, T., Kragh, M., Christiansen, P., Karstoft, H., Jørgensen, R. N., & Rückert, U. (2018). Multi-Modal Detection and Mapping of Static and Dynamic Obstacles in Agriculture for Process Evaluation. *Frontiers in Robotics and AI*, *5*, 28. Advance online publication. DOI: 10.3389/frobt.2018.00028 PMID: 33500915

Koul, R. M. (2023). Convergence of machine learning and IoT: Towards intelligent sensing and decision-making. IJEETE Journal of Research, 1(1).

Krishnababu, M. E., Devi, B. R., Soni, A., Panigrahi, C. K., Sudeepthi, B., Rathi, A., & Shukla, A. (2024). A Review on Precision Agriculture Navigating the Future of Farming with AI and IoT. *Asian Journal of Soil Science and Plant Nutrition*, *10*(2), 336–349. DOI: 10.9734/ajsspn/2024/v10i2291

Kuli, B. K., Debnath, J., Sheikh, A., Das, S., & Balai, P. S. (2025). Smart Farming Revolution: AI, IoT, and Robotics in Precision Agriculture and Soil Conservation. *International Journal of Scientific Research in Science, Engineering and Technology*, *12*(2), 688–706. DOI: 10.32628/IJSRSET25122193

Kum, S., Oh, S., & Moon, J. (2024). Edge AI framework for large scale smart agriculture. In *2024 27th Conference on Innovation in Clouds, Internet and Networks (ICIN)* (pp. 143–147). IEEE. DOI: 10.1109/ICIN60470.2024.10494451

Kumar Reddy, C. K., Reddy, P. S., Pilly, A., & Doss, S. (2025). Transformative Effects of Smarter Chatbots. 333-350. https://doi.org/DOI: 10.1002/9781394234028.ch18

Kumar Stephen, V. M. V., Palanisamy, R., & Antonio Rutaf Manalang, M. (2024). Emergency medical services system. In *2024 1st International Conference on Innovative Engineering Sciences and Technological Research (ICIESTR)* (pp. 1-6). IEEE. DOI: 10.1109/ICIESTR60916.2024.10798196

Kumar, B. M., Kumar, K. K., Sasikala, P., Sampath, B., Gopi, B., & Sundaram, S. (2023). Sustainable green energy generation from wastewater: IoT and ML integration. In Mishra, B. (Ed.), *Sustainable science and intelligent technologies for societal development* (pp. 440–463). IGI Global Scientific Publishing., DOI: 10.4018/979-8-3693-1186-8.ch024

Kumari, K., & Nafchi, A. M. (2024). Sustainable agriculture with AI, machine learning, deep learning, and IoT for future farming. 2024 ASABE Annual International Meeting, 2400906. DOI: 10.13031/aim.202400906

Kumar, K., & Kumar, V. (2023, November). Integration of artificial intelligence and machine learning for Internet of Things. *In 2023 International Conference on Sustainable Communication Networks and Applications (ICSCNA)* (pp. 491–497). IEEE. DOI: 10.1109/ICSCNA58489.2023.10370106

Kumar, M., Kumar, A., Sajjan Kumar, S., Patle, G. T., Kumar, S., & Tanwar, A. (2024). Role of cutting-edge technology in horticulture crops for shaping the future smart farming ecosystem: A comprehensive review. *Environment and Ecology*, *42*(4B), 1933–1942. DOI: 10.60151/envec/KSOK1056

Kumar, N. M., & Mallick, P. K. (2018). The Internet of Things: Insights into the building blocks, component interactions, and architecture layers. *Procedia Computer Science*, *132*, 109–117. DOI: 10.1016/j.procs.2018.05.170

Kumar, R., & Singh, A. (2023). Artificial intelligence tools for the agriculture value chain. *Electronics (Basel)*, *13*(22), 4362. DOI: 10.3390/electronics13224362

Kute, S., Tyagi, A. K., Sahoo, R., & Malik, S. (2022). Building a smart healthcare system using Internet of Things and machine learning. In *Big data management in sensing* (pp. 159–177). River Publishers. DOI: 10.1201/9781003337355-11

Lakhiar, I. A., Yan, H., Zhang, C., Wang, G., He, B., Hao, B., Han, Y., Wang, B., Bao, R., Syed, T. N., Chauhdary, J. N., & Rakibuzzaman, M. (2024). A Review of Precision Irrigation Water-Saving Technology under Changing Climate for Enhancing Water Use Efficiency, Crop Yield, and Environmental Footprints. *Agriculture*, *14*(7), 1141. DOI: 10.3390/agriculture14071141

Lal, R. (2011). Soil health and climate change: an overview. *Soil health and climate change*, 3-24.

Lal, R. (2016). Feeding 11 billion on 0.5 billion hectare of area under cereal crops. *Food and Energy Security*, *5*(4), 239–251. DOI: 10.1002/fes3.99

Lasker, A. (2024). Exploring ethical considerations in generative ai. *International Journal of Advanced Research*, *12*(4), 531–535. Advance online publication. DOI: 10.21474/IJAR01/18578

Lavelle, P., Decaëns, T., Aubert, M., Barot, S., Blouin, M., Bureau, F., Margerie, P., Mora, P., & Rossi, J. P. (2006). Soil invertebrates and ecosystem services. *European Journal of Soil Biology*, *42*, S3–S15. DOI: 10.1016/j.ejsobi.2006.10.002

Lee, H., & Park, J. (2023). Computer vision in smart agriculture and precision farming. *Artificial Intelligence in Agriculture*, *5*(1), 100234. DOI: 10.1016/j.aiia.2024.06.004

Lee, M. H., Yao, M. H., Kow, P. Y., Kuo, B. J., & Chang, F. J. (2024). An Artificial Intelligence-Powered Environmental Control System for Resilient and Efficient Greenhouse Farming. *Sustainability (Basel)*, *16*(24), 10958. DOI: 10.3390/su162410958

Lee, S. (2019). A novel design and implementation of data acquisition and preprocessing system for precision agriculture. In *Agriculture and environment perspectives in intelligent systems* (Vol. 24, pp. 52–71). IOS Press., DOI: 10.3233/AISE190004

Lesk, C., Rowhani, P., & Ramankutty, N. (2016). Influence of extreme weather disasters on global crop production. *Nature*, *529*(7584), 84–87. DOI: 10.1038/nature16467 PMID: 26738594

Li, T., Xie, F., Xiong, Y., & Feng, Q. (2023). Multi-Arm Robot Task Planning for Fruit Harvesting Using Multi-Agent Reinforcement Learning. *arXiv preprint arXiv:2303.00460*. https://arxiv.org/abs/2303.00460 DOI: 10.1109/IROS55552.2023.10341822

Liakos, K. G., Busato, P., Moshou, D., Pearson, S., & Bochtis, D. (2018). Machine Learning in Agriculture: A Review. *Sensors (Basel)*, *18*(8), 2674. DOI: 10.3390/s18082674 PMID: 30110960

Li, D., Nanseki, T., Chomei, Y., & Kuang, J. (2023). A review of smart agriculture and production practices in Japanese large-scale rice farming. *Journal of the Science of Food and Agriculture*, *103*(4), 1609–1620. DOI: 10.1002/jsfa.12204 PMID: 36069313

Lin, K., Xu, X., & Xiao, F. (2022). MFFusion: A multi-level features fusion model for malicious traffic detection based on deep learning. *Computer Networks*, *202*, 108658. DOI: 10.1016/j.comnet.2021.108658

Li, S., Liu, J., Yang, Y., Shen, R., & Jiang, J. (2025). Thinking as human: Self-reflective reinforcement learning framework for fertilization decision-making. *Smart Agricultural Technology*, *10*, 100841. DOI: 10.1016/j.atech.2025.100841

Liu, J., ... (2023). "Exploring the Integration of Digital Twin and Generative AI in Agriculture," 2023 15th International Conference on Intelligent Human-Machine Systems and Cybernetics (IHMSC), Hangzhou, China, pp. 223-228, DOI: 10.1109/IHMSC58761.2023.00059

Liu, E., Gold, K. M., Combs, D., Cadle-Davidson, L., & Jiang, Y. (2022). Deep semantic segmentation for the quantification of grape foliar diseases in the vineyard. *Frontiers in Plant Science*, *13*, 978761. DOI: 10.3389/fpls.2022.978761 PMID: 36161031

Liu, J., Li, J., Chen, Y., Lian, S., Zeng, J., Geng, M., Zheng, S., Dong, Y., He, Y., Huang, P., Zhao, Z., Yan, X., Hu, Q., Wang, L., Yang, D., Zhu, Z., Sun, Y., Shang, W., Wang, D., & Chen, X. M. (2023). Multi-scale urban passenger transportation CO2 emission calculation platform for smart mobility management. *Applied Energy*, *331*, 120407. DOI: 10.1016/j.apenergy.2022.120407

Liu, Y., & Wang, H. (2023). Enhancing global agricultural monitoring systems for climate-smart agriculture. *Climate Services for Agriculture*, *2*(1), 100037. DOI: 10.1016/j.csag.2024.100037

Li, W., Chai, Y., Khan, F., Jan, S. R. U., Verma, S., Menon, V. G., Kavita, , & Li, X. (2021). A Comprehensive Survey on Machine Learning-Based Big Data Analytics for IoT-Enabled Smart Healthcare System. *Mobile Networks and Applications*, *26*(1), 234–252. DOI: 10.1007/s11036-020-01700-6 PMID: 40477400

Li, X., & Zhang, Y. (2023). Artificial intelligence and its role in soil microbiology and agricultural sustainability. *Microbiology*, *12*, 100006. DOI: 10.1016/bs.mim.2024.05.006

Li, X., Zhou, Y., Liu, J., Wang, L., Zhang, J., & Fan, X. (2022). The detection method of potato foliage diseases in complex background based on instance segmentation and semantic segmentation. *Frontiers in Plant Science*, *13*, 899754. DOI: 10.3389/fpls.2022.899754 PMID: 35865287

Li, Z., Chen, P., Shuai, L., Wang, M., Zhang, L., Wang, Y., & Mu, J. (2022). A copy paste and semantic segmentation-based approach for the classification and assessment of significant rice diseases. *Plants*, *11*(22), 3174. DOI: 10.3390/plants11223174 PMID: 36432903

Llisterri Giménez, N., Monfort Grau, M., Pueyo Centelles, R., & Freitag, F. (2022). On-device training of machine learning models on microcontrollers with federated learning. *Electronics (Basel)*, *11*(4), 573. DOI: 10.3390/electronics11040573

Lloret, J. (2022). Edge Computing in Precision agriculture. 1-1. .DOI: 10.1109/FMEC57183.2022.10062622

Long, S. P. (2025). Needs and opportunities to future-proof crops and the use of crop systems to mitigate atmospheric change. Philosophical Transactions B, 380(1927), 20240229. DOI: 10.1098/rstb.2024.0229

López, I. D., Grass, J. F., Figueroa, A., & Corrales, J. C. (2023). A proposal for a multi-domain data fusion strategy in a climate-smart agriculture context. *International Transactions in Operational Research*, *30*(4), 2049–2070. DOI: 10.1111/itor.12899

Lowder, S. K., Sánchez, M. V., & Bertini, R. (2021). Which farms feed the world and has farmland become more concentrated? *World Development*, *142*, 105455. DOI: 10.1016/j.worlddev.2021.105455

Luna, D. F. O., & Ruiz, P. A. M. (2019). Automation and control of greenhouse implemented technologies: a review. *Visión electrónica, 2*(2), 381-394.

Luo, L., Liu, B., Chen, M., Wang, J., Wei, H., Lu, Q., & Luo, S. (2025). DRL-enhanced 3D detection of occluded stems for robotic grape harvesting. *Computers and Electronics in Agriculture*, *229*, 109736. DOI: 10.1016/j.compag.2024.109736

Lu, Y., Chen, D., Olaniyi, E., & Huang, Y. (2022). Generative adversarial networks (GANs) for image augmentation in agriculture: A systematic review. *Computers and Electronics in Agriculture*, *200*, 107208. DOI: 10.1016/j.compag.2022.107208

Lytridis, C., Kaburlasos, V. G., Pachidis, T., Manios, M., Vrochidou, E., Kalampokas, T., & Chatzistamatis, S. (2021). An Overview of Cooperative Robotics in Agriculture. *Agronomy (Basel)*, *11*(9), 1818. DOI: 10.3390/agronomy11091818

Majeed, Y., Fu, L., & He, L. (2024). Editorial: Artificial intelligence-of-things (AIoT) in precision agriculture. *Frontiers in Plant Science*, *15*, 15–2024. DOI: 10.3389/fpls.2024.1369791 PMID: 38344185

Majumder, S., Khandelwal, Y., & S. K, (2024). "Computer vision and Generative AI for yield prediction in Digital Agriculture," 2024 2nd International Conference on Networking and Communications (ICNWC), Chennai, India, pp. 1-6, .DOI: 10.1109/ICNWC60771.2024.10537337

Makondo, N., Kobo, H. I., Mathonsi, T. E., & du Plessis, D. P. D. (2024). Implementing an efficient architecture for latency optimisation in smart farming. *IEEE Access : Practical Innovations, Open Solutions*, *12*, 140502–140526. DOI: 10.1109/ACCESS.2024.3466994

Malhotra, M., Ghosh, R., Kumar, N., & C, K. K. R. (2025). AI-Powered Harmony. In *Advances in social networking and online communities book series* (pp. 95-126). https://doi.org/DOI: 10.4018/979-8-3373-0543-1.ch004

Martínez, L., & Pérez, R. (2023). Transforming agriculture with machine learning, deep learning, and IoT. *Smart Agriculture*, *4*(2), 66–78. DOI: 10.1007/s44279-024-00066-7

Martin, P. (2021). A future-focused view of the regulation of rural technology. *Agronomy (Basel)*, *11*(6), 1153. DOI: 10.3390/agronomy11061153

Mavridou, E., Vrochidou, E., Papakostas, G., Pachidis, T., & Kaburlasos, V. (2019). Machine Vision Systems in Precision Agriculture for Crop Farming. *Journal of Imaging*, *5*(12), 89. DOI: 10.3390/jimaging5120089 PMID: 34460603

McKinney, S. M., Sieniek, M., Godbole, V., Godwin, J., Antropova, N., Ashrafian, H., Back, T., Chesus, M., Corrado, G. S., Darzi, A., Etemadi, M., Garcia-Vicente, F., Gilbert, F. J., Halling-Brown, M., Hassabis, D., Jansen, S., Karthikesalingam, A., Kelly, C. J., King, D., & Shetty, S. (2020). International evaluation of an AI system for breast cancer screening. *Nature*, *577*(7788), 89–94. DOI: 10.1038/s41586-019-1799-6 PMID: 31894144

McLandsborough, A. (2023). Nothing Runs like a (Autonomous) Deere: How Government-Sponsored Reskilling Can Cushion the Blow of Technological Advancement for Kansan Farm Workers. *Kan. JL & Pub. Pol'y*, *33*, 197.

Mehta, C. R., Chandel, N. S., & Dubey, K. (2023). Smart Agricultural Mechanization in India—Status and Way Forward. In *Smart Agriculture for Developing Nations: Status, Perspectives and Challenges* (pp. 1–14). Springer Nature Singapore. DOI: 10.1007/978-981-19-8738-0_1

Mehta, S., & Patel, D. (2023). *Sustainable artificial intelligence solutions for agricultural efficiency*. IntechOpen., DOI: 10.5772/intechopen.112996doi.org

Mendhurwar, S., & Mishra, R. (2018). Emerging synergies between Internet of Things and social technologies. *Journal of Global Information Technology Management*, *21*(2), 75–80. DOI: 10.1080/1097198X.2018.1462918

Milella, A., Reina, G., & Nielsen, M. (2019). A multi-sensor robotic platform for ground mapping and estimation beyond the visible spectrum. *Precision Agriculture*, *20*(2), 423–444. DOI: 10.1007/s11119-018-9605-2

Milioto, A., Lottes, P., & Stachniss, C. (2018). Real-time Semantic Segmentation of Crop and Weed for Precision Agriculture Robots Leveraging Background Knowledge in CNNs. *arXiv [Cs.CV]*. Retrieved from http://arxiv.org/abs/1709.06764 DOI: 10.1109/ICRA.2018.8460962

Mishra, H., & Mishra, D. (2024). *Artificial Intelligence and Machine Learning in Agriculture: Transforming Farming Systems* (Vol. I). Research Trends in Agriculture Science.

Mittal, U., Sai, S., Chamola, V., & Sangwan, D. (2024). A Comprehensive Review on Generative AI for Education. *IEEE Access : Practical Innovations, Open Solutions*, *1*, 142733–142759. Advance online publication. DOI: 10.1109/ACCESS.2024.3468368

Miyato, T., Kataoka, T., Koyama, M., & Yoshida, Y. (2018). Spectral normalization for generative adversarial networks. In *International Conference on Learning Representations (ICLR).* https://openreview.net/forum?id=B1QRgziT-

Mizik, T. (2023). How can precision farming work on a small scale? A systematic literature review. *Precision Agriculture*, *24*(1), 384–406. DOI: 10.1007/s11119-022-09934-y

Mohanty, S. P., Hughes, D. P., & Salathé, M. (2016). Using deep learning for image-based plant disease detection. *Frontiers in Plant Science*, *7*, 215232. DOI: 10.3389/fpls.2016.01419 PMID: 27713752

Molotoks, A., Smith, P., & Dawson, T. P. (2021). Impacts of land use, population, and climate change on global food security. *Food and Energy Security*, *10*(1), e261. DOI: 10.1002/fes3.261

Monchusi, B. B., Kgopa, A., & Mokwana, T. (2024). Integrating IoT and AI for Precision Agriculture: Enhancing Water Management and Crop Monitoring in Small-Scale Farms. *International Conference on Intelligent and Innovative Computing Applications*. 2024. 151-158. DOI: 10.59200/ICONIC.2024.017

Moshayedi, A. J., Khan, A., Yang, Y., Hu, J., & Kolahdooz, A. (2024). *Robots in Agriculture: Revolutionizing Farming Practices.* EAI Endorsed Transactions on AI and Robotics., DOI: 10.4108/airo.5855

Movahed, A. B., Movahed, A. B., Aliahmadi, B., & Nozari, H. (2024). Green and Sustainable Supply Chain in Agriculture 6.0. In Oskounejad, M., & Nozari, H. (Eds.), *Advanced Businesses in Industry 6.0* (pp. 32–45). IGI Global Scientific Publishing., DOI: 10.4018/979-8-3693-3108-8.ch003

Mowat, K. M. (2025). A Case Study on Readiness of Engineers in Agricultural Technology Start-Ups (Doctoral dissertation, The University of Nebraska-Lincoln).

Muhammed, D. (2025). *Artificial Intelligence of Things (AIoT) for Smart Agriculture: Architectures and Solutions* [Phdthesis, Sorbonne Université, Faculté des Sciences et Ingénierie]. https://hal.science/tel-05046536

Muhammed, D., Ahvar, E., Ahvar, S., Trocan, M., Montpetit, M. J., & Ehsani, R. (2024). Artificial Intelligence of Things (AIoT) for smart agriculture: A review of architectures, technologies and solutions. *Journal of Network and Computer Applications*, *228*, 103905. DOI: 10.1016/j.jnca.2024.103905

Munawar, M., Singh, T. M., Kishor Kumar Reddy, C., & Doss, S. (2025). Synergies Between Artificial Intelligence and Haptic Systems in Advancing Medical Technology. In Mamodiya, U., Kalam, A., Bossoufi, B., Awad, M., & Somwanshi, D. (Eds.), *Integrating AI With Haptic Systems for Smarter Healthcare Solutions* (pp. 517–542). IGI Global Scientific Publishing., DOI: 10.4018/979-8-3373-2307-7.ch023

Murphy, K. P. (2012). *Machine learning: A probabilistic perspective*. MIT Press.

Musalamadugu, T. S., & Hemachandran, K. (2023). *Generative AI for medical imaging analysis and applications*. Future Medicine AI. DOI: 10.2217/fmai-2023-0004

Mzoughi, O., & Yahiaoui, I. (2023). Deep learning-based segmentation for disease identification. *Ecological Informatics*, *75*, 102000. DOI: 10.1016/j.ecoinf.2023.102000

Nag, B. (2024). Navigating Ethical Dilemmas in Generative AI. *Advances in Computational Intelligence and Robotics Book Series*, 189–222. DOI: 10.4018/979-8-3693-9173-0.ch007

Naidoo, P., & Sibanda, M. (2024). Emerging trends and future directions of the industrial internet of things. *From Internet of Things to Internet of Intelligence*, 91–110.

Narvekar, C., Shinde, S., Mishra, V., & Kshirsagar, D. (2023). Productivity improvement with generative AI framework for data enrichment in agriculture. *International Journal on Recent and Innovation Trends in Computing and Communication*, *11*(9), 679–684. DOI: 10.17762/ijritcc.v11i9.8859

Nautiyal, C., Nautiyal, P., Papnai, G., Mittal, H., Agrawal, K., Shivani, , Vishesh, , & Nandini, R. (2024). Importance of Smart Agriculture and Use of Artificial Intelligence in Shaping the Future of Agriculture. *Journal of Scientific Research and Reports*, *30*(3), 129–138. DOI: 10.9734/jsrr/2024/v30i31864

Neves, M. F., Casagrande, B. P., Cambaúva, V., de Oliveira Teixeira, G., & Toledo, P. J. F. (2023). Agriculture 6.0: A new proposal for the future of agribusiness. Rev. De Gestão Soc. *Engenharia Ambiental*, *17*, e04004.

Neves, M., Casagrande, B., Cambaúva, V., de Oliveira Teixeira, G., & Toledo, P. (2023). Agriculture 6.0: A New Proposal for the Future of Agribusiness. *Revista de Gestão Social e Ambiental.*, *17*(9), e04004. DOI: 10.24857/rgsa.v17n9-021

Ng, A. Y., & Jordan, M. I. (2002). On discriminative vs. generative classifiers: A comparison of logistic regression and naive Bayes. In *Advances in Neural Information Processing Systems* (pp. 841–848). https://proceedings.neurips.cc/paper/2001/file/1f71e393b3809197ed63b7091b20f0e4-Paper.pdf

Ngadi, N., Zaelany, A. A., Latifa, A., Harfina, D., Asiati, D., Setiawan, B., Ibnu, F., Triyono, T., & Rajagukguk, Z. (2023). Challenge of agriculture development in Indonesia: Rural youth mobility and aging workers in agriculture sector. *Sustainability (Basel), 15*(2), 922. DOI: 10.3390/su15020922

Nigar, N. (2024). *AI in remote patient monitoring*. arXiv. https://arxiv.org/abs/2407.17494

Nijhuis, S., & Edari, A. (2024). Advancements in smart cities, Internet of Things, and Internet of Nature: A brief overview of applications and best practices in the Netherlands. *Al Edari, 169*, 14–32.

Nikhil, U. V., Pandiyan, A. M., Raja, S. P., & Stamenkovic, Z. (2024). Machine Learning-Based Crop Yield Prediction in South India: Performance Analysis of Various Models. *Computers, 13*(6), 137. DOI: 10.3390/computers13060137

Nkwo, M., Ikwunne, T., Adejoro, C., & Anuyah, O. (2025). Exploring Usability of AI Systems in the Global South—Toward Responsible Human-Centered AI for Sustainable Cities and Communities. In *Usability for the World: Building Better Cities and Communities* (pp. 125–165). Springer. DOI: 10.1007/978-3-031-88326-2_5

O'Connor, D., & Smith, J. (2023). AIIRA: AI Institute for Resilient Agriculture. *AI Magazine, 44*(1), 51. DOI: 10.1002/aaai.12151

Obaid, M. K., Abood, B. S. Z., Alazzai, W. K., & Jasim, L. (2024). From Field to Fork: The Role of AI and IoT in Agriculture. E3S Web of Conferences, 491, 02006. DOI: 10.1051/e3sconf/202449102006

Oliveira, R., & Silva, R. D. S. (2023). Artificial Intelligence in Agriculture: Benefits, Challenges, and Trends. *Applied Sciences (Basel, Switzerland), 13*(13), 7405. Advance online publication. DOI: 10.3390/app13137405

Onteddu, A. R., Kundavaram, R. R., Kamisetty, A., Gummadi, J. C. S., & Manikyala, A. (2025). Enhancing Agricultural Efficiency with Robotics and AI-Powered Autonomous Farming Systems. *Malaysian Journal of Medical and Biological Research, 12*(1), 1.

Osasona, F., Amoo, O. O., Atadoga, A., Abrahams, T. O., Farayola, O. A., & Ayinla, B. S. (2024). Reviewing the ethical implications of AI in decision making processes. *International Journal of Management & Entrepreneurship Research, 6*(2), 322–335. DOI: 10.51594/ijmer.v6i2.773

Padmaja, P., Sophia, I. J., Hari, H. S., Kumar, S. S., & Somu, K.. (2021). Distribute the Message over the Network Using another Frequency and Timing Technique to Circumvent the Jammers. *J Nucl Ene Sci Power Generat Techno, 10*, 9.

Pal, D., Mondal, R., Saha, S., & Chatterjee, S. (2025). INTEGRATION OF ADVANCED TECHNOLOGIES LIKE AI AND IOT INTO AGRICULTURAL EXTENSION.

Pallottino, F., Violino, S., Figorilli, S., Pane, C., Aguzzi, J., Colle, G., Nerio Nemmi, E., Montaghi, A., Chatzievangelou, D., Antonucci, F., Moscovini, L., Mei, A., Costa, C., & Ortenzi, L. (2025). Applications and perspectives of Generative Artificial Intelligence in agriculture. *Computers and Electronics in Agriculture*, *230*, 109919. DOI: 10.1016/j.compag.2025.109919

Patel, D., Gandhi, M., Shankaranarayanan, H., & Darji, A. D. (2022). Design of an Autonomous Agriculture Robot for Real Time Weed Detection using CNN. *arXiv preprint arXiv:2211.12077*. https://arxiv.org/abs/2211.12077

Patel, K., & Desai, M. (2023). Artificial intelligence in sustainable vertical farming. *arXiv*. https://arxiv.org/abs/2312.00030arXiv

Patil, M. A., & Manohar, M. (2022). Enhanced radial basis function neural network for tomato plant disease leaf image segmentation. *Ecological Informatics*, *70*, 101752. DOI: 10.1016/j.ecoinf.2022.101752

Patil, P. (2022). Applications of deep learning in traffic management: A review. *International Journal of Business Intelligence and Big Data Analytics*, *5*(1), 16–23.

Patil, Y. H., Patil, R. Y., & Ozer, O. (2024). Next-Gen Agriculture 6.0: A Smart Solution Pioneering the Future of Farming. In Reddy, C. K. K., Doss, S., Khan, S. B., & Alqhatani, A. (Eds.), *Evolution and Advances in Computing Technologies for Industry 6.0* (pp. 167–187). CRC Press. DOI: 10.1201/9781003503934-7

PayPal. (n.d.). *Machine learning fraud detection technologies*. https://www.paypal.com/us/brc/article/payment-fraud-detection-machine-learning

Pietrantoni, L., Favilla, M., Fraboni, F., Mazzoni, E., Morandini, S., Benvenuti, M., & De Angelis, M. (2024). Integrating collaborative robots in manufacturing, logistics, and agriculture: Expert perspectives on technical, safety, and human factors. *Frontiers in Robotics and AI*, *9*, 1342130. DOI: 10.3389/frobt.2024.1342130 PMID: 39687349

Pingali, P. L. (2012). Green revolution: Impacts, limits, and the path ahead. *Proceedings of the National Academy of Sciences of the United States of America*, *109*(31), 12302–12308. DOI: 10.1073/pnas.0912953109 PMID: 22826253

Pintus, M., Colucci, F., & Maggio, F. (2025). Emerging Developments in Real-Time Edge AIoT for Agricultural Image Classification. *IoT*, *6*(1), 13. DOI: 10.3390/iot6010013

Polly, R., & Devi, E. A. (2024). Semantic segmentation for plant leaf disease classification and damage detection: A deep learning approach. *Smart Agricultural Technology*, *9*, 100526. DOI: 10.1016/j.atech.2024.100526

Pratapagiri, S. et al. (2021). "Early Detection of Plant Leaf Disease Using Convolutional Neural Networks," 2021 3rd Interna-tional Conference on Electronics Representation and Algorithm (ICERA), pp. 77-82, doi: DOI: 9.1109/ICERA53111.2021.9538659

Praveen, R., Mittal, M., Parida, P., Begum, R., Kumar, Y., & Nijhawan, G. (2025). Deep Learning Applications for Detecting Crop Diseases from Image Data. *2025 International Conference on Computational, Communication and Information Technology (ICCCIT)*, 566–571. DOI: 10.1109/ICCCIT62592.2025.10928061

Preanto, S. A., Ahad, M. T., Emon, Y. R., Mustofa, S., & Alamin, M. (2024). A Semantic Segmentation Approach on Sweet Orange Leaf Diseases Detection Utilizing YOLO. *arXiv preprint* arXiv:2409.06671.

Prema Kumar, M. et al. (2024). "Enhanced Plant Disease Detection in Precision Agriculture Using Autoencoder and Transfer Learning," 2024 Third International Conference on Artificial Intelligence, Computational Electronics and Communication System (AICECS), MANIPAL, India, pp. 1-6, DOI: 10.1109/AICECS63354.2024.10957233

Puttanapura, J., Singh, T. M., Reddy, C. K. K., & Doss, S. (2024). Improvised prediction of rainfall using random forest classifier over Taiwan. In *Proceedings of the 2024 4th International Conference on Soft Computing for Security Applications (ICSCSA)* (pp. 152–156). IEEE. DOI: 10.1109/ICSCSA64454.2024.00031

Qazi, S., Khawaja, B. A., & Farooq, Q. U. (2022). IoT-Equipped and AI-Enabled Next Generation Smart Agriculture: A Critical Review, Current Challenges and Future Trends. *IEEE Access : Practical Innovations, Open Solutions*, *10*, 21219–21235. DOI: 10.1109/ACCESS.2022.3152544

Qi, H., & Altinakar, M. S. (2011). Simulation-based decision support system for flood damage assessment under uncertainty using remote sensing and census block information. *Natural Hazards*, *59*(2), 1125–1143. DOI: 10.1007/s11069-011-9822-8

Qi, Z., Du, L., Huo, R., & Huang, T. (2023). Predictive maintenance based on identity resolution and transformers in IIoT. *Sensors (Basel)*, *16*(9), 310.

Qu, J., Cui, G., Li, Z., Fang, S., Zhang, X., Liu, A., Han, M., Liu, H., Wang, X., & Wang, X. (2024). Advanced Flexible Sensing Technologies for Soft Robots. *Advanced Functional Materials*, *34*(29), 2401311. DOI: 10.1002/adfm.202401311

Qu, J., Zhang, Z., Qin, Z., Guo, K., & Li, D. (2024). Applications of autonomous navigation technologies for unmanned agricultural tractors: A review. *Machines (Basel)*, *12*(4), 218. DOI: 10.3390/machines12040218

Radford, A., Metz, L., & Chintala, S. (2016). Unsupervised representation learning with deep convolutional generative adversarial networks. arXiv preprint arXiv:1511.06434

Radford, A., Narasimhan, K., Salimans, T., & Sutskever, I. (2018). *Improving language understanding by generative pre-training*. OpenAI. https://cdn.openai.com/research-covers/language-unsupervised/language_understanding_paper.pdf

Rahimi Azghadi, M., Olsen, A., Wood, J., Saleh, A., Calvert, B., Granshaw, T., Fillols, E., & Philippa, B. (2024). Precise robotic weed spot-spraying for reduced herbicide usage and improved environmental outcomes: A real-world case study. *arXiv preprint arXiv:2401.13931*. https://arxiv.org/abs/2401.13931

Rahmadian, R., & Widyartono, M. (2020). Autonomous Robotic in Agriculture: A Review. *2020 Third International Conference on Vocational Education and Electrical Engineering (ICVEE)*, 1–6. DOI: 10.1109/ICVEE50212.2020.9243253

Rahman, R. A., & Shah, B. (2016, March). Security analysis of IoT protocols: A focus in CoAP. In 2016 3rd MEC international conference on big data and smart city (ICBDSC) (pp. 1-7). IEEE. DOI: 10.1109/ICBDSC.2016.7460363

Rahman, S. M., Ibtisum, S., Podder, P., & Hossain, S. M. (2023). Progression and challenges of IoT in healthcare: A short review. arXiv preprint arXiv:2311.12869.

Rahman, I. (2023). AI-powered Personalized Treatment Recommendation Framework for Improved Healthcare Outcomes. *Journal of Computational Social Dynamics*, *8*(11), 42–51.

Rahman, M., & Chowdhury, H. (2023). Blockchain and artificial intelligence for ensuring the authenticity of organic legumes. *Legume Research*, *46*(5), 786. DOI: 10.18805/LRF-786

Rahman, Z. U., Asaari, M. S. M., Ibrahim, H., Abidin, I. S. Z., & Ishak, M. K. (2024). Generative adversarial networks (GANs) for image augmentation in farming: A review. *IEEE Access : Practical Innovations, Open Solutions*, *12*, 179912–179943. DOI: 10.1109/ACCESS.2024.3505989

Rai, N., Zhang, Y., Ram, B. G., Schumacher, L., Yellavajjala, R. K., Bajwa, S., & Sun, X. (2023). Applications of deep learning in precision weed management: A review. *Computers and Electronics in Agriculture*, *206*, 107698. DOI: 10.1016/j.compag.2023.107698

Rai, S., & Goyal, N. (2020). Waste management through smart bin. [IJERT]. *International Journal of Engineering Research & Technology (Ahmedabad)*, *9*(09), 2278–0181.

Raj, L., Gaurav, S., Kumar, A., Kumar, H., & Rampal, A. (March 06, 2025). AI-Powered Crop Health Monitoring System. Available at *SSRN*: https://ssrn.com/abstract=5175088 or http://dx.doi.org/DOI: 10.2139/ssrn.5175088

Rajwar, K., Pant, M., & Deep, K. (2024). Application of machine learning in agriculture: Recent trends and future research avenues. *arXiv preprint* arXiv:2405.17465.

Rakshit, S., Gorai, S. K., Das, B., Sahu, S., Gills, R., Ray, P., & Mukherjee, A. (2025). Empowerment of Rural Women and Youth Retention as Drivers of the Agricultural Economy. In Sharma, N. K., Rai, P. K., & Rai, D. C. (Eds.), *Indian Agriculture: Challenges, Priorities and Solutions* (pp. 149–181). Springer Nature., DOI: 10.1007/978-981-96-5273-0_8

Ramadugu, R., & Doddipatla, L. (2022). The role of AI and machine learning in strengthening digital wallet security against fraud. *Journal of Big Data and Smart Systems*, *3*(1).

Rana, R., & Bhambri, P. (2025). The Role of IoT in Shaping the Future of Geospatial AI. In *Recent Trends in Geospatial AI* (pp. 177–216). IGI Global Scientific Publishing.

Rane, J., & Kaya, Ö., & Mallick, S., & Rane, N. (2024). Smart farming using artificial intelligence, machine learning, deep learning, and ChatGPT: Applications, opportunities, challenges, and future directions. .DOI: 10.70593/978-81-981271-7-4_6

Rangarajan, D., Rangarajan, A., Reddy, C. K. K., & Doss, S. (2025). Exploring the next-gen transformations in healthcare through the impact of AI and IoT. In *Intelligent Systems and IoT Applications in Clinical Health* (pp. 73–98). IGI Global.

Rani, N. S., Krishna, A. S., Sunag, M., Sangamesha, M. A., & Pushpa, B. R. (2024). Infield disease detection in citrus plants: Integrating semantic segmentation and dynamic deep learning object detection model for enhanced agricultural yield. *Neural Computing & Applications*, *36*(35), 22485–22510. DOI: 10.1007/s00521-024-10451-4

Rani, R., Sahoo, J., Bellamkonda, S., Kumar, S., & Pippal, S. K. (2023). Role of Artificial Intelligence in Agriculture: An Analysis and Advancements With Focus on Plant Diseases. *IEEE Access : Practical Innovations, Open Solutions*, *11*, 137999–138019. DOI: 10.1109/ACCESS.2023.3339375

Rashid, A. B., & Kausik, M. A. K. (2024). AI revolutionizing industries worldwide: A comprehensive overview of its diverse applications. *Hybrid Advances*, *7*, 100277. DOI: 10.1016/j.hybadv.2024.100277

Redhu, N. S., Thakur, Z., Yashveer, S., & Mor, P. (2022). Artificial intelligence: A way forward for agricultural sciences. In Sharma, P., Yadav, D., & Gaur, R. K. (Eds.), *Bioinformatics in Agriculture* (pp. 641–668). Academic Press., DOI: 10.1016/B978-0-323-89778-5.00007-6

Revathi, S., Ansari, A., Susmi, S. J., Madhavi, M., Gunavathie, M. A., & Sudhakar, M. (2024).Integrating Machine Learning-IoT Technologies Integration for Building Sustainable Digital Ecosystems.In Multidisciplinary Applications of Extended Reality for Human Experience (pp. 259-291).IGI Global. DOI: 10.4018/979-8-3693-2432-5.ch012

Riedelbauch, D., Höllerich, N., & Henrich, D. (2023). Benchmarking teamwork of humans and cobots—An overview of metrics, strategies, and tasks. *IEEE Access : Practical Innovations, Open Solutions*, *11*, 43648–43674. DOI: 10.1109/ACCESS.2023.3271602

Rithika, A., Singh, T. M., Reddy, C. K. K., & Doss, S. (2025). Leveraging generative AI for designing next-generation haptic systems: Advancements, challenges, and applications. In Mamodiya, U., Kalam, A., Bossoufi, B., Awad, M., & Somwanshi, D. (Eds.), *Integrating AI with haptic systems for smarter healthcare solutions* (pp. 327–350). IGI Global., DOI: 10.4018/979-8-3373-2307-7.ch015

Rizvi, S. M. H., Naseer, A., Rehman, S. U., Akram, S., & Gruhn, V. (2024). Revolutionizing Agriculture: Machine and Deep Learning Solutions for Enhanced Crop Quality and Weed Control. *IEEE Access : Practical Innovations, Open Solutions*, *12*, 11865–11878. DOI: 10.1109/ACCESS.2024.3355017

Roy, P., Padhiary, M., Hoque, A., Sahu, B., Roy, D., & Kumar, K. (2025). Machine Learning for Precision Agriculture and Crop Yield Optimization: Techniques and Applications. In *The Role of Artificial Intelligence in Advancing Applied Life Sciences* (pp. 189–234). IGI Global Scientific Publishing. DOI: 10.4018/979-8-3693-9208-9.ch007

Roy, S., Ansal, D. K., & Kumar, M. (2025). Soil Texture Prediction: Advances in Remote Sensing, Image Analysis, and Machine Learning. *2025 Fourth International Conference on Smart Technologies, Communication and Robotics (STCR)*, 1–4. DOI: 10.1109/STCR62650.2025.11020387

Sagar, S., & N Birje, M. 2025. Precision Agriculture Using Internet of Things and Cloud Computing:Review, A. (•••).. . *SN Computer Science*, *6*(4), 1–18.

Sai, S., Kumar, S., Gaur, A., Goyal, S., Chamola, V., & Hussain, A. (2025). Unleashing the power of generative AI in Agriculture 4.0 for smart and sustainable farming. *Cognitive Computation*, *17*(1), 1–18. DOI: 10.1007/s12559-025-10420-6

Salimans, T., Goodfellow, I., Zaremba, W., Cheung, V., Radford, A., & Chen, X. (2016). Improved techniques for training GANs. In *Advances in Neural Information Processing Systems* (pp. 2234–2242). https://proceedings.neurips.cc/paper_files/paper/2016/file/8a3363abe7921c41837d905f2b29721f-Paper.pdf

Sapkota, R., & Karkee, M. (2025). Generative AI in agriculture: Creating image datasets using DALL·E's advanced large language model capabilities. *arXiv*. https://arxiv.org/abs/2307.08789

Sarkar, D., Bali, R., Sharma, T., Sarkar, D., Bali, R., & Sharma, T. (2018). Machine learning basics. Practical Machine Learning with Python: A Problem-Solver's Guide to Building Real-World Intelligent Systems, 3-65.

Sarkar, D., & Banerjee, S. (2023). Sustainable development of AI applications in agriculture: A review. *Procedia Computer Science*, *219*, 350–359. DOI: 10.1016/j.procs.2023.10.350

Savitha, C., & Talari, R. (2025). Evaluating the performance of random forest, support vector machine, gradient tree boost, and CART for improved crop-type monitoring using greenest pixel composite in Google Earth Engine. *Environmental Monitoring and Assessment*, *197*(4), 437. DOI: 10.1007/s10661-025-13880-3 PMID: 40108031

Schaffers, H., Komninos, N., &Pallot, M. (2012).Smart cities as innovation ecosystems sustained by the future internet.

cheau, M. C., Achim, M. V., Găbudeanu, L., Brici, I., & Vîlcea, A. L. (2022). Legal, economic and cyber security framework considerations for drone usage. *Applied Sciences (Basel, Switzerland)*, *12*(9), 4663. DOI: 10.3390/app12094663

Schizas, N., Karras, A., Karras, C., & Sioutas, S. (2022). TinyML for ultra-low power AI and large-scale IoT deployments: A systematic review. *Future Internet*, *14*(12), 363. DOI: 10.3390/fi14120363

Schwalbert, R. A., Amado, T., Corassa, G., Pott, L. P., Prasad, P. V. V., & Ciampitti, I. A. (2020). Satellite-based soybean yield forecast: Integrating machine learning and weather data for improving crop yield prediction in southern Brazil. *Agricultural and Forest Meteorology*, *284*, 107886. DOI: 10.1016/j.agrformet.2019.107886

Sehrawat, D., & Gill, N. S. (2019, April). Smart sensors: Analysis of different types of IoT sensors. In 2019 3rd International Conference on Trends in Electronics and Informatics (ICOEI) (pp. 523-528). IEEE.

Selinger, E., & Durant, D. (2021). Amazon's Ring: Surveillance as a slippery slope service. *Science as Culture*, *31*(1), 1–15. DOI: 10.1080/09505431.2021.1983797

Senathipathi, S., Abirami, N., Narayanasamy, P., Mathankumar, S., Lotus, A. A., & Muthulekshmi, M. (2025). Smart Robotic Crop Replanting and Recovery Using CNNs and Edge Computing. *2025 3rd International Conference on Disruptive Technologies (ICDT)*, 1349–1354. DOI: 10.1109/ICDT63985.2025.10986524

Seth, A. (2023). *Modeling postoperative delirium risk: Decoding electronic health data with artificial intelligence* (Doctoral dissertation, Johns Hopkins University). Johns Hopkins University Repository.

Seto, K. C., Güneralp, B., & Hutyra, L. R. (2012). Global forecasts of urban expansion to 2030 and direct impacts on biodiversity and carbon pools. *Proceedings of the National Academy of Sciences of the United States of America*, *109*(40), 16083–16088. DOI: 10.1073/pnas.1211658109 PMID: 22988086

Shaheb, M. R., Sarker, A., & Shearer, S. A. (2022). *Precision agriculture for sustainable soil and crop management*. IntechOpen., DOI: 10.5772/intechopen.101759

Shaikh, F. K., Memon, M. A., Mahoto, N. A., Zeadally, S., & Nebhen, J. (2022). "Artificial Intelligence Best Practices in Smart Agriculture," in IEEE Micro, vol. 42, no. 1, pp. 17-24, 1 Jan.-Feb, .DOI: 10.1109/MM.2021.3121279

Shaikh, T. A., Rasool, T., & Lone, F. R. (2022). Towards leveraging the role of machine learning and artificial intelligence in precision agriculture and smart farming. *Computers and Electronics in Agriculture*, *199*, 107119. DOI: 10.1016/j.compag.2022.107119

Shailaja, K., Srinivasulu, B., Thirupathi, L., Gangula, R., Boya, T. R., & Polem, V. (2023). An Intelligent Deep Feature Based Intrusion Detection System for Network Applications. *Wireless Personal Communications*, *129*(1), 345–370. DOI: 10.1007/s11277-022-10100-w

Shamshiri, R. R., Kalantari, F., Ting, K. C., Thorp, K. R., Hameed, I. A., & Weltzien, C. (2018). Advances in greenhouse automation and controlled environment agriculture: A transition to plant factories and urban agriculture. *International Journal of Agricultural and Biological Engineering*, *11*(1), 1–22. DOI: 10.25165/j.ijabe.20181101.3210

Sharma, K., & Nandal, R. (2019). A literature study on machine learning fusion with IoT. 2019 3rd International Conference on Trends in Electronics and Informatics (ICOEI), 1440–1445. https://doi.org/DOI: 10.1109/ICOEI.2019.8862656

Sharma, K., & Shivandu, S. (2024). Integrating artificial intelligence and Internet of Things (IoT) for enhanced crop monitoring and management in precision agriculture. Scientific Intelligence, Advance online publication. DOI: 10.1016/j.sintl.2024.100292

Sharma, H., Sidhu, H., & Bhowmik, A. (2025). Remote Sensing Using Unmanned Aerial Vehicles for Water Stress Detection: A Review Focusing on Specialty Crops. *Drones (Basel)*, *9*(4), 241. DOI: 10.3390/drones9040241

Sharma, M., Kumar, C. J., Singh, T. P., Talukdar, J., Sharma, R. K., & Ganguly, A. (2023). Enhancing disease region segmentation in rice leaves using modified deep learning architectures. *Archiv für Phytopathologie und Pflanzenschutz*, *56*(20), 1555–1580. DOI: 10.1080/03235408.2024.2310326

Sharma, P., Berwal, Y. P. S., & Ghai, W. (2020). Performance analysis of deep learning CNN models for disease detection in plants using image segmentation. *Information Processing in Agriculture*, *7*(4), 566–574. DOI: 10.1016/j.inpa.2019.11.001

Shen, Z., Wang, Z., Chew, J., Hu, K., & Wang, Y. (2025). Artificial intelligence empowering robo-advisors: A data-driven wealth management model analysis. *International Journal of Management Science Research*, *8*(3), 1–12. DOI: 10.53469/ijomsr.2025.08(03).01

Shinde, N. K., Seth, A., &Kadam, P. (2023).Exploring the synergies: a comprehensive survey of blockchain integration with artificial intelligence, machine learning, and iot for diverse applications. Machine Learning and Optimization for Engineering Design, 85-119.

Shoaib, M., Hussain, T., Shah, B., Ullah, I., Shah, S. M., Ali, F., & Park, S. H. (2022). Deep learning-based segmentation and classification of leaf images for detection of tomato plant disease. *Frontiers in Plant Science*, *13*, 1031748. DOI: 10.3389/fpls.2022.1031748 PMID: 36275583

Sibona, F., Harrison, N., Wallace, N., Stevens, C., Sukkarieh, S., Raubenheimer, D., Simpson, S. J., & González, L. A. (2025). A Framework to Enable Eco-Cyber-Physical Systems for Robotics-Focused Digital Twins in Smart Farming. *Journal of Field Robotics*, rob.22562. DOI: 10.1002/rob.22562

Singh, A., & Kumar, P. (2023). Leveraging imitation learning in agricultural robotics. *Frontiers in Robotics and AI*, *10*, 1441312. DOI: 10.3389/frobt.2024.1441312 PMID: 39483488

Singh, T. M., Reddy, C. K. K., & Lippert, K. (2024). The revolution and future of blockchain technology in cybersecurity: A comprehensive analysis. In *Artificial intelligence for blockchain and cybersecurity powered IoT applications* (1st ed., p. 28). CRC Press., DOI: 10.1201/9781003497585-5

Singh, T., Kundroo, M., & Kim, T. (2024). WSN-Driven Advances in Soil Moisture Estimation: A Machine Learning Approach. *Electronics (Basel)*, *13*(8), 8. Advance online publication. DOI: 10.3390/electronics13081590

Skendžić, S., Zovko, M., Živković, I. P., Lešić, V., & Lemić, D. (2021). The impact of climate change on agricultural insect pests. *Insects*, *12*(5), 440. DOI: 10.3390/insects12050440 PMID: 34066138

Smith, P., Poch, R. M., Lobb, D. A., Bhattacharyya, R., Alloush, G., Eudoxie, G. D., & Hallett, P. (2024). Status of the World's Soils. *Annual Review of Environment and Resources*, •••, 49.

Sonmez, D., & Cetin, A. (2024). "An End-to-End Deployment Workflow for AI Enabled Agriculture Applications at the Edge," 2024 6th International Conference on Computing and Informatics (ICCI), New Cairo - Cairo, Egypt, pp. 506-511, DOI: 10.1109/ICCI61671.2024.10485167

Soussi, A., Zero, E., Sacile, R., Trinchero, D., & Fossa, M. (2024). Smart sensors and smart data for precision agriculture: A review. *Sensors (Basel)*, *24*(8), 2647. DOI: 10.3390/s24082647 PMID: 38676264

Spagnuolo, M., Todde, G., Caria, M., Furnitto, N., Schillaci, G., & Failla, S. (2025). Agricultural Robotics: A Technical Review Addressing Challenges in Sustainable Crop Production. *Robotics (Basel, Switzerland)*, *14*(2), 9. DOI: 10.3390/robotics14020009

Speer, B., Miller, M., Schaffer, W., Gueran, L., Reuter, A., Jang, B., & Widegren, K. (2015). *Role of smart grids in integrating renewable energy (No.NREL/TP-6A20-63919). National Renewable Energy Lab*. NREL. DOI: 10.2172/1215177

Stanghellini, C. (2013, June). Horticultural production in greenhouses: efficient use of water. In *International Symposium on Growing Media and Soilless Cultivation 1034* (pp. 25-32).

Steele, R., & Lo, A. (2013). Telehealth and ubiquitous computing for bandwidth-constrained rural and remote areas. *Personal and Ubiquitous Computing*, *17*(3), 533–543. DOI: 10.1007/s00779-012-0506-5

Strauß, P., Schmitz, M., Wöstmann, R., & Deuse, J. (2018). Enabling of predictive maintenance in the brownfield through low-cost sensors, an IIoT-architecture and machine learning. *Proceedings of the 2018 IEEE International Conference on Big Data (Big Data)*, 1474–1483. https://doi.org/DOI: 10.1109/BigData.2018.8622076

SwathiSree. M., Reddy, C. K. K., Vaishnavi, K., &Harika, V. (2025). Informatics and Internet of Things uses in clinical medicine. In Intelligent systems and IoT applications in clinical health (pp. 99-126).IGI Global.

Szabó-Szentgróti, E., Rámháp, S., & Kézai, P. K. (2023). Systematic review of cashierless stores (just walk out stores) revolutionizing the retail. *Management & Marketing*, *18*(s1), 427–448. DOI: 10.2478/mmcks-2023-0023

Taesi, C., Aggogeri, F., & Pellegrini, N. (2023). COBOT applications—Recent advances and challenges. *Robotics (Basel, Switzerland)*, *12*(3), 79. DOI: 10.3390/robotics12030079

Tagarakis, A. C., Dordas, C., Lampridi, M., Kateris, D., & Bochtis, D. (2021). A smart farming system for circular agriculture. *Engineering Proceedings*, *9*(1), 10.

Talaviya, T., Shah, D., Patel, N., Yagnik, H., & Shah, M. (2020). Implementation of artificial intelligence in agriculture for optimization of irrigation and pesticide spraying. *Artificial Intelligence in Agriculture*, *4*, 58–73. DOI: 10.1016/j.aiia.2020.04.002

Tao, W., Zhao, L., Wang, G., & Liang, R. (2021). Review of the internet of things communication technologies in smart agriculture and challenges. *Computers and Electronics in Agriculture*, *189*, 106352. DOI: 10.1016/j.compag.2021.106352

Thamaraikannan, N., & Manju, S. (2023). "Classification of Satellite Agriculture Imagery to Identify Crop Type Using Deep Learning Techniques - A Review," *2023 International Conference on Computer Communication and Informatics (ICCCI)*, Coimbatore, India, pp. 1-5, DOI: 10.1109/ICCCI56745.2023.10128228

Thampi, S. M., Thomas, T., & Mukherjee, P. (Eds.). (2025). *Securing the Connected World: Exploring Emerging Threats and Innovative Solutions*. Springer Nature. DOI: 10.1007/978-3-031-82826-3

Thilakarathne, N. N., Bakar, M. S. A., Abas, P. E., & Yassin, H. (2025). Internet of things enabled smart agriculture: Current status, latest advancements, challenges and countermeasures. *Heliyon*, *11*(3), e42136. DOI: 10.1016/j.heliyon.2025.e42136 PMID: 39959477

Thirupathi, L., Rajesh, A., & Sandeep, R. (2023). Internet of Lighting for Smart Cities. In Tiwari, M., Ismail, Y., Verma, K., & Garg, A. K. (Eds.), *Optical and Wireless Technologies. OWT 2021. Lecture Notes in Electrical Engineering* (Vol. 892). Springer., DOI: 10.1007/978-981-19-1645-8_41

Thirupathi, L., Srinivasulu, B., Khanapurkar, U., Rambabu, D., & Preeti, C. M. (2023). Heart Disease Prediction in Healthcare Communities by Machine Learning Over Big Data. In Bhateja, V., Sunitha, K. V. N., Chen, Y. W., & Zhang, Y. D. (Eds.), *Intelligent System Design. Lecture Notes in Networks and Systems* (Vol. 494). Springer., DOI: 10.1007/978-981-19-4863-3_38

Tong, T., Ye, F., Zhang, Q., Liao, W., Ding, Y., Liu, Y., & Li, G. (2024). The impact of labor force aging on agricultural total factor productivity of farmers in China: Implications for food sustainability. *Frontiers in Sustainable Food Systems*, *8*, 1434604. DOI: 10.3389/fsufs.2024.1434604

Tsolakis, N., Bechtsis, D., & Bochtis, D. (2019). AgROS: A Robot Operating System based emulation tool for agricultural robotics. *Agronomy (Basel)*, *9*(7), 403. DOI: 10.3390/agronomy9070403

Tsouros, D. C., Bibi, S., & Sarigiannidis, P. G. (2019). A Review on UAV-Based Applications for Precision Agriculture. *Information (Basel)*, *10*(11), 349. DOI: 10.3390/info10110349

Tzafestas, S. G. (2018). Synergy of IoT and AI in modern society: The robotics and automation case. *Robot. Autom. Eng. J*, *31*(5), 1–15. DOI: 10.19080/RAEJ.2018.03.555621

Urrea, C., & Kern, J. (2025). Recent Advances and Challenges in Industrial Robotics: A Systematic Review of Technological Trends and Emerging Applications. *Processes (Basel, Switzerland)*, *13*(3), 3. Advance online publication. DOI: 10.3390/pr13030832

van der Burg, S., Giesbers, E., Bogaardt, M. J., van der Veen, M., & van Est, R. (2024). Ethical aspects of AI robots for agri-food: A relational approach based on four case studies. *AI & Society*, *39*(2), 541–555. DOI: 10.1007/s00146-022-01429-8

Van Dijk, M., Morley, T., Rau, M. L., & Saghai, Y. (2021). A meta-analysis of projected global food demand and population at risk of hunger for the period 2010–2050. *Nature Food*, *2*(7), 494–501. DOI: 10.1038/s43016-021-00322-9 PMID: 37117684

van Hilten, M., & Wolfert, S. (2022). 5G in agri-food—A review on current status, opportunities and challenges. *Computers and Electronics in Agriculture*, *201*, 107291. DOI: 10.1016/j.compag.2022.107291

Vaswani, A., Shazeer, N., Parmar, N., Uszkoreit, J., Jones, L., Gomez, A. N., . . . Polosukhin, I. (2017). Attention is all you need. In *Advances in Neural Information Processing Systems* (pp. 5998–6008). https://proceedings.neurips.cc/paper_files/paper/2017/file/3f5ee243547dee91fbd053c1c4a845aa-Paper.pdf

Vaswani, A., Shazeer, N., Parmar, N., Uszkoreit, J., Jones, L., Gomez, A. N., & Polosukhin, I. (2023). Attention Is All You Need. *arXiv [Cs.CL]*. Retrieved from http://arxiv.org/abs/1706.03762

Venkata Reddy, P. S., Nandini Prasad, K. S., & Puttamadappa, C. (2024). "Conversational AI bot based on IoT Knowledgebase for Smart Agriculture," *2024 International Conference on Knowledge Engineering and Communication Systems (ICKECS)*, Chikkaballapur, India, pp. 1-7, DOI: 10.1109/ICKECS61492.2024.10616582

Verma, S. K., & Verma, A. (2023). Precision irrigation management: An innovative technique for efficient water management. *Indian Farming*, *73*(4), 21–24.

Vieira, S., Pinaya, W. H. L., & Mechelli, A. (2020). Main concepts in machine learning. In *Machine learning* (pp. 21–44). Academic Press. DOI: 10.1016/B978-0-12-815739-8.00002-X

Villamil, S., Hernández, C., & Tarazona, G. (2020). An overview of internet of things. *Telkomnika*, *18*(5), 2320–2327. DOI: 10.12928/telkomnika.v18i5.15911

Visconti, P., Rausa, G., Del-Valle-Soto, C., Velázquez, R., Cafagna, D., & De Fazio, R. (2024). Machine learning and IoT-based solutions in industrial applications for smart manufacturing: A critical review. *Future Internet*, *16*(11), 394. DOI: 10.3390/fi16110394

Vizniuk, A., Diachenko, G., Laktionov, I., Siwocha, A., Xiao, M., & Smoląg, J. (2025). A comprehensive survey of retrieval-augmented large language models for decision making in agriculture: Unsolved problems and research opportunities. *Journal of Artificial Intelligence and Soft Computing Research*, *15*(2), 115–146. DOI: 10.2478/jaiscr-2025-0007

Von Stamm, B., & Trifilova, A. (Eds.). (2009). *The future of innovation*. Gower Publishing, Ltd.

Wang, C. (2024). Intelligent Agricultural Greenhouse Control System Based on Internet of Things and Machine Learning. *arXiv preprint arXiv:2402.09488*. https://arxiv.org/abs/2402.09488

Wang, S., Xu, D., Liang, H., Bai, Y., Li, X., Zhou, J., Su, C., & Wei, W. (2025). Advances in Deep Learning Applications for Plant Disease and Pest Detection: A Review. *Remote Sensing (Basel)*, *17*(4), 698. DOI: 10.3390/rs17040698

Wang, T., Xu, X., Wang, C., Li, Z., & Li, D. (2021). From Smart Farming towards Unmanned Farms: A New Mode of Agricultural Production. *Agriculture*, *11*(2), 145. DOI: 10.3390/agriculture11020145

Watanabe, K., & Tanaka, Y. (2023). Towards autonomous mapping in agriculture: A review of supportive technologies. *Robotics and Autonomous Systems*, *12*(1), 104514. DOI: 10.1016/j.robot.2023.104514

Wolfert, S., Ge, L., Verdouw, C., & Bogaardt, M. J. (2017). Big data in smart farming – A review. *Agricultural Systems*, *153*, 69–80. DOI: 10.1016/j.agsy.2017.01.023

Xu, J., Ren, C., Zhang, X., Wang, C., Wang, S., Ma, B., He, Y., Hu, L., Liu, X., Zhang, F., Lu, L., Li, S., Zhang, J., Zhu, Y.-G., Vitousek, P., & Gu, B. (2025). Soil health contributes to variations in crop production and nitrogen use efficiency. *Nature Food*, *6*(6), 1–13. DOI: 10.1038/s43016-025-01155-6 PMID: 40234681

Yadav, P., Rathwad, G., & Jain, J. (2024). *Generative AI: Shaping the Future While Disrupting the Present*. International Journal For Multidisciplinary Research. DOI: 10.36948/ijfmr.2024.v06i05.28085

Yerebakan, M., & Hu, B. (2024). Human–Robot Collaboration in Modern Agriculture: A Review of the Current Research Landscape. *Advanced Intelligent Systems*, *6*(7), 2300823. Advance online publication. DOI: 10.1002/aisy.202300823

Yin, S., Xi, Y., Zhang, X., Sun, C., & Mao, Q. (2025). Foundation models in agriculture: A comprehensive review. *Agriculture*, *15*(8), 847. DOI: 10.3390/agriculture15080847

Yuan, X., Li, S., Chen, J., Yu, H., Yang, T., Wang, C., Huang, S., Chen, H., & Ao, X. (2024). Impacts of global climate change on agricultural production: A comprehensive review. *Agronomy (Basel)*, *14*(7), 1360. DOI: 10.3390/agronomy14071360

Zaki, M. J. MeiraJr, W., &Meira, W. (2020). Data mining and machine learning: Fundamental concepts and algorithms. Cambridge University Press.

Zaman, Q. (2024). The role of artificial intelligence in early disease detection: Transforming diagnostics and treatment. [MJH]. *Multidisciplinary Journal of Healthcare*, *1*(2), 43–54.

Zhang, D., Malkin, N., Liu, Z., Volokhova, A., Courville, A., & Bengio, Y. (2022, June). Generative flow networks for discrete probabilistic modeling. In *International Conference on Machine Learning* (pp. 26412-26428). PMLR.

Zhang, L., & Chen, Y. (2023). Trends in the sustainable use of artificial intelligence: A bibliometric analysis. *Sustainable Computing : Informatics and Systems*, *5*(2), 1222–1235. DOI: 10.1007/s43621-025-01222-9

Zhang, X., Wang, C., & Li, Y. (2022). Practices, challenges, and future of digital transformation in smallholder agriculture: Insights from a literature review. *Agriculture*, *14*(12), 2193.

Zhang, Y. (2019). The role of precision agriculture. *Resource Magazine*, *26*(6), 9–9.

Zhao, K., Nguyen, M., & Yan, W. (2024). "Evaluating Accuracy and Efficiency of Fruit Image Generation Using Generative AI Diffusion Models for Agricultural Robotics," 2024 39th International Conference on Image and Vision Computing New Zealand (IVCNZ), Christchurch, New Zealand, pp. 1-6, DOI: 10.1109/IVCNZ64857.2024.10794462

Zhao, S., Song, J., & Ermon, S. (2017). *Towards deeper understanding of variational autoencoding models*. arXiv preprint arXiv:1702.08658

Zhao, C., Liu, B., Piao, S., Wang, X., Lobell, D. B., Huang, Y., Huang, M., Yao, Y., Bassu, S., Ciais, P., Durand, J.-L., Elliott, J., Ewert, F., Janssens, I. A., Li, T., Lin, E., Liu, Q., Martre, P., Müller, C., & Asseng, S. (2017). Temperature increase reduces global yields of major crops in four independent estimates. *Proceedings of the National Academy of Sciences of the United States of America*, *114*(35), 9326–9331. DOI: 10.1073/pnas.1701762114 PMID: 28811375

Zhao, X., & Liu, H. (2023). Artificial intelligence in agricultural technology: Enhancing crop harvest prediction. *International Journal of Agricultural Sciences*, *10*(2), 22240. DOI: 10.21275/SR24825222240

About the Contributors

C Kishor Kumar Reddy, currently working as Associate Professor, Dept. of Computer Science and Engineering, Stanley College of Engineering and Technology for Women, Hyderabad, India. He has research and teaching experience of more than 13 years. He has published more than 100 research papers in National and International Conferences, Book Chapters, and Journals indexed by Scopus and others. He is an author for 2 text books and 135 +edited books. He is the member of ACM< ISTE, CSI, IAENG, UACEE, IACSIT

Binju Saju is a distinguished Assistant Professor with over ten years of academic experience, currently teaching at the Adi Shankara Institute of Technology and Engineering in Kalady, India in the Department of AI and Data Science. An expert in Artificial Intelligence, Machine Learning, Image Processing, Generative and Explainable AI. She has a Ph.D. from Christ University, Bangalore. She is a prolific researcher with numerous publications and patents, and she actively mentors students in research and patent writing. With a strong foundation in Artificial Intelligence and a passion for education, Dr. Saju is dedicated to advancing knowledge and fostering innovation in her field.

Asegul Hulus holds the positions of assistant professor and lecturer in the field of computing and is also acknowledged as a fellow of the Higher Education Academy (FHEA). She is a distinguished researcher and published author with expertise in the field of S.T.E.A.M., encompassing areas such as Artificial Intelligence (AI), Human-Computer Interaction (HCI), User Experience (UX) and Interaction Design, Metaverses, and Web3 (Blockchain, Digital Assets). She is a member of prestigious organizations such as the Association for Computing Machinery (ACM) and the Institute of Electrical and Electronics Engineers (IEEE), among others. In addition, her skill set includes graphic design, video editing, audio production, and video game proficiency. Dr. Hulus actively fulfils the roles of chair, chief editor, and peer reviewer for a diverse array of journals, conferences, and publication houses.

Moreover, she maintains a prominent position as one of the founding authorities in her field. Additionally, she assumes the role of a mentor for the Women4Cyber Foundation and holds several leadership positions within professional computing committees. Among these committees is the Council on Women in Computing (ACM-W) of the Association for Computing Machinery, where she serves as an investigative journalist and is one of the four main committee members for the Professional Chapters Committee of ACM-W. For her significant contributions to AI, Dr. Hulus has been appointed honorary convener for women's empowerment activities within the AI Foundation Trust's technical advisory board. She is the inaugural female member of the board, specializing in both AI technology and the encouragement of women in the field. Her commitment to S.T.E.A.M. education and diversity initiatives, amongst her other affiliations, has earned her recognition. She is celebrated for her prolific writing and public speaking on the intersection of these areas. Her work particularly focuses on advocating for gender diversity and addressing systemic challenges within the computing sector

Tatjana Sibalija (BSc, MSc, PhD in Mechanical Engineering, from Faculty of Mechanical Engineering, University of Belgrade) is Full Professor at the Faculty of Computing, Union University, Belgrade. Her research interest includes artificial intelligence, data science, manufacturing and industrial engineering, process modelling, design and optimisation. She is the editor of 3 books; author of 6 books, 12 book chapters, 130+ papers published in peer-reviewed international journals / presented at refereed international conferences (h-index=20, i10-index=30), including several invited keynote speeches. She has served as a chair of many conference sessions, a member of 50+ scientific program committees of international conferences, a member of the editorial board and guest editor of several international scientific journals. She has been awarded eight awards and honours. Since 2009, she has been appointed FP7 / Horizon 2020 / Horizon Europe (EU Research Framework Programmes) expert / advisor on the projects review, selection and progress monitoring

Haïfa Nakouri is a Machine Learning specialist and currently holds the Invited Professor of Computer Science position at the University of Quebec at Chicoutimi (UQAC). She is also an Assistant Professor in Business Computing at the Higher School of Digital Economy, University of Manouba, Tunisia. Dr. Nakouri obtained her Business Computing Bachelor, M.Sc., and PhD degrees from the Higher Institute of Management, University of Tunis, Tunisia (ISG Tunis) in 2007, 2009, and 2016, respectively. She is associated with the LARODEC laboratory at ISG Tunis and focuses on research topics such as Machine learning, Responsible/ Secure AI, Computer Vision, and Image Processing. Her developed expertise in analyzing the vulnerabilities and shortcomings of Machine Learning models and designing solutions to enhance their resilience. Dr. Nakouri is a member of the UQAC Cybersecurity Research Chair, where she collaborates with experts in the field to advance knowledge in learning model security. She is actively involved in various international research projects and collaborations and has published over 25 research papers. Additionally, she has contributed to Program Committees, Organizing Committees, and chaired several reputable international conferences and workshops.

***

Angamuthu A was born in Puducherry, India on June 17, 1999. He received the B.Tech. degree in Information Technology from Christ Institute of Technology, Puducherry in 2020 and the M.Tech. degree in Network and Information Security from Pondicherry University, Puducherry in 2025. He is currently working as an Assistant Professor at Chennai Institute of Technology, Chennai, India. His research interests include information security, cloud computing, and cyber forensics.

Dhivyabharathi A.J completed M.Tech in Advanced Computing in 2014. she is currently pursuing Ph.D in the Department of Computer Science and Engineering in School of computing at SASTRA Deemed University-India. Her research focusses on Computer Vision and Deep Learning. She is a reviewer of a journal (igi-global). She qualified the GATE 2012 examination with an 87 percentile demonstrating her academic excellence.

Ravi Aavula working as an Associate Professor in the Department of Data Science, Anurag University, has about 21 years of teaching experience. He received his B.Tech degree in Computer Science and Engineering with distinction and M.Tech. degree in Software Engineering with distinction from JNTUH University, HYDERABAD. He received Ph.D degree in Computer Science and Engineering from GITAM University, Andhra Pradesh state. He has published 20 research papers in refereed International journals and 10 research papers in the proceedings of various International conferences. He has received several best paper awards for his research papers at various International conferences. His areas of research include Machine Learning, Artificial Intelligence, Data Mining, Computer Networks, Deep Learning, Cloud Computing. He is an active member of CSE, ISTE.

Shoeab Akhter is a passionate researcher and soon-to-be final-year Practicum student, B.Sc., in the Agriculture Department of the College of Agricultural Sciences at IUBAT—International University of Business Agriculture and Technology,

Dhaka 1230, Bangladesh. He is doing an internship at the Bangladesh Atomic Energy Commission. His research title is 'Green Synthesis of Nanoparticles for Agricultural Use.' Loved by nature and with a passion for the environment, his pursuits include areas of interest like agricultural sciences, food security and food science, environmental science, waste management, green energy, AI, IoT-based farming, and agritourism, among others. Along with his objective of creating an impact through writing and research, he combines new, innovative, suitable technologies and nature conservation with real-world issues. He has contributed book chapters (Emerald, CABI, Bentham Science, Springer, Cambridge Scholars, IGI Global) and has participated in various major projects on research. With the hope of bringing about a change in the world, he aspires to contribute to global knowledge and create a lasting impact on sustainability in society.

Sabyasachi Bhattacharya is an academic and professional associated with the Government of Assam, known for his involvement in technical education. He has held positions at several educational institutions including Barak Valley Engineering College and Residential Girls' Polytechnic, Golaghat, both located in Assam. His work contributes to the advancement of technical education and the empowerment of students, particularly in engineering and polytechnic disciplines.

Bontha Mamatha is an Assistant Professor in the department of Computer Science and Engineering at Keshav Memorial Engineering College, Osmania University, Hyderabad . her research interests include IOT, Computer Networks and Artificial Intelligence. She has around 15 journal and conference paper publications. She has served has reviewer for various conferences .

Suman Dabbu earned his Bachelor of Engineering and Master of Engineering in Biomedical Engineering from the Department of Biomedical Engineering, University College of Engineering, Osmania University. He began his career as a knowledge scientist at Scitech Patent Arts Pvt Ltd before joining the Department of Biomedical Engineering as an Assistant Professor in 2007. He was promoted to Associate Professor in 2020. Dr. Dabbu has 17 years of teaching and 10 years of research experience. He received his PhD in Biomedical Engineering from Osmania University, specializing in biomedical signal processing. His research interests include the analysis of physiological signals related to various diseases, development of indigenous medical instruments, and the study of movement disorders. Dr. Dabbu has published over 10 papers in reputed journals and presented 10 papers at various national and international conferences. He has received research grants to the tune of 50.Lakhs from funding agencies such as SERB-DST, DST-PURSE, TSCOST, UGC, and TEQIP-Govt of India. He was also awarded the SERB-TARE fellowship from the Government of India at IICT Hyderabad for a period of three years from 2023-26. Dr. Dabbu has won three best paper awards at international conferences, including ICNB-2 in 2011, IEEE-INDICON in 2015, and ICSDP in 2022. He serves as the Director of the Biomedical Instrumentation Centre, which undertakes consultancy assignments for various hospitals in Telangana state. He played a key role in developing calibration protocols for medical equipment according to the standards set by the Medical Device Directives, Government of India. He served as the Head of the Department of Biomedical Engineering from February 2013 to 2015 and from February 2021 to 2023. He is currently the Chairperson of the Board of Studies in Biomedical Engineering at Osmania University. Dr. Dabbu has held several administrative positions at the university, including Additional Controller of Examinations from 2019 to 2021 and Joint Director of ITE&C since 2023. Additionally, Dr. Dabbu is an Associate NCC Officer at 2(T) AIR SQN (TECH), NCC. He attended the pre-commissioning course at Officers Training Academy, Kamptee, and Air Force Station, Tambaram, and was commissioned as a Flying Officer in November 2018

Basudha Dewan is an Assistant Professor in Electronics and Communication Engineering Department at Manipal University, Jaipur. She holds a Ph.D. in Electronics & Communication Engineering from MNIT Jaipur, specializing in TFET biosensors, steep subthreshold devices, TCAD simulation, and nano-scale device physics. She has a strong academic background with an M.Tech. from Punjab Engineering College, Chandigarh, and a B.Tech. from Jaypee University of Information Technology, Waknaghat. Her research focuses on the modelling and simulation of advanced biosensors, leading to multiple publications in prestigious journals such as Materials Science & Engineering B, Micro and Nanostructures, Applied Physics A, and Silicon. Dr. Dewan has also contributed to book chapters and conference proceedings and has patents related to biosensor technology. She has received funding from SERB for organizing the 1st International Conference on Sustainable Computing and Intelligent Systems (ICSCIS 2025). Additionally, she serves as a reviewer for reputed journals like Physica Scripta, Springer's Silicon, and Scientific Reports (Nature). Dr. Dewan has experience teaching subjects such as Digital Electronics, Data Structures & Algorithms, and Computer Architecture. She has also been associated with IIIT Una as a faculty member. Her technical expertise includes TCAD simulation tools (Silvaco, Sentaurus), Verilog, VHDL, and PSpice. With a commitment to advancing semiconductor research and biosensor technology, she continues to contribute significantly to academia and industry.

Shugufta Fatima is a Research Scholar in the department of Computer Science and engineering, Annamalai University, Chidambaram, India and currently working as an Assistant Professor in the Department of Computer Science and Engineering at Stanley College of Engineering and Technology for Women, Hyderabad, India, with 8 years of teaching experience. She holds a Bachelor's and Master's degree in Computer Science and Engineering from Osmania University, earning multiple Academic Excellence awards. Her interests include Machine Learning, Deep Learning, and Artificial Intelligence, areas in which she has published research in reputable journals and holds a patent. She is also a book chapter reviewer with IGI Global and has organized workshops, conferences, and mentoring sessions to foster learning and innovation. Dedicated to creating an inclusive and engaging environment, she inspires intellectual curiosity and critical thinking in her students through innovative teaching methods.

G. Muthupandi was born in Tamilnadu, India on 20th March 1990. He received B.E. degree in Computer Science and Engineering and M.E. degree in the discipline of Embedded and Real Time System, He is currently working as an Assistant Professor in Computer Science and Engineeering at Chennai Institute of Technology, Chennai, India. He is a researcher with more than Ten years of teaching experience and expertise in the field of the Internet of Things (IoT) and Embedded Systems. Driven by a passion for innovation and a dedication to pushing the boundaries of design and technology, I am a highly accomplished inventor and researcher. My portfolio showcases a commitment to impactful solutions, evidenced by the granting of 10 design patents, reflecting a keen eye for aesthetics and functionality. Beyond design, I hold 20 published utility patents, demonstrating a deep understanding of practical problem-solving. My expertise is further validated by contributions to the academic community, including 8 publications at prestigious IEEE conferences and 2 peer-reviewed journal papers. This blend of practical invention and scholarly contribution positions me as a versatile and accomplished professional ready to tackle complex challenges and create meaningful advancements.

Ria Ghosh is the Co-founder and Managing Director of Forency LLP, specializing in digital forensics for India's premier investigative agencies. A frequent contributor to academic and professional publications, her work explores the intersection of AI, cybersecurity, and forensic science across fields like global security, personalized medicine, and environmental sustainability.

Israt Jahan is a research assistant at CHOWDHURY AI AND CYBERSECURITY RESEARCH LAB, located in Adelaide, South Australia. She obtained her B.Sc. degree in Computer Science and Engineering with a specialization in Data Science from East West University, Dhaka, Bangladesh. Her primary research interests revolve around multimodal learning, trustworthy AI, medical image analysis, and natural language processing (NLP). She has contributed chapters to various proceedings of the Springer and IEEE conferences and Scopus International journals on advanced diagnostic systems, explainable AI techniques, text and image combined representations for modality transparency, and diagnostic accuracy. She is inclined to work on research that bridges the gap between state-of-the-art research and real-world application, either in aiding patient outcomes or in creating trustworthy AI in medicine.

A. J. M. Sirajul Karim was born on January 15, 1954 in Feni. He earned a B.Sc.Ag. in 1977 and an M.Sc.Ag. (Soil Science) in 1979 from Bangladesh Agricultural University (BAU). Joining the Bangladesh Agriculture Research Institute (BARI) in 1979, he became Senior Scientific Officer in 1981. He received a USAID scholarship for an M.S. at Colorado State University in 1983. He joined IPSA Gazipur Agriculture University (GAU) as Assistant Professor in 1987. He earned a Ph.D. in 1995 and Postdoc in 2000 from Kyushu University, Japan. He supervised 17 Ph.D. and 37 M.S. students, published over 115 papers, served Gazipur Agriculture University (GAU) in key roles, and led a national project on Hill Soils from 2013 to 2018. At present he is Chair and Professor, College of Agricultural Sciences, International University of Business Agriculture and Technology (IUBAT).

Shikha Khullar is an accomplished academic and researcher with a Ph.D. in Data Mining and Artificial Intelligence. She currently serves as an Associate Professor at Poornima University, where she is recognized for her dedication to academic excellence and her dynamic involvement in radical research. With a deep-rooted passion for innovation and discovery, Dr. Khullar actively engages in interdisciplinary research that bridges technology and real-world applications, particularly in areas such as intelligent systems, smart data analytics, and sustainable digital solutions. Her work is characterized by a forward-thinking approach, where she not only explores theoretical models but also contributes to practical implementations that address pressing societal and industrial challenges. At Poornima University, she has been instrumental in mentoring young minds, guiding Ph.D. scholars, and contributing to curriculum development in emerging areas such as Artificial Intelligence, Blockchain, and Smart Agriculture Technologies. She is the member of IAENG and ERDA

Shohidullah Miah is Professor and Director of the College of Agricultural Sciences, IUBAT. Dr. Miah did his B.Sc. and Master's in Fisheries from Bangladesh Agricultural University. He was awarded his Ph.D. in Aquaculture at the University of Malaya, Malaysia. Dr. Miah has also been a Post-Doctoral Fellow at Japan Society for the Promotion of Science (JSPS) and the Center for Excellence (COE) Japan. He has professional experience of over 33 years in fisheries research and environmental water pollution control. His special interests are: wastewater treatment, fish population dynamics, mass culture of animal food, biochemical analysis, and food preservation. Dr Miah has published more than 40 academic papers and is the author of several technical reports.

R. Madana Mohana is an experienced academician with over 20 years in teaching and research. He earned his Ph.D. in Computer Science & Engineering from Sri Venkateswara University, Tirupati. He has published 66 research papers and holds 24 patents (including national and international). His research interests include AI, Machine Learning, Deep Learning, Data Science, and Outcome-Based Education. He is a Senior Member of IEEE, Fellow of IETE, and Life Member of ISTE.

Madiha Munawar, a second-year Computer Science and Engineering (CSE) student with a keen interest in software engineering and research. I have contributed to book chapters on emerging technologies and have also participated as a reviewer, critically reviewing book chapters, and am looking forward to conducting more research and publishing chapters, conference papers and journal articles. With a deep passion for coding, I actively explore diverse programming languages and innovative problem-solving techniques. I have participated in International conference and presented my paper in that, also in coding competitions and contributed to book chapters on emerging technologies. Driven by an insatiable thirst for knowledge, I aspire to create impactful software solutions that address real-world challenges with efficiency and creativity.

Rathinavel Subramanian was born in India on May 7, 1994. He received the B.E. degree in Electrical and Electronics Engineering in 2015 and the M.E. degree in Control and Instrumentation Engineering in 2017. He is a researcher with more than eight years of teaching experience and expertise in the field of the Internet of Things (IoT). He is currently working as an Assistant Professor at Chennai Institute of Technology, Chennai, India.

N. Sasikaladevi received Ph. D. Degree in Computer Science in 2013. She has published more than 42 papers in reputed International journals and conferences including publications in SCI-Indexed Journals. She is a reviewer of more than a dozen of reputed journals including IEEE transactions on Services Computing and IEEE Journal of Internet of Things. s. Her research focus includes the design of machine learning strategies to solve Discrete Logarithm Problem. She published several books and chapters in reputed publisher including Prentice Hall of India, Lambert Academic Publisher, IGI Global, Springer and Science Direct. She has received fund from Department of Science and Technology, Government of India to carry out projects in the domain of Services computing and Security System. She has received young scientist award and women scientist award from DST, India. She is also involved in the security enriched payment system design as a part of Digital India Initiative. Her current research interests include information security, digital authentication, security of wireless sensor networks, computer vision, deep learning and security of vehicular ad-hoc networks.

Anindya Nag obtained an M.Sc. in Computer Science and Engineering from Khulna University in Khulna, Bangladesh, and a B.Tech. in Computer Science and Engineering from Adamas University in Kolkata, India. He is currently a lecturer in the Department of Computer Science and Engineering at the Northern University of Business and Technology in Khulna, Bangladesh. His research focuses on health informatics, medical Internet of Things, neuroscience, and machine learning. He serves as a reviewer for numerous prestigious journals and international conferences. He has authored and co-authored about 43 publications, including journal articles, conference papers, book chapters, and has co-edited nine books.

Arya Nambiar is a pre-final year student pursuing a Bachelor of Fashion Technology (BFTech) at the National Institute of Fashion Technology (NIFT), Kannur. She has a keen interest in the intersection of technology and fashion, particularly in areas such as inventory management, automation, and sustainable production. Arya is a detail-oriented and motivated learner, consistently exploring innovative solutions to enhance efficiency in the apparel and retail industries.

Shyam Sunder Pabboju is an Assistant Professor in the Department of Computer Science and Engineering, Mahatma Gandhi Institute of Technology, Hyderabad, Telangana, India. He has over 20 years of teaching experience.He has received a Ph.D. degree in Computer Science & Engineering from Osmania University, Hyderabad, Telangana, India in January 2024.He has published over 30 research articles/papers in various reputed International Journals and Conference Proceedings published by reputed publishers. My interests are Big Data, Parallel Computing, Soft Computing, Distributed and Cloud Computing.

Ayona Padhi is a pre-final year student pursuing a Bachelor of Fashion Technology (B.F.Tech) with a specialization in Fashion Management. She is currently engaged in her Graduation Project, which allows her to apply academic knowledge to real-world industry practices. Her work focuses on exploring innovative approaches within the fashion business domain, emphasizing strategic thinking, market analysis, and sustainable practices

Manvitha Pola is a dedicated undergraduate student currently pursuing a BTech in Artificial Intelligence and Machine Learning at Sreenidhi Institute of Science and Technology, India. With a strong foundation in AI and ML, she is actively involved in both academic and research activities. At Sreenidhi, Manvitha has distinguished herself through her exceptional academic performance and her commitment to advancing her field of study. Her passion for machine learning has driven her to explore various research areas, including data science and natural language processing. Manvitha is actively involved in research, currently working on a paper focused on the applications of machine learning in healthcare revolution. Her research explores the potential of AI/ML technologies to drive innovations and improve healthcare practices, leveraging advanced techniques to analyze complex medical data and develop intelligent decision support systems.

Jagadeshwari Puttanapura is a dedicated Computer Science and Engineering student at Stanley College of Engineering and Technology for Women, India. With a keen interest in software development, machine learning, and data science, she have consistently demonstrated strong academic performance and a passion for technology. She has actively participated in several coding competitions, securing top positions in contests. Currently working as a Research Intern at University of South Alabama

Deva Rajashekar is an accomplished academician and researcher with a strong background in engineering and software systems. He earned his Bachelor's degree in Engineering and Master's degree in Software Engineering from the prestigious Jawaharlal Nehru Technological University (JNTU) Hyderabad, Telangana, India, in 2006 and 2009, respectively. Demonstrating a deep commitment to advancing his knowledge and expertise, he further pursued and successfully completed his Ph.D. from the same University, JNTU Hyderabad.Since 2009, Dr. Rajashekar has been actively contributing to the field of education as an Assistant Professor at various engineering institutes. His dedication to teaching and mentoring students has made him a respected figure in the academic community. Over the years, he has played a pivotal role in shaping the minds of future engineers and researchers, imparting both theoretical knowledge and practical skills.His research interests lie primarily in the cutting-edge domains of Machine Learning and Deep Learning. His work in these areas reflects his passion for exploring innovative solutions to complex problems and leveraging artificial intelligence to drive technological advancements. Through his research, he has contributed to the growing body of knowledge in these fields, earning recognition from peers and academic circles.

C Kishor Kumar Reddy, currently working as Associate Professor, Dept. of Computer Science and Engineering, Stanley College of Engineering and Technology for Women, Hyderabad, India. He has research and teaching experience of more than 10 years. He has published more than 60 research papers in National and International Conferences, Book Chapters, and Journals indexed by Scopus and others. He is an author for 2 text books and 2 co-edited books. He acted as the special session chair for Springer FICTA 2020, 2022, SCI 2021, INDIA 2022 and IEEE ICCSEA 2020 conferences. He is the corresponding editor of AMSE 2021 conferences, published by IoP Science JPCS. He is the member of ISTE, CSI, IAENG, UACEE, IACSIT.

Rupayan Roy is an Assistant Professor at the National Institute of Fashion Technology (NIFT), Kannur. He holds a Ph.D. in Textile Engineering from the Indian Institute of Technology (IIT) Delhi, with extensive academic and research experience in smart textiles, wearable technology, sustainable manufacturing, and ergonomics. Dr. Roy has authored over 50 research articles and book chapters in reputed publications, and has actively participated in more than 10 national and international conferences. His current research interests include the application of AI, IoT, and automation in apparel production, textile waste management using HTL, and intelligent air filtration media. He is also engaged in curriculum development and is known for his interdisciplinary contributions bridging fashion, technology, and sustainability.

Jawagar S was born in Puducherry, India, on November 20, 2001. He received the B. Tech. degree in Computer Science and Engineering from Christ College of Engineering and Technology, Moolakulam, Puducherry, India, in 2023, and the M.Tech. degree in Computer Science and Engineering from Pondicherry University, Puducherry, India, in 2025. He is currently working as an Assistant Professor at Chennai Institute of Technology, Chennai, India. His research interests include deep learning, nature-inspired algorithms, and evolutionary techniques.

Ishu Sharma is a distinguished academician and researcher currently associated with the Chandigarh Group of Colleges (CGC), Jhanjeri, located in Mohali, Punjab, India. She is also connected with Chandigarh Engineering College, part of the CGC group, known for its focus on engineering, technology, and applied sciences. With a strong background in engineering and higher education, Dr. Sharma has contributed significantly to teaching, research, and academic leadership. Her areas of interest likely include emerging technologies, innovation in education, and industry-academia collaboration. At CGC Jhanjeri, she plays a vital role in shaping the academic curriculum and mentoring students toward technical excellence and professional growth.

Ehsan Sheybani is a respected academic and researcher affiliated with the University of South Florida (USF) in Tampa, USA. He serves as a faculty member in the field of engineering and computer science, with expertise spanning areas such as artificial intelligence (AI), data analytics, machine learning, and wireless sensor networks. Dr. Sheybani has authored numerous research publications and has been involved in interdisciplinary projects, contributing to both theoretical advancements and practical applications in technology. His work often focuses on real-world solutions, including healthcare analytics, smart systems, and cybersecurity.

Thakur Monika Singh holds an M.Tech in Computer Science and Engineering from Osmania University and is currently pursuing a Ph.D. in the same field at KL University. Since 2017, she has been a faculty member in the Computer Science and Engineering Department at Stanley College of Engineering and Technology for Women. With a rich background in teaching and research, she is also a certified project-based learning mentor by Wipro and holds a TalentNext certification in Java Full Stack development. Her research interests focus on Machine Learning, and she has published several papers in national and international journals. Additionally, she holds a patent to her name.

Mahima Tasnim is currently a third-semester MBA student at Underwood University, USA. Her academic pursuit began soon after she graduated with an honors degree in Bachelor of Science in Agriculture (BSAg) from the International University of Business Agriculture and Technology. Her love of business and agriculture is still helping steer her education in both aspects. With the hope of bringing about a change in the world, he aspires to contribute to global knowledge and create a lasting impact on sustainability in society.

Lingala Thirupathi works as Assistant Professor, Dept. of CSE (Artificial Intelligence and Machine learning), Sreenidhi Institute of Science and Technology, Hyderabad, India. He has research and teaching experience of more than 18 years. He has published more than 50 research papers at national and international conferences, book chapters, and journals indexed by Scopus and others. He is the author of 3 textbooks. He acted as the reviewer for Springer journals and other international conferences. He is a member of IAENG.

Index

A

Agricultural Automation 116, 118, 119, 120, 122, 126, 127, 128, 129, 130
Agricultural Robotics 3, 23, 52, 61, 63, 67, 76, 87, 90, 110, 134, 147, 338
AI-Driven Automation 260
AIoT 54, 85, 180, 181, 182, 183, 184, 185, 186, 187, 188, 189, 190, 191, 192, 193, 194, 196, 197, 199
AI-Powered Machinery 126
Automation 11, 14, 16, 18, 26, 28, 31, 32, 34, 35, 36, 37, 40, 46, 50, 51, 55, 62, 63, 64, 74, 83, 89, 91, 104, 111, 113, 116, 117, 118, 119, 120, 121, 122, 123, 124, 125, 126, 127, 128, 129, 130, 131, 132, 133, 135, 142, 144, 146, 147, 148, 152, 153, 154, 155, 170, 171, 172, 176, 184, 187, 188, 189, 197, 211, 218, 226, 227, 229, 230, 231, 232, 233, 238, 239, 243, 244, 245, 246, 258, 260, 264, 268, 271, 272, 274, 275, 276, 278, 342, 343, 344, 347, 349, 350, 353, 355, 356, 357, 358, 362, 363, 366, 367, 369, 371

B

Bilateral Filtering 281, 282, 284, 285, 286, 292, 293, 294, 299, 304, 307

C

CLAHE 281, 282, 284, 285, 286, 292, 293, 294, 296, 299, 304, 307
Cloud Computing 14, 38, 42, 144, 172, 180, 199, 229, 355, 356, 360
CNN 21, 54, 69, 70, 83, 231, 281, 282, 290, 292, 293, 294, 299, 304, 305, 306, 308, 310, 322, 323, 349
Convolutional Neural Networks 4, 22, 69, 83, 146, 152, 167, 177, 320, 324, 330, 332
Crop Yield Optimization 67, 74, 76, 86, 313
Cybersecurity 207, 211, 228, 229, 233, 237, 239, 241, 244, 246, 251, 256, 260, 273, 275, 276, 279, 371

D

Data Ownership 1, 5, 17, 19, 78, 106, 129, 163, 193, 195, 315, 331
Digital Transformation 15, 131, 139, 142, 145, 166, 233, 259, 367

E

Edge AI 3, 49, 50, 73, 91, 94, 97, 101, 109, 196, 259, 321, 332, 343, 352, 364, 367
eFishery 185, 188
Energy Efficiency 226, 235, 274, 347, 361

G

GANs 11, 25, 27, 29, 30, 31, 50, 53, 89, 90, 94, 95, 96, 97, 98, 104, 105, 107, 109, 203, 204, 205, 207, 208, 209, 210, 211, 212, 213, 215, 216, 219, 223
Generative AI 1, 2, 3, 5, 6, 7, 12, 17, 18, 19, 21, 22, 23, 25, 26, 27, 29, 50, 89, 90, 94, 95, 96, 97, 98, 99, 100, 101, 103, 104, 105, 106, 107, 109, 163, 203, 204, 205, 206, 207, 208, 210, 211, 213, 214, 215, 216, 217, 218, 219, 220, 221, 222, 223, 225, 226, 256, 279, 370

I

Industrial Automation 226, 243, 245, 344, 347, 355, 356, 358, 362, 366
IoT sensors 6, 15, 42, 44, 46, 117, 120, 136, 148, 151, 154, 155, 156, 157, 159, 160, 174, 175, 177, 184, 186, 201, 226, 234, 236, 243, 244, 245, 251, 264, 265, 266, 269, 270, 271, 334, 344, 347, 357, 359, 371

M

Multi-Agent Systems 61, 62

P

Precision Agriculture 3, 14, 21, 23, 30, 37, 44, 51, 52, 53, 54, 55, 57, 61, 65, 74, 82, 83, 84, 85, 86, 94, 98, 99, 108, 109, 110, 116, 118, 119, 120, 125, 126, 131, 135, 136, 137, 138, 139, 141, 142, 145, 146, 147, 149, 152, 156, 165, 167, 168, 172, 173, 178, 181, 185, 186, 189, 198, 199, 201, 276, 278, 283, 293, 301, 308, 309, 313, 315, 320, 330, 337
Precision Farming 17, 27, 32, 37, 51, 65, 94, 101, 120, 132, 137, 142, 143, 144, 145, 148, 158, 187, 195, 226, 229, 233, 258, 268, 337
Predictive Analytics 30, 38, 42, 49, 50, 57, 73, 123, 127, 144, 145, 147, 151, 153, 161, 173, 174, 177, 188, 195, 242, 245, 246, 249, 261, 266, 268, 318, 329, 335, 342, 344, 353, 356, 360
Privacy 16, 38, 47, 48, 50, 51, 78, 79, 97, 101, 128, 143, 162, 163, 179, 190, 192, 194, 203, 205, 214, 215, 216, 219, 226, 228, 233, 239, 242, 244, 246, 248, 250, 251, 258, 260, 261, 268, 273, 275, 276, 330, 331, 342, 352, 355, 356, 357, 359, 360, 364, 365, 367, 368
Probabilistic Modeling 206, 223

R

Real-time Data Collection 28, 141, 164, 225, 226, 229, 361
Regenerative Intelligence 1
Robotics 3, 4, 13, 14, 17, 18, 19, 20, 23, 26, 28, 35, 37, 46, 47, 49, 50, 51, 52, 54, 55, 57, 58, 60, 61, 62, 63, 64, 67, 72, 75, 76, 77, 79, 80, 81, 83, 84, 85, 86, 87, 89, 90, 95, 107, 110, 112, 119, 124, 130, 133, 134, 135, 145, 146, 147, 171, 172, 187, 194, 207, 223, 227, 230, 264, 315, 320, 332, 333, 336, 337, 338, 339, 348, 351, 371

S

Semantic Segmentation 54, 108, 281, 282, 283, 288, 309, 310
Sensor Fusion 73, 90, 92, 332
Sensor Networks 67, 76, 108, 144, 185, 187, 321, 361
Smart Agriculture Technologies 135, 184
Smart Cities 145, 146, 166, 168, 226, 227, 229, 231, 232, 233, 249, 251, 253, 254, 256, 258, 263, 276, 279, 342, 343, 344, 349, 350, 353, 356, 358, 362, 365, 367
Smart Devices 227, 249, 355
Smart Farming 7, 9, 11, 12, 19, 25, 28, 30, 31, 39, 42, 49, 51, 52, 55, 56, 84, 87, 90, 94, 97, 108, 109, 124, 129, 136, 138, 139, 141, 142, 143, 145, 146, 147, 148, 149, 150, 154, 155, 158, 159, 161, 162, 163, 164, 165, 166, 167, 168, 169, 171, 172, 173, 174, 180, 182, 186, 187, 190, 191, 194, 196, 197, 198, 200, 201, 258, 268, 269, 277, 314, 334, 337
Smart Healthcare 241, 359, 360, 368, 370
Smart Irrigation 11, 15, 100, 120, 128, 141, 144, 149, 150, 156, 159, 165, 170, 178, 189, 198, 258, 268, 269
Smart Solutions 13, 74, 177, 251, 257, 258, 260
Soil Health 19, 26, 53, 57, 59, 60, 62, 63, 64, 65, 66, 67, 70, 71, 72, 73, 74, 75, 78, 80, 87, 98, 115, 118, 122, 131, 152, 226, 313, 314, 315, 316, 317, 318, 319, 321, 322, 323, 326, 328, 335
Soil Monitoring 33, 147, 237, 318, 319, 320, 321, 322, 335
Sustainability 5, 14, 17, 18, 19, 20, 21, 26, 27, 28, 33, 34, 36, 43, 46, 49, 51, 58, 60, 61, 74, 76, 80, 81, 83, 91, 94, 101, 106, 108, 112, 113, 116, 120, 123, 131, 133, 134, 135, 136, 137, 139, 142, 143, 144, 145, 147, 148, 151, 155, 156, 158, 161, 165, 166,

171, 172, 184, 186, 188, 189, 194, 197, 198, 201, 226, 228, 233, 235, 236, 237, 238, 240, 244, 245, 248, 250, 253, 254, 260, 264, 267, 268, 269, 313, 314, 316, 321, 324, 326, 329, 331, 334, 335, 337, 338
Sustainable Agriculture 35, 42, 52, 72, 82, 108, 130, 135, 146, 167, 194, 195, 198, 225, 268, 314, 317, 325, 337, 338
Sustainable Farming 37, 43, 50, 63, 65, 67, 78, 109, 112, 147, 150, 177, 195, 313, 315, 319, 326, 327

V

VAEs 25, 27, 29, 30, 31, 50, 90, 95, 96, 97, 107, 203, 204, 205, 208, 209, 210, 211, 213, 219

www.ingramcontent.com/pod-product-compliance
Lightning Source LLC
LaVergne TN
LVHW080910300926
849162LV00003B/629